Adolf Heschl

The Intelligent Genome

Springer

Berlin
Heidelberg
New York
Barcelona
Hong Kong
London
Milan
Paris
Tokyo

Adolf Heschl

The Intelligent Genome

On the Origin of the Human Mind by Mutation and Selection

With 20 Drawings (*Homo sapiens*) by Herbert Loserl

 Springer

ADOLF HESCHL
Executive Manager
Konrad Lorenz Institute
for Evolution and Cognition Research
Adolf Lorenz Gasse 2
3422 Altenberg, Austria
adolf.heschl@kla.univie.ac.at

Translator

ALBERT EDWARD FULFORD
Weimarer Straße 16
67459 Böhl-Iggelheim
Germany

Title of the German edition: Adolf Heschl, Das intelligente Genom
ISBN 3-540-64202-1 © Springer-Verlag Berlin Heidelberg New York 1998

Cover photo by courtesy of Roger D. Kornberg, reprinted with permission
from SCIENCE, Vol. 292, No.5523, p.1876
© 2001 American Association for the Advancement of Science

ISBN 3-540-67166-8 Springer-Verlag Berlin Heidelberg New York

Library of Congress Cataloging-in-Publication Data
Heschl, Adolf, 1959-
  [Intelligente Genom. English]
  The intelligent genome : on the origin of the human mind by mutation and selection/
Adolf Heschl.
      p. cm.
  Includes bibliographical references and index.
  ISBN 3540671668 (hardcover)
    1. Intellect–Genetic aspects. 2. Brain–Evolution. 3. Genetic psychology. 4. Behavior
genetics. I. Title. QP398 .H54513 2001 612.8'2–dc21

Springer-Verlag Berlin Heidelberg New York
a member of BertelsmannSpringer Science+Business Media GmbH
http://www.springer.de

© Springer-Verlag Berlin Heidelberg 2002
Printed in Germany

Cover design: D&P, Heidelberg
Typesetting: Camera-ready by the author
SPIN 10725432   31 /3130 - 5 4 3 2 1 0 - Printed on acid-free paper

To my daughter Katharina and my two sons Oliver and Benjamin,
who have systematically convinced me of the fact
that I cannot teach them much,
in fact, that I cannot teach them anything at all.

Contents

Preface

The essential ideas reported here were published for the first time in 1990 in an article for the *Journal of Theoretical Biology,* under the title "L = C. A simple Equation with Astonishing Consequences" (Vol. 145, 13—40).

The differentiation of the body cells was prepared by modifications in the molecular structure of the germ cell and we are on the right track if we consider that all differentiations occurring in the course of ontogeny depend on the chemical and physical molecular structure of the germ cell, in other words on the nuclear substance of the germ cell.

August Weismann

Preface

This book is the result of a series of personal experiences and encounters which caused the attention of my genome to be directed onto a completely unexpected path, namely to realize that the totality of all these particular experiences, in some coded way or other, must have already been known to it in advance. To arrive finally at this seemingly paradoxical conclusion, I had to pass through several distinct ontogenetic stages every one of which contributed substantially to the continuation of the direction not chosen once for all, but one continually genetically controlled.

At the age of 13, I became irreversibly imprinted on Lorenz' beloved "Companion in the bird's environment" (1935/70) which caused in me an equally irreversible fixation on the organisms described therein in all the details of their behavior (birdwatcher syndrome). A little bit later Johann Gepp showed me how to recognize and scientifically determine the many diverse kinds of insects in the field. During my university studies, I had to learn how at least some species of insects have developed a quite unusual method of communicating with each other. Under the supervision of Herbert Heran, a former assistant of Karl v. Frisch, I was given the unique opportunity of repeating the majority of all the ingenious experiments on the particular mechanisms involved in the bee dance. Uwe Krebs convinced me that even the rather cold-blooded and hence often quite lethargic reptiles, for example, in the form of warans, can be highly interesting subjects for comparative ethological considerations. Reinhart Schuster taught me in great detail the morphological and behavioral taxonomy of the fascinating group of spiders and, by doing so, he showed me in a convincing manner how far one can push pure, that means technically "unarmed" systematic thinking. Ernst Topitsch helped me lose every respect for traditional philosophy by demonstrating how much farther the critical analysis of different world view systems, and be it that incredibly impressive one developed by Immanuel Kant, can bring the skeptical thinker. Johann Götschl fascinated me with his conviction that it must be possible to elaborate a both comprehensive and consistent one-world-theory which is thought to finally integrate into one picture the best concepts from the most varied scientific disciplines.

During my stay in Geneva Sylvain Dionnet and Anastasia Tryphon let me participate at short range in what Jean Piaget so promisingly had termed "genetic epistemology" where children of all ages, from the baby up to the young adult, are studied in their respective individually unique ways of constructing step by step by ontogeny a more and more purely mental, i.e., increasingly flexible and abstract model of the world. Ariane Etienne introduced me to the newly created field of cognitive ethology which has set itself the very ambitious goal of elaborating a consistent synthesis between the purely ontogenetic study of cognition and a systematic phylogenetic comparison of species.

Back to Austria Rupert Riedl, my academic mentor and furtherer, taught me among many other things that a very special, i.e. typically Viennese combination of abstract theorizing and empirical verification—its unwritten motto goes: "If the data don't (immediately) fit the theory, all the worse for the data" (famous examples: Freud, Boltzmann, Mach, Gödel, Popper, Weiss, Schrödinger, Kammerer, Bertalanffy, etc.)—is one of the best prerequisites for at least occasionally doing a little bit more than just "science as usual." In fact, there exists no theory in the world which would have absolutely no empirical value and be it a purely subjective, ideological and/or scientific one (usually, it is a variable mixture of all three). In addition, Manfred Wuketits showed me that even the sometimes rather dry matter of evolutionary epistemology need not necessarily be boring. Laco Kovac gave me much personal assistance by repeatedly urging me to further elaborate what I had begun to develop and, finally, there was yet a short letter from Manfred Eigen in which he let me know that he "basically agrees" with the view expressed in this book.

Monika Kickert, the former "soul" of the Konrad Lorenz institute, taught me how to better understand the intricate ethological relationships among the different generations of the Lorenz school. Last but not least, a very special thank you to "Guna" Gunatilaka and, most of all, Silvia, my wife, both of whom have never ceased to encourage me to continue with my work.

Adolf Heschl
Altenberg a. d. Donau, September 2001

Introduction: Nature Explains Nurture

By asking the question: what is it that
makes man a man? I am saying on
the one hand, that there is his culture
and, on the other, his genome, that is
clear. However, what are the genetic
limits of culture? We have absolutely
no idea. And that is a pity, because
this is the most fascinating problem,
the most fundamental one that
exists.

Jacques Monod

In this book it will not be proved, as has already been done a hundred times
before, that man is not much more than an ape, although possibly slightly more
intelligent or, as the zoologist Desmond Morris demonstrated with a convincing cover picture some years ago, a largely naked ape (Morris 1967). On the
contrary, we will accept such a viewpoint simply as given and instead will address far more difficult and, I hope in the long run, far more exciting questions
about the real nakedness of the natural being which we call *Homo sapiens*. For
the average zoologist as well as for the scientifically interested layman it has
indeed been clear for some time that we humans are biologically related not
only to some of these more hairy creatures, such as the bonobo, chimpanzee,
gorilla, and orangutan, but of course in the same manner also to the rest of the
large family of primates consisting of over 200 different species (overview in
Diamond 1992; Kenyon and Moraes 1997; for the technique of transplanting
mitochondria, see Osusky, Kissova and Kovac 1997). The degree of purely external similarity is simply too striking to allow us to doubt the fact. For the case
of the bonobos or pygmy chimpanzees (*Pan paniscus*), which are at least genetically as closely related to humans as the common chimpanzee (*Pan troglodytes*), Sue Savage Rumbaugh, one of the most experienced experts on the
behavior of this species, set out the following arguments:

They are not chimpanzees of any size or shape. They are more like persons with small brains
and extra-long body hair. ... In particular, when you see a bonobo walking on two legs, as
humans do, you get a strong impression of what the human ancestor would have looked like
(Sue Savage-Rumbaugh and Roger Lewin 1994, p. 93/97).

Likewise, during the last few decades so many additional scientific data and proofs have been found which demonstrate this connection that it requires fundamentalist intransigence or at least considerable resistance to the flood of information available in modern society to permanently insulate oneself against the evolutionary worldview. So it was to a great extent the merit of the increasing numbers of biologists who succeeded in making themselves heard and thus promoting a slow, but nevertheless general change in thinking. Of course, all this began with the heretical insight of the meanwhile world-famous Englishman and world traveler Charles Darwin, who strove to spread his ideas more than a hundred years ago. In the meantime, the theory of evolution is considered as one of the best corroborated theories within the natural sciences, a fact which gives justified rise to the hypothesis that even man can be integrated into the concept of a common development of the living world. Concerning his purely physical or, strictly speaking, biological characteristics, there already exists a quite extensive agreement about the idea that we have many of them in common with other organisms on this planet and, in the meantime, even considerable parts of human behavior are seen by an increasing number of researchers as conditioned by our biology. Hence the human animal seems to behave in many respects in ways a zoologist would expect, but only if we are ready to include the exact phylogenetic tree—as far as it is known today—into our considerations.

This book, however, is intended to do a little bit more than just join the large choir filled with famous voices, who herald the scientific Song of Songs of the evolutionary dependence of man. Since if we hear again and again that the species *Homo sapiens* still carries in itself some important biologically determined residues of its physical and even behavioral constitution which should be taken into consideration, but at the same time excluding specifically human characters such as thinking, consciousness, language, morality, in short, the capability for cultural tradition, from biological evolution, then we must ask ourselves whether it is not so much a scientific boundary which is touched here but rather an ideological taboo. It is concerned with, as we can easily see, the widespread conviction that man holds a special position within the animal kingdom, so pleasantly comforting and therefore regularly emphasized. In the meantime, such a seemingly liberal position is even held by the Pope, who only recently accepted that the "bare theory" of a physical relationship between humans and animals can be seen as a "fact". However, he did this not without simultaneously mentioning the not insignificant restriction that the immortal soul still needs to be personally inspired by God (Holden 1996). We will see that these and similar opinions, which occasionally—perhaps in an unconscious manner—are still defended even by some biologists, can no more be maintained if we consider our current state of knowledge. Rather, we will detect, why the entire human psyche—"soul" is just a much more attractive word for it—

is nothing else than a quite autonomous product of our own biological evolution.

Regardless of the numerous differences of opinion at the political-ideological level, we find ourselves, as members of the same species, to be unanimous in our indiscriminate acceptance of a uniquely privileged position in evolution. Surprisingly, even the majority of biologists agree in this respect and, correspondingly, show themselves eager to furnish reliable evidence in favor of the special position of the human species within the entire animate world. Even Desmond Morris who, by his remarkable popularity, contributed a great deal to the scientific demystification of the picture of the human being to the public, at the end of his often quite unconventional considerations arrives at the very personal conclusion that our species, as the only one among the many thousand others on earth, would have succeeded in developing into, what he calls a "wonder child of evolution" (Morris 1995).

One could ask in a somewhat grandiloquent form whether nature actually disregard humans, as did the Swiss biologist Adolf Portmann (1897—1982), who by his morphologic studies on the "living shape" of organisms arrived at almost creationist-like conclusions. In the words of Portmann this sounds as follows: "But in man a completely new design was tried" (Portmann 1971). Of course, one could say that man occupies a privileged position as long as one concentrates only superficially on some of his special characteristics—for the embryologist Portmann (1967) for example these consisted of a type of "secondary nidicolousness", coupled to an "extra-uterine first year". There is no doubt, man as constituting a physiological premature birth holds a special position within the entire animal kingdom, with corresponding modifications in the area of his behavior (Hassenstein 1987). Nevertheless, he only holds a special position in the same manner as does every other animal species, since after a more or less closer examination of any given species one can easily detect again and again specific traits, which are typical for that species alone. In fact, it is for exactly this reason that zoologists and animal taxonomists have, for quite a long time and with legitimate scientific justification spoken of the existence of species which are all different to each other and hence singular in a way. The refined tactics of the domestic cuckoo to first let another species of bird incubate its eggs and then raise its young, certainly lends it a privileged position among all birds. However, at the same time every other bird species on this planet has at least one unique property. The superbly simple and nevertheless very successful method used by bloodsucking mites of landing on their usually extremely mobile and, for a small mite, not easily caught hosts—by letting themselves drop from a higher vantage point with admirable timing, makes this species likewise into an eccentric of evolution. However, the same holds true for every other randomly chosen species of mite. Likewise with our next relatives within the animal kingdom, namely the bonobo and the chimpanzee, they all have their specific characteristic of emotional expression, of locomotion and many additional details in their outer appearance, which they have acquired completely independently of our own species and thus give them the right for a privileged position like every other species.

What then is the main concern of our investigation? It is to determine the actual range of validity of the Darwinian theory of evolution, of course, but understood in a far more fundamental way than has been the case up to now. It is not unusual features which are of interest, since the whole field of biology is full of them, but it is the much more difficult question of whether we humans really represent an integral part of biological evolution. This includes the far more delicate question about whether we are subject to exactly the same general evolutionary mechanisms of mutation and selection as are all other existing species. Since it is exactly in this area where, in sharp contrast to the tremendous variability of the specific individual features, which separate the different species, we see a quite unusual uniformity, from which as far as we know no single species could really escape.

Interestingly enough, man prefers to see himself with regard to the latter question as the great exception from biological evolution, as some kind of a fantastical "yes, but"-mutant within the whole remaining realm of organisms. Now we are at the point where I want to sound the warning which I announced at the beginning of this introduction. That is to say I will not primarily attempt to illustrate the fact that the „human ape", as one would expect, has many characters in common with his close biological relatives and many others not so near, but, going far beyond that, I would like to advance both theoretical considerations as well as numerous empirical proofs of the fact that even our seemingly exceptional species with all its supposed and real particularities—be they of morphological, ethological or cognitive nature—can be fully integrated into the fundamental mechanisms of evolution and hence is still subject to continuing biological evolution.

Such a procedure will lead us—as the reader in the course of the volume will soon notice—to rather unusual and even frighteningly unorthodox conclusions, which nevertheless are particularly interesting just because of their apparent impossibility. By doing so, the most provocative insight will consist of discovering that as a matter of principle, i.e., for both strong evolutionary and epistemological reasons, we, as multicellular organisms, must be strictly excluded from the possibility of cognitive gain during our individual lives. Expressed in a somewhat more technical way this means that evolutionarily relevant modifications, that is the acquisition of any new trait, can take place only in the phylogeny (germ line) of multicellular organisms, in contrast to ontogeny (individual development) which must remain excluded from progress. It can also be shown that, in an indirect way, this is the logical extrapolation and generalization respectively of what, after all, Weismann had in mind when he

argued against Lamarck's famous theory of the inheritance of acquired characteristics (cf. chap. 11). As a consequence, it is exactly these paradoxically so-
feared genes which get the last word in this persistent and still continuing
game of evolution. Even worse, this influence goes far beyond that always quite
limited area which, until today, has been attributed to those "bad" genes by any
challenger of the Conditio humana, from Jacques Monod (1973) and Hans
Eysenck (1973) to Edward Wilson (1975) and Richard Dawkins (1976), the inventor of genetic "selfishness". Hence, our final result will not be a deceitfully
appeasing "not in our genes" (Lewontin, Rose and Kamin 1985) a position
which E. O. Wilson once characterized by the following words: "It is a remarkable fact that in recent years the most effective opposition to the study of human evolution has come not from religious fundamentalists and the political
right, but from biologists and other scientists who identify themselves with the
radical left. The focus of these critics was on the inheritance of intelligence..."
(Wilson 1978). Likewise we will not adopt a further interactive, even though
well-meant compromise as offered for example by researchers such as Robert
Boyd and P. J. Richerson (1976, 1985), Luca and Francesco Cavalli-Sforza (1994),
and, still more recently, by a group around Jeffrey Elman in a book with the
promising title "Rethinking Innateness" from 1996:

Where does knowledge come from? This was the question we began with. As we suggested
at the outset, the problem is not so much that we do not know what the sources of knowledge are. The problem is rather in knowing how these sources combine and interact. The
answer is not Nature or Nurture, it is Nature and Nurture (Elman et al. 1996, p. 357).

The problem with that pragmatic statement is that the real issue consists of
exactly that which is not recognized at all as a special problem here, namely,
the problem of the *origin* of knowledge. With regard to this fundamental problem, our answer will read quite different, namely in the sense that nature explains nurture: It can be shown that the entire knowledge of the (multicellular) individual must already preexist in its genome, i.e., the totality of genetic
information comprising genes and other coding or otherwise semantic units
of DNA or RNA. Hence we want to do precisely, what so far no biologist has
ever dared, that is to assert that man is completely pre-programmed with regard to his attainable knowledge. Accordingly, we maintain at the same time to
have solved definitively the apparently so tricky nature/nurture-problem.

We will see that living systems are operationally and hence cognitively closed systems which can only change their identity through a random-like breaking off of their metastable internal dynamic structures. And that is called
evolution. That does not mean, however, as is often considered and consciously
misunderstood by many people, that all the details of the concrete fate of an
individual must be already written down in the embryo. The crucial point here

consists of understanding that the necessary knowledge to get along with the challenges of the environment must be already available in all its necessary details in the fertilized egg cell, but that this, however, does not require that both the concrete manner as well as the succession of the encounters with an often very variable environment should be foreseeable.

Consequently the whole matter also has nothing to do with the long since superseded idea of the so-called "innate" according to Lorenz (1937), which is commonly thought must already be present at birth ready-made for use. There is no doubt that we make an impression of being really badly incomplete at birth—and even worst before —, but this is exactly what constitutes our so far proven life-style. It is not a compromise between rigidity and preformation of genetically preprogrammed "eidylons" (*Greek*: "eidolo" = [internal] "image") versus unlimited flexibility and structural plasticity of genetically indeterminate "xenidrins" (*Greek*: "xenos" = [influenced] "from outside") as still Lumsden and Wilson's (1981, 1983) seemingly provoking (Lewontin 1981), yet soon rejected (Maynard Smith and Warren 1982) "coevolutionary" thesis of genes and culture tried to make us believe which is the topic, but simply the question whether we humans obtain the necessary information for the successful organization of our special kind of life-cycle from the environment or from inside ourselves.

For that it is essential to understand that no information in the sense of meaning is present in the inorganic environment, but only in itself completely senseless physical states and arrangements, which have nothing to say at all to the organism, but instead very often threaten it in a highly dangerous way. Or, has anybody among us ever experienced that a falling rock warned him/her just in time before it would have destroyed him or her? Even if we were lucky enough to perceive its buzzing sound in time, those acoustic waves which meet our ear would keep silent in the true sense of the word about their threatening "side effect". If it were not for our extremely well-designed ear with its—hopefully—sufficiently fast connections over the so-called *Heschl gyrus*, the primary auditory cortex, into one of our cortical motor areas, we would indeed have difficulty in surviving.

Even in the case where we apparently have "learned" through the experience of several mountain tours, to avoid slopes prone to rock falls it is still our brain alone, which already knows how to link specific past perceptions with later events to prevent the worst. We in fact have no possibility of simply taking up the necessary information in any directed way from the environment, if, in the course of our long phylogeny, this information has not already entered our genome by mutation and selection. Since, if we assume that the first possibility would work, then of course there would be nothing left for us to fear because we could easily solve every acute problem the very same moment we were confronted with it. Hence things are going to get even a little bit worse: In future, we will have to interpret even the phenomenon of learning in a completely differently manner than today, namely not as the acquisition of new knowledge about the world, but as the application of the already pre-existent knowledge in all the necessary detail to those often very special problems of our increasingly complicated primate world. What seduces us to make such a, at first sight, completely absurd statement? During each individual step of any "learning" process, no matter how complicated it is, our brain knows exactly at every single moment what it has to do or not do. Since if it suddenly no longer knew how to react for any reason, then woe betide us for natural selection would not have pity on us.

With regard to the relevant history of science, in this case biology, we will rely primarily not so much on the great selection theorist Charles Darwin, but rather on the, at least for a wider public, somewhat "forgotten" evolutionist August Weismann (1834—1914), who together with and nevertheless completely independent of Gregor Mendel (1822—1884), prepared the development of modern genetics (Löther 1990). However, whereas Weismann was still forced to develop his famous "non-hereditability of acquired characters" by proceeding in a strictly casuistic-empirical manner, i.e., step by step ("… I do not say that the effects of use and nonuse cannot and may not be heritable, I only think they are not." 1895, p. 61), it has in the meantime become possible to theoretically deduce the universal validity of the so-called "Weismann's dogma" directly from the theory of evolution itself. Consequently, this insight must also be valid for all kinds of behavior, as will be shown. Viewed from this perspective, the environment influences permanently us in a purely physical way, there is no doubt about that, but it cannot instruct us in any directed way. In this evolutionary play, that can be already said in advance for the reassurance of the reader, we all are very active players, without, however, at the same time being ever able to foresee its future outcome.

My intention here consists, and I invite every open-minded reader to join in the attempt to view the human race from the absolutely disrespectful perspective of a merciless Drosophila researcher who erroneously—let us say due to the absentmindedness of the scientific assistant who is known for regularly making bad mistakes—starts to investigate some of those wingless and naked specimens of *Homo sapiens*. It is known that the particular attractiveness of such an everyday mistake often leads to unexpected insights, insights which, under different, i.e., conventional conditions of thinking, one would consider to be absolutely nonsensical and absurd in advance. The bare attempt, I mean, is worth doing and it requires, against all conventional expectations, neither any special knowledge-based prerequisites nor any purely technical insights reserved for the privileged clique of so-called scientific experts. A quite recent situation report from the world of university administration may serve here to encourage the non-specialist reader:

Seen pragmatically, one will have to be glad, if our universities succeed in avoiding the production of one-sidedly trained and personally distorted "idiot savants" (Sigurd Höllinger 1992, p. 73).

Thus, it may probably be sufficient to be able to differentiate a blue tit from a mole at first sight in one's own garden, even though both occasionally or permanently live in "caves". It is the simple which often hides the provocative, and in any case it must be provocative, since otherwise it would not be so simple.

1 The Myth of a Wonder of Nature

What is most difficult thing to
achieve?—To get wise to oneself.
Wilhelm Busch

Despite great efforts to understand the complete human being within the context of modern evolutionary theory, something of a myth, a particularly persistent one, has remained at the forefront of most theoretical approaches. What we mean here is not that the sometimes vehement denial of the theory of evolution as a whole or of its validity with regard to humans by those many different traditions in philosophy and the humanities, not to mention modern day fundamentalism. Instead we mean that fascinating phenomenon that natural scientists, among them not a few biologists, in fact, even many evolutionary biologists, despite Darwin, continue to stick to the myth that man holds a special position. A statement on this subject made by Konrad Lorenz is only one among many:

For the purpose of this book (note: "Behind the Mirror") ... the categorical difference between man and all other living beings is essential—the "hiatus", as Nicolai Hartmann termed it, that great gulf between two levels of real being, which has originated through the fulguration [lat.: "flash"] of the human mind. ... Hence it is no exaggeration to say that the mental life of man represents a new kind of living (Konrad Lorenz 1973/77, p. 168/172).

This strange respect for our own species is even more surprising since everyday empirical research is constantly delivering such a huge number of new results which are increasingly restricting the field of mere speculation. Nonetheless, it seems also to be the theory itself and the conclusions which, so far, have been deduced from evolutionary theory, which explain why up to this day nobody has really ventured to formulate at least a provisional attempt at a consistent extrapolation into the specifically human area. That is not to say that we do not have "evolutionary" approaches to the human phenomenon—far from it, in the meantime, we have hundreds of them—, but most of them are still purely metaphorical. Even though, the basic leitmotiv, as can be easily verified, was prepared by Darwin himself in his quite unambiguous hint at a gradual but continuous mental evolution in man. However, it could be that it is just by following that advice that we would potentially have to call into question some of our most cherished and hence indispensable beliefs about ourselves.

What then transforms, or it would be better to say seems to transform man into this unique exception to biological evolution? Basically, it is again and again the same three features, which are regularly thought to be taken as a valid justification for the particular status of being an exception to the laws of evolution: the human capacity for learning, language and culture. In addition, true self-consciousness allows the human being to be the only creature capable of reflecting upon his own position in nature and hence able to escape the laws of nature, at least at this purely mental level. Closely associated with this traditional picture we find anew the philosophically time-honoured body/ soul-problem as well as, in a new more "scientific" reformulation, the very popular concept of an ultimately undissoluble brain/mind-dualism. However, looked at more closely the problem of dualism unveils itself to be an intellectual creation of a special kind. Whereas for all other organs of the human body the discoveries of modern medicine have succeeded in showing that their functions are identical with the underlying material, i.e., biochemical processes, one still continues to insist with astonishing persistence to keep the human mind clean from all earthly contaminations. In this way nobody has ever seriously asked the question about the "real nature" of the functioning of the liver or the bladder, even though, viewed from the problem itself, we have to contend with the same situation as it is in the case of that mysterious interaction between brain and mind, which obstinately resists any resolution.

In fact, it seems indeed that the human brain is the very first biological organ which is ashamed of its ignoble origin. Perhaps, a conceivable evolutionarily adaptive function of this remarkable arrogance of a behaviorally important control mechanism over the rest of the body, as the brain to a certain degree undoubtedly is, could consist of disclosing as little as possible about that important control. The almost perfect neuronal control over the whole body, which to this extent is lacking in other animal species, seems to confirm this view. This means that it is simply not true, as it is generally described, that the mind—or even "its" brain—as something abstract as opposed to the body. On the contrary, there is no species, in which the mind exercises such a strong and nearly perfect control over the whole body. Accordingly, we see only in *Homo sapiens* the fantastic possibility of influencing to a remarkable degree the functioning of many internal organ systems (e.g., digestion, heart beat, peripheral blood circulation) by means of conscious thinking (e.g., autogenous training). This ability, however, does not demonstrate an example of a mysterious mind/body-interaction, but instead is based on a well-defined collaboration of two distinct neuronal fiber systems, the existence of which has obviously stood the test of time in the course of our evolution:

The *vegetative* (or: *autonomous*) and *somatic* nervous systems work hand in hand. Their neuronal morphological substrates can no longer be differentiated in the central areas, particularly in the upper brainstem, the hypothalamus and the telencephalon (Wilfrid Jänig 1993, p. 349).

In principle, this psychosomatic relationship was already known by the animist physician G. E. Stahl (1660—1734) who was convinced that the "soul" has a very powerful influence on one's own health and illness. However, let us now continue with the usual interpretation of the wonder being which is man. In this perspective, human language by being founded upon individual self-consciousness leads to the first breakthrough of the limits set by one's own personal experience and hence to the formation of a proper social field of knowledge and abilities. Finally, within the various cultural traditions of various national groups some sort of a preliminary high point of a unique development is reached which one could describe as the emancipation of man from his own biological limitations. Together with the transmission of information from one member of a group or an organized society to the next one, a particular stage in human evolution seems to have been achieved in which the sharp contrast to the still primarily genetically based, i.e., Darwinian evolution becomes more than clear. In the words of a prominent American evolutionary biologist this sounds as follows:

Cultural evolution has progressed at rates that Darwinian processes cannot begin to approach. ... Human cultural evolution, in strong opposition to our biological history, is Lamarckian in character. What we learn in one generation, we transmit directly by teaching and writing (Steven J. Gould 1979).

Even though this description is of course a somewhat simplistic picture and, in the meantime, has been qualified to a certain degree (cf. the discussion in Callebaut 1993, p. 423—431), it nevertheless very nicely describes the common cliché which still today guides the current thinking about this subject. So if we accept Gould's view, all of a sudden, all that what otherwise has had to be strictly forbidden becomes possible for the majority of the species. In addition, for the first time, a true transmission of acquired characteristics becomes feasible, something from which, interestingly enough, the French naturalist Jean Baptiste de Lamarck (1744—1829) and, as will be explained a little later, to a certain degree even Darwin himself, erroneously expected to represent the actual mechanism of behavioral evolution. In contrast to purely biological development, which usually takes place over thousands of years and proceeds only by very small gradual steps, a completely new possibility arises of achieving much faster progress than before. Hence man is the first and in fact the only living creature on this planet which has managed to emancipate himself so to speak once and for all from his own biology through the invention of culture and tradition. However, all that happened, as one can assume, only after a long and tough biological evolution, in the course of which *Homo sapiens* finally succeeded in dissociating himself from his many ancestors and relatives who are still ape-like today. What was left over and to what many evolutionists regularly point out with great emphasis only concerns rudiments, that means more or less small behavioral fragments of our biological descent and phylogeny, the effects of which we should take into account if we are really intent on building a more humane future for our race. The result of such considerations is an entirely bizarre hybrid, a strange human animal, driven by its biologically rooted instincts, doing its best to dismiss this rather ignoble heritage and to advance within an all encompassing human culture.

If we claim that most biologists and evolutionary theorists have adopted such a view of the matter, we should nevertheless specify this in more detail. It is of course not really true that there are no attempts at all to give a biologically and evolutionarily oriented interpretation of complex human behavior. On the contrary, the last decades have shown how far Darwin's theory has already advanced in this respect and how different in the meantime some of the behavioral patterns which have, in the main preoccupied philosophers of the various conflicting schools, are on the way to being explained in a strictly causal-material manner based on results from scientific empirical research. Thus, modern behavioral ecology as the conceptual continuation of classical ethology (1973 Nobel Prize winners and founders: Karl v. Frisch, Niko Tinbergen, Konrad Lorenz) has meanwhile succeeded in explaining in a convincing way the biological fitness value of a great variety of behavioral patterns, from comparatively simple perception processes up to complete life cycle strategies of very varied animal species by simply relying on the validity of evolutionary theory. Consequently, there is no reason to reproach biologists with professional incompetence or even, as it is done with impressive regularity, to think that the concept of evolution could not maintain its initial promise, or to rejoice that evolutionary theory in general has lost its scientific validity.

It is also not really the case that one would have purposely left out the human race with the intent to award our species a particularly privileged position within the natural sciences. No, even that is only partly true if one looks at what has already been done with the interpretation of human behavior by the ambitious group of sociobiologists. In fact, here the exact opposite seems to be true since sociobiology as put forward by Edward O. Wilson (1975) is nothing more than the first and, in a certain sense, already completely overdue attempt to describe the enormous variability of both animal and human social behavior within the framework of truly applied evolutionary thinking (for a recent vindication after years of often groundless dismissal, see Alcock 2001). The fact that, since its foundation in the early 1970s, this discipline has already earned not just one important accolade makes it as provocative as the first still "shocking" results of classical ethology were many decades ago—consider those endless debates around an innate aggression instinct (for a harsh and to some extent legitimate criticism of Lorenz's too simplistic hydraulic model of drive accumulation, see Pilz and Moesch 1975). It goes without saying that in this way the picture of the formerly predominantly descriptive-comparative method in the behavioral sciences has changed decisively.

Let us consider, for example, one of the most important changes in theory concerning the unit of selection. Whereas the still quite anthropomorphic concept of the conservation of the species dominated most earlier discussions and controversies, which often were caused by the—simply erroneous—concept itself, the idea of physically separable units such as identifiable genes and genomes led to far more stimulating impulses both for the further refinement of behavioral theory as well as its consistent transformation into empirical research projects. Interestingly enough, this development also reinstated in some sense Darwin's early idea of the individual organism as the fundamental unit of evolution , but as a reaction against the assumed "selfish" concept of evolution nevertheless both forgotten as well as rejected (Heschl 1994a). Yet, even though both sociobiology and modern behavioral ecology (Krebs and Davies 1991) as the currently most important evolutionary approaches to the behavior of living organisms have already succeeded in including even very complex behavior patterns of the undoubtedly most intelligent of all apes into their considerations, however, the position of the species *Homo sapiens* as an exception itself remained widely untouched. What we mean here is his unusual learning capacity, which ultimately, as every clear-thinking person can easily accept, is not at all easy to explain simply by biology. For that purpose it is sufficient to take a short look at the history of only some of the more relevant inventions of the last few centuries, for example, firearms, steam engines, artificial electricity, internal combustion engines, computers, etc., and we rapidly understand that such an accelerated development cannot be explained by that very slow movement which is called evolution. The latter can certainly be held responsible for some striking morphological differences between humans and dinosaurs, but in the end is unable to explain those numerous miracles which the human mind has produced in such a short time span.

In the face of such impressive comparisons, almost everybody instantaneously surrenders and even confirmed sociobiologists such as Richard Dawkins feel obliged to introduce a new concept of cultural evolution based on the reproduction of so-called memes to at least metaphorically bind the Spiritus hominis, the escaping human mind to the theory of evolution. However, looked at more closely, Dawkins' memes are nothing more than just thoughts or ideas which like a virus seem to be transferred from one person to another. The true origin of this concept called mimetics is not difficult to recognize: it is the very attractive and hence still very popular metaphor of the contagious idea (Lynch 1996). What comes out is a strangely parallel and yet considered fair race between archaic staid phylogeny and high-flying cultural history, with all those well known possibilities of placing your bets on the victory of one or other of the opponents (Campbell 1975). Suddenly, one is reminded those gladiatorial contests in ancient Rome with the difference that today bets on secret genetic hints have to contend with bets on cultural innovations. As a rule the rescue for our permanently endangered society and culture is of course expected to come from the latter category even though the geneticists seem to be on the point of moving into the fast lane. How will this race finally end or, what interests us much more in this context, can such a race exist at all? Since if we really accept the existence of Dawkins' memes as replicators which are taken to be completely independent of organisms and their genes, do we then not have the much more difficult problem of risking that the whole content of already well-corroborated evolutionary theory will ultimately flow out through a tiny, but nevertheless very effective leak? To call this leak "mimetics" instead of "culture" is undoubtedly a very clever trick but it does not resolve the existing conceptual contradictions.

2 Evolutionary Epistemology or the Difficulties of Getting Started

If anyone could understand the baboon he would do more for philosophy than Locke.

Charles Darwin

Even Darwin himself anticipated that, ultimately, his theory of a continuous evolution of life must also be valid for the exceptional characters of man, even including a completely new epistemology of our species, and he was very clear in this regard (cf. Engels 1989). Nevertheless, the numerous social controversies which he initiated with his radically new views on the origin of man and his evolutionary change in the course of phylogeny did not—as was in fact to be expected—lead to a really comprehensive integration of the human species into his theory of evolution. Rather, they left behind a very special compromise which primarily had the purpose of defusing some of the potentially very strong threats at that time—and not only then—and of establishing self-awareness among people. Through this compromise, the thesis was recognized that at least the body/organic evolution of man has proceeded according to the Darwinian principles of natural selection and that this process of purely morphological change will probably also take place in the future. At the same time, however, the idea that the particular mental capacities of *Homo sapiens* could be understood sufficiently within the frame of selection theory was explicitly excluded. In addition, this restriction was not understood as only being provisional and perhaps caused by some unknown methodological limits, but viewed with regret by Darwin and many of his more fervent supporters (e.g., Spencer, Haeckel, Boltzmann, Mach) as a fundamental theoretical impossibility. This may sound like a long-outdated story of the natural sciences but in fact it is exactly here that we have to search for the last still active bastion of the myth of a special position of man in nature. It runs in a very similar manner as do big political or even, if war is thought of as diplomacy by other means, military decisions in the history of humankind. Once both the winners and losers are defined and a treaty is agreed on, usually for a quite long time nobody is in a hurry to question the new doctrine (for those who are married: compare wedding and divorce). It is not before significant changes take place in the respective structures of power of the parties involved or a change in the validity of their legitimations, i.e., arguments for or against a case, that a given political state is allowed to be modified. A very similar phenomenon occurs in the field of science, for which the physicist and subsequently a philosopher of science Thomas Kuhn (1962) developed his own theory of paradigmatic change describing the replacement of one existing scientific doctrine by a new and often quite different one. His more concrete ideas about the famous "path of discovery" (cf. Bloom 2000 with together with Makous 2000) will be dealt with in one of the later chapters.

However, only one century after Darwin, i.e., in the course of the second half of the 20th century, several attempts were made to question seriously the quite unanimously established ideological compromise between Darwin's "dangerous ideas" (Dennett 1996) and older and obviously "not so dangerous" views held by various conservative institutions, among which the Catholic church with its immortal "soul" as a very efficient spearhead of restoration. The main idea behind all these rather cautious attempts was to both elaborate and propagate a first naturalistic interpretation of human mental capacities. Even so, the results were strikingly modest, regardless of which scientific discipline they originated from. What came out each time was a living being abounding which with many surprising corresponding characteristics and similarities with the other higher primates, but nevertheless a living being who was still condemned to carry on representing a sort of prodigy and who, from time to time, even seemed to accelerate with regard to his unusual exceptionality. One of the very first semi-scientific, semi-philosophical attempts to undermine somehow the special position of man in relation to the rest of evolution gave itself the impressive name of "evolutionary epistemology" the basic axioms of which, nowadays, have been frankly copied and are now live on in a more natural science-oriented "evolutionary psychology" (cf. Krebs 1996; a major part of the criticism we address here to evolutionary epistemology also helds for evolutionary psychology which simply has replaced the outdated Lorenzian "instincts" by a variety of more modern sounding Chomskyan "modules"; for a critical discussion of the radically modular or „ecological" approach to animal learning and cognition, see MacPhail and Bolhuis 2001; for a reasonable compromise between hierarchical and modular architecture of human intelligence, see Velichkovsky 1994).

Donald Campbell (1974) deserves special merit for coining the term at a time when "evolutionary" still had very little to do with matters of cognition. Interestingly enough, one of the most committed adherents of this approach did not stem from biology, nor even from any other discipline of the natural sciences. On the contrary, it was the recently deceased epistemologist and philosopher of science Karl Popper (1902—1994) who was one of the major co-founders of evolutionary epistemology (for other important developments in the

German speaking area, see Leinfellner 1976, Lorenz 1973, Oeser 1974, Riedl 1980, Vollmer 1975, Wagner 1987, Wuketits 1984; for a comparable and partly even preceding development in the US, see Campbell 1960; for the Anglo-Saxon area representative is Plotkin 1982, 1994). Popper was convinced that even the very successful institution of modern science with its numerous different disciplines should proceed by following the example of animate nature if it is really interested in achieving new insights. "From the amoeba to Einstein", that was his catchphrase, one and the same successful *method*, namely that of trial and error, is applied, because it succeeds in bringing their applicants closer to the absolute ideal of objective truth with a probability that is at least non-negative.

According to Popper the sole difference between animals and man lies in the fact that only the latter, by means of his special mental abilities such as thinking, imagination, reason and logic, who has in the meantime become capable of evading the direct impact of natural selection. Popper argued that instead of losing their lives with every minor, but nevertheless potentially lethal error, humans usually prefer to sacrifice their ideas and conjectures about the world. Such a procedure has the inestimable advantage of allowing a theoretically unlimited number of repetitions which of course must lead to an enormous acceleration of the whole process of cognitive development. In such a way, it gets progressively easier to understand why biological and cultural evolution display such different rates of change and, ultimately, this situation also explains man's special position within nature. At the least, Popper who, in this respect, argues in a somewhat similar way to Dawkins, seems to thrust the conventional separation between the mind creature man and the mere biological organism animal into the background in favor of a rather more homogeneous mode of development. The criticism which one of course can still apply and which would soon collapse the whole "evolutionary" model consists of showing that we are dealing here with only very superficial analogies, if not purely external similarities, between both kinds of evolution and hence we have to be very careful about assuming any deeper causal relationship. On the contrary, it is more than questionable if, for example, a truly random mutation in a biological-genetic sense can correspond to a conjecture or hypothesis in a human-creative sense in any thinkable way.

A further important evolutionary approach to the explanation of the peculiarities of the human mind came from a real biologist who, however, very early in his life, felt himself attracted towards philosophy. The Swiss zoologist who became world famous later as a developmental psychologist and epistemologist, Jean Piaget, attempted to explain the puzzle of the origin of the human mind through its emergence and continuous development in the course of the individual's life. For him, perceiving as well as true cognition was not something biologically preordained but the result of an extremely complicated and protracted interaction between the growing child and its environment. This kind of permanent interaction, in which at least at the beginning of the child's development the exact contributions of subjective versus objective world cannot to be separated from each other (adualism), was thought to lead to a constant expansion of the intellectual capacities of the individual. However, Piaget's idea of an initial adualistic cognitive starting point of the young child is disputed as the perception ecologist J. J. Gibson makes clear:

> The assertion that children are able to understand the difference between true and imagined not before their intellect has developed, is mentalistic nonsense. The growing child comprehends increasing parts of its reality, and this the more the more places of its natural habitat are contacted by it (retranslated from James Gibson 1982, p. 277).

That the assumption of an initial inseparability of inside and outside is in fact nonsensical is shown by some recent models of neuronal self-reference, in which the body on its own always "knows" exactly what it or what the outer world is doing (Wolpert, Ghahramani and Jordan 1995; Berthoz 1995). Be it as may be, the important point here is that the active manipulation of the young child in its still limited environment represents the motor for the development of increasingly more intelligent behavioral patterns. In this way, the growing infant elaborates in a playful manner both the things and objects of its environment as well as the corresponding concepts and strategies , allowing him, step by step, to solve his problems better.

One of the best examples of what Piaget (1936) meant by this kind of developmental change is given by the, as he termed it, *circular reactions* (variants: primary, secondary, tertiary) which are easy-to-observe and slightly, but constantly changing repetitions of single behavior patterns with the sole purpose of increasing the understanding of a possibly existing causal relationship between separate phenomena. Now, all parents will know very well what we are talking about. Consider a young child less than one year old in front of a so-called "mobile" which is perhaps the most universal toy of our time (and perhaps, in Neanderthals and *Cro-Magnon* times as well, consisting of chiming mammoth teeth): every time the child sets either the whole of the toy or a part of it in motion, the mobile moves and makes a pleasant noise as well. Interestingly enough, the child does not usually stop playing after two or three trials, when one would expect it to have seen enough of this simple procedure. Far from it, every child spends minutes if not hours seriously "investigating" the fascinating relationship between its own body movements and the equally changing views and events in the world out there represented by the funny mobile. Similar stereotyped behavior occurs everywhere during human development and, in fact, it seemed to be one of the particular characteristics of our species. However, the important point here is that there is a constant variation inherent in it which causes the result to be not always a purely quantitative one, but instead from time to time also leads to a qualitative change from one cognitive strategy to solve a given problem to another.

Thus, Piaget was one of the first to stress the indispensable necessity of the autonomous activity of the child for the development of its intelligence. In his view, the environment fulfils a rather more passive role and, in the best case, only puts the necessary material at the child's disposal. Correspondingly, an unattractive and not very stimulating environment can even lead to severe deficits in cognitive development. For Piaget, however, this does not mean that the information underlying mental development is to be searched in the stimuli of the milieu, but rather that the actual origin of human cognitions lies in a specific construction made by the subject, i.e., the child itself. With such a surprisingly clear position in favor of the cognitive system concerned, in the present case the developing subject, Piaget came very close to a proper biological interpretation of human intelligence (Piaget 1967). This relationship has now been rediscovered in quite a number of scientific disciplines, and this even allows us to speak of a kind of renaissance of Piaget's genetic epistemology which can be understood as a developmental theory of knowledge. Even though the term "genetic" in Piaget's context has little if anything to do with the actual biological technical term, at least it demonstrates a highly interesting conceptual connection between epistemology and biological theory.

Finally, a further important pioneer of current evolutionary epistemology comes from one of the more qualified disciplines of the natural sciences with respect to the subject. In a similar manner to Piaget, his famous Swiss colleague, the Austrian physician and behavioral scientist Konrad Lorenz likewise started his academic career with a strong personal predilection to go in a certain philosophical direction. However, whereas Piaget was strongly influenced by rather Lamarckian inspired thinkers such as the French writer Henri Bergson (1859—1941), Lorenz was much more interested in the well-known German philosopher whose work seemed to be much easier to combine with a Darwinian-biological interpretation of man. Immanuel Kant's famous *a priori* categories of human reason (Kant 1781), understood as indispensable logical preconditions of perception and thinking, appeared to be optimally suited for an at least partially biological interpretation of man's special position in the animal world. Accordingly, Lorenz published during World War II—besides some frankly Social Darwinian works (Lorenz 1940a, 1940b) in the tradition of Herbert Spencer (1893)—an article with the symptomatic title "Kant's doctrine of the *a priori* in the light of contemporary biology" (Ger.: "Kant s Lehre vom *Apriorischen* im Lichte gegenwärtiger Biologie." *Blätter für Deutsche Philosophie* 15, 94—125). In this article, Lorenz arrived at the conclusion that Kant's theory of cognition represents exactly the right "interface" between traditional philosophy and modern evolutionary biology. Looked at more closely,

Lorenz argued, Kant's categories were to be seen as being nothing more than the phylogenetically acquired, hence—in contrast to Piaget—truly genetic preconditions of human reasoning. That insight opened for the first time the possibility of considering, even at this stage, purely mental capacities of man within the framework of Darwinian thinking. However, since natural selection is necessarily restricted to genetically transmitted characteristics, only innate components of any behavior can be subject to biological evolution. In consequence, inborn aspects that researchers thought to have discovered in both animal and human behavioral patterns were over-emphasized for understandable reasons and, finally, led to the very popular, but at the same time also controversial term "study of instinct." The great field scientist and ecologist Niko Tinbergen was the first person to introduce this term into behavioral research when he published a very influential textbook in 1951. He did not anticipate the controversy that he would initiate by doing so. Even though there were negative reactions, his main working hypothesis was a very plausible one and, viewed from an evolutionary standpoint, did not lack scientific legitimacy. The behavior of animals always takes place under certain and often very selective environmental conditions, and hence it appears to be quite reasonable if the ethologist, as the professional investigator of behavior, makes the assumption that the specific kind of observed behavioral pattern can only be understood if taken as a biologically imposed adaptation to exactly those conditions. Any other assumption would it render considerably more difficult or even mysterious to clarify the actual function of behavior.

Obviously, the concentration of a whole field of research together with several generations of scientists involved in that one single important characteristic of innateness also had some considerable disadvantages. In particular, an almost dogmatically fixed opposition developed to all those behavioral features which were usually considered as being learnt or otherwise acquired, and this was accompanied by a corresponding hardening of the frontiers between the two competing scientific communities on this and the other side of the Atlantic. At this point, it was certainly not by coincidence that the best known apostles of the innate came primarily from Europe, whereas the great learning theorists as well as the most convinced learning dogmatists and pedagogues were and are still to be found across the ocean. The influential American behaviorists, such as Burrhus Skinner (1904—1990), for example, would have had great problems in reaching the position comparable to the one Lorenz did in a Central Europe where history was always a too heavy burden to allow one to accept readily ideas of a practically unlimited plasticity of behavior as officially proclaimed in classic behaviorism.

Nevertheless, the lover of science and non-biologist Popper was already on the right road to equating even the highest mental achievements of human beings, that is its grandiose scientific theories and discoveries, with the evolution of living organisms. Only in the most favorable case, is the result of this permanent scientific selection process by means of Popperian falsification or refutation mental progress in as much as our knowledge can, paradoxically, only increase step by step out of our ignorance, from one successful refutation of one scientific hypothesis or an assumption of the next one. Because, as must be stressed again and again in the sense of Popper, all those not yet refuted hypotheses may only be provisionally seen as being somewhat truer as those many errors which have already been banished to the garbage dump of the history of science. Consequently, each new day and each new scientific experiment can make or break, at one stroke, all that wonderful knowledge which has been obtained over years of laboriously detailed work. Even whole worldviews may suddenly collapse often due to some completely subsidiary observation or consideration. This happened for example with the Ptolemaic system, in which the sun still had to circle around the earth, with the myth of creation, where Adam and Eve still had to engender the entire human race, with the phlogiston theory, according to which with the burning process was connected to the production of a certain mysterious substance (*phlogiston*; "heat substance", postulated by the chemist G. E. Stahl during the first half of the 18th century, but refuted by A. L. Lavoisier in 1775), as well as with many other ideas, which in their time have been regarded as being true and correct. There is much to be said for Popper's critical theory of science, and this also explains much of its persistent popularity. Far beyond that, Popper's spirit clearly dominates the whole of western scientific activity. There is no researcher nor an author of an important textbook, no society critic and even no influential politician, who would dare to question this incredibly successful method of scientific progress. Obviously, the fascination of this approach must to be investigated in the particularly instructional style which Popper always practiced with regard to his masterly instructions on the correct method of success. Poppers stipulations concerning the correct functioning of science might therefore be the answer thereby to an apparently fundamental wish for some kind of primeval trust, a wish, which for various reasons, even exists in the sciences. Here, the subliminal fear of possibly ending up with a completely false method of cognitive gain as a chaotic borderline of the scientific enterprise led to a surprisingly orthodox conformity within the otherwise so revolutionary and society-changing self-understanding of human culture. Hence it is no miracle to see Popper figuring as one of, if not *the* most important philosopher of science of the whole 20th century. His easily comprehensible instructions for the right way to cognitive gain, which we could circumscribe as: "Rely only on that, what really works", builds trustworthy ideological, but also technological confidence.

Is it possible to seriously criticize Popper's position or even to have serious misgivings about it? Let us attempt to do it here briefly—a more substantial criticism follows in chapter 15—, by concentrating on the proper core of the problem, namely of cognitive gain, more precisely, how do really new ideas and hypotheses originate. As paradoxical as this may sound, this problem was never seriously tackled by the great philosopher himself. For Popper the important thing was only how to deal with *already existing* ideas, in other words how to sift the wheat from the chaff most effectively. The solution to the much more difficult problem of how exactly these new ideas emerge or could emerge was tacitly presupposed. It seems perhaps a little bit daring, but we can show that it is exactly at this point that biology, or more precisely its subdiscipline evolutionary biology, is in fact able to offer a completely new answer, even though an answer which, as will be shown, is not so easy to accept subjectively. It was Paul Feyerabend (1976), the great "anti-Methodist" and declared Anti-Popperian among the contemporary epistemologists, who became world-famous with his battle cry "anything goes" which severely disconcerted not a few simpler philosophical minds. Today, these words sound like the long-desired emancipation from a by far too normative diction of the Popperian schoolmasters who, for a very long time, constituted the tacit majority among the elitist class of "serious" philosophers. Even so, how much is actually possible, but, at the same time, how little can be real progress—this can only be demonstrated with Darwin's theory.

Compared with Popper, a universal intellect of research Jean Piaget came much closer to our actual subject, that is a consistent synthesis of "biology and knowledge" (Piaget 1971). His emphasis on the mental activity of the subject who, in his view, was not only a merciless selector of the few correct from those all too many false hypotheses, but a creative producer of new ideas is centers itself around the really interesting part of our investigation. The only reproach, with which one now would have to confront him, consists of having overshot the mark somewhat, since in Piaget's perspective there no longer exist any limits of human cognition. If the fact or rather the "mystery" of physical aging were not there (Theimer 1983), coupled with the constant decrease of memory and, above all, the drama of the individual death, we would not even be able to stop growing more and more experienced, more intelligent, and more and more perfect. If our personal mental life, as was also thought by Popper, in fact represented a Darwinian process, during which ever better adapted ideas follow each other, then Piaget would be right and we would then automatically have the power to freely develop into a boundless infinity of an increasingly more complex mental being. If, however, we as individuals—and as we will see, much speaks for it—represent only a small fragment out of the whole scenario of organic evolution, we have to reckon with a rather modest contribution to the development of our species.

Finally, it was Konrad Lorenz as one of the rare trained biologists among the many self-appointed evolutionary epistemologists who presented a very disputed and at the same time almost popular version of a biologist's interpretation of the rational creature called man. Our thinking, so he tries to make us believe, is, to a great extent born within us, but nevertheless we can surmount these our archaic thinking instincts to a certain degree through learning and communication processes. As can be easily shown in comparative ethology, there seems to be, in fact, something like different degrees of behavioral freedom regarding the species under consideration, whereby of course only man and no other animal deserves the lion's share. However, there exists quite a severe, but nevertheless regularly overlooked problem of this almost too clear and hence commonly accepted dichotomy concept and this concerns the exact type of transition from one faculty to the other, that is, from instinct over to learning.

To explain it in a somewhat simplified way: A human or animal individual wants to learn something new. To start up, it necessarily needs a first concrete instruction, for example in the form of "take this or that for true and react in such a way or just not react to it". This first instruction must have been already available and therefore also in some way innately pregiven or, properly speaking, inherited, because how else this could still perfectly unsuspecting subject understand any instruction given by the environment? However, now inevitably comes the next step in the learning process and so the whole problem begins anew. Who says *now* to the subject, what it has to do in the next moment? This kind of question-and-answer game, which Plato the great epistemologist mastered perfectly,, can be continued in principle for an arbitrarily long time. As can be easily seen, a useful distinction between innate and learned cannot be resolved in such a way. There is no doubt that this distinction still has a certain empirical value, that is, for example to distinguish certain arbitrarily definable categories of behavioral patterns—as there are such trivial differences such as "available at the moment of birth" versus "not available", "rigid" versus "flexible" motor co-ordinations, "short term" versus "long-term" behavioral programs—but eventually, it has little if anything to do with that typical difference in origin, i.e., phylogeny or ontogeny, with which it is usually associated. Nonetheless it is exactly at this point where, up to this day, the roads part, which then obviously culminates in the very crucial question about the true reach of our supposed personal liberties.

Very often, Konrad Lorenz still has a name for being the one who emphasized most strongly the instinctive character of human behaviour. However, looked at more closely, the exact opposite is the case. By formulating a learning concept which, at least at the beginning of ethology, was diametrically opposed to the innate he just made possible the biological miracle of man and by doing so he put himself into the first row of those not so rare biologists who repeatedly stress the unexplainable special position of our species within biology. By contrast, Darwin himself was always concerned with the abolishment of any such biological miracle. Thus it seems that we can only come closer to a new solution of the *Causa humana* if we approach the case in a much more impartial way than we have been used to doing up till now. That means nothing more than that, as a methodological principle, we should not care at all how the result will finally appear. Of course that should not prevent anyone from making a subjective judgment of the value of our result at the end of our investigation. However, for purely methodological reasons it is advisable for us to employ once again the refreshingly uninhibited approach of a professional zoologist, but now to push it further into an area which still represents an unspoken taboo, namely the area of human epistemology itself, man's undoubtedly most exclusive and, at the same time, most controversial domain (for examples of philosophical criticism of traditional evolutionary epistemology, see Löw 1983; Köchler 1983; Lütterfelds 1987; Hösle 1988). For this purpose we will now first ignore Hilary Putnam's unshakable conviction that "reason can't be naturalized" (Putnam 1983)—he erroneously thinks that epistemology must be purely normative—as well as deliberately ignore the well-meant advice of David Hull, a special philosophical connoisseur of evolutionary attempts at explanation, who still quite recently declared:

... the fundamentals of evolutionary theory are currently in a state of flux. Now is not the time to take a particular interpretation of biological evolution and apply it uncritically to social evolution (David Hull 1981, p. 73).

However, for David Hull's sake let us nevertheless start once again from the beginning and let us scrutinize more closely than ever before man the animal with the help of a zoological magnifying glass. If, however, as has already happened too often in a very miraculous way, man again tries to avoid such a strict biological investigation, we will have to re-examine one of his most reliable escape routes more closely this time, namely his apparently so exceptional capacity to acquire new knowledge. However, to successfully execute something demanding like that it still calls for a little bit of biological theory.

3 To Be or Not to Be

Nothing is easier than to admit in words the truth of the universal struggle for life, or more difficult at least I have found it so than constantly to bear this conclusion in mind. Yet unless it be thoroughly engrained in the mind, I am convinced that the whole economy of nature, with every fact on distribution, rarity, abundance, extinction, and variation, will be dimly seen or quite misunderstood.

Charles Darwin

The theory of evolution has a close association with the well-known quotation from Shakespeare's *Hamlet*. The term "evolution" in its modern sense as a development which can only be understood by an extended historical consideration was first used by the astronomer Sir John F. W. Herschel (1792—1871) who tried to arrange the existing diversity of stars into certain common developmental lines. Darwin himself stated his opinion of Herschel as the model philosopher of the time with the following revealing words:

... Sir John Herschel's "Introduction to the Study of Natural Philosophy" aroused in me a burning enthusiasm; at least, to make the most insignificant contribution to the edifice of science. No single or a dozen other books influenced me as strongly as this one (retranslated from Charles Darwin in F. Darwin 1887, Vol. 1, p. 55).

As a matter of fact Darwin's famous-notorious "struggle for life" is exactly about—*nomen est omen*—being or not being, even though not always in such a theatrical manner as one expects from some all too popular accounts of evolutionary theory. On the contrary, very often the most important processes are those which can take place in a very cooperative form viewed externally (Dugatkin 1997) or which are even completely hidden from the naked eye (see plants) without any actual physical fight or obvious struggle. The survival of the fittest, as evolution has often been called since Herbert Spencer (1820—1903), describes the simple and easily comprehensible fact that differently equipped organisms in a limited habitat with limited resources will have accordingly different chances for survival and reproduction (for the absurdity of the corresponding tautology argument, see Mayr 1988, Oeser 1974b, and especially Vollmer 1985).

Today, differential reproduction, that is, the quantitatively different rates of multiplication for distinct individuals, represents a more modern and, above all, substantially better term for it , since it is semantically more neutral and thus easier to manage (for an elaborate molecular treatment, see Eigen 1971). If resources and habitats are scarce—and in reality they are always limited—, then this must necessarily lead to any kind of competition between the organisms concerned. Thereby, competition can refer both to individuals of the same (*intraspecific*) or different (*interspecific selection*) species, depending in each case on the respective ecological relationships we are looking at. Generally, competition manifests itself most intensely between members of the same species, because we can expect that their individuals are already adapted to a quite similar ecological niche through both their comparatively uniform morphology and associated survival strategies. Conversely, the rivalry between individuals can be expected to decrease markedly if organisms are not as closely related to each other as within a given species, because then the points of contact for possible Darwinian competition consequently become fewer and fewer.

This does not exclude, however, that additional causal relationships exist even between rather distantly related biological species, e.g., between rats and humans, which in the long run will prove to be a perhaps a very significant competition in the future. If one regards the habitat of earth globally , one will eventually have to arrive at the opinion that all present-day organisms are in a kind of association dependent on each other to a varying degrees. A currently very popular, even though a little bit exaggerated example may illustrate this situation. The species Homo sapiens itself is on the way to changing the once quite varied living conditions on the globe in such a manner that biodiversity will be reduced with only comparatively few animal and plant species left over for future evolution if the current trend should continue. Even a still more extreme scenario is conceivable, in which the whole of mankind inadvertently, but nevertheless surprisingly simultaneously blows itself up together with the remaining species on Noah's ark by pressing a single atomic button. Even if this is not so easy as it may seem to be at first sight—in fact, every single living organism of the world would have to be destroyed—such a theoretical consideration at least shows, how close the global connections theoretically can be. With this as a background it is, however, quite legitimate to consider each serious rival to our own, obviously slightly "deranged" species as an important and above all real competitor for our continuing existence on the globe (e.g., mice, rats, insects, viruses).

Thus the competition for both naked survival as well as potentially eternal existence in the form of our own descendants in a limited habitat must necessarily lead to a simultaneous evolutionary modification of living systems, whereby those better or at least not more unfavorably adapted organisms automatically maintain the upper hand. Since Darwin, the totality of all mechanisms which together drive this process, is called natural selection and it stresses the simple fact that nature, so to speak, does its best to sift the chaff from the wheat. At this point, however, it is appropriate to make some critical notes since many of these central Darwinian terms are taken from the human domain of colloquial language and hence repeatedly give rise to some all too popular misinterpretations. First of all "better" means only that, for any reason, a certain type of organism—this can be both a certain species as a whole as well as a group of individuals from within a species—is on the point of replacing other forms of life in purely quantitative terms. This statement, and this is forgotten time and time again in some simplified accounts of evolutionary theory, has absolutely nothing to do with an evaluation in the sense of human value systems (good/bad, better/worse etc..).

What is meant here is purely and simply the quantitative determination of physically real existence (e.g., this worm still lives, this rodent still burrows, this strangely intelligent ape still runs around, etc.), whereby at least for the theory all currently existing forms of life are to be treated as evolutionarily equivalent. Simply because their mere existence—that is the only point where perhaps even Heidegger's mysterious "Being-in-the-world" (1978), an only for particularly devoted insiders understandable state of "factual existence", would make sense, if at all—is the only relevant criterion for evolution. Even if certain clearly negative statistical trends are already visible—we know for example from a growing number of species that are unable to cope with a completely new environment, i.e., our technical civilization and they will probably become extinct—the only significant factor right up till the last minute is the mere survival of any individuals, be it a single pair (or one individual in asexually reproducing organisms). It is not until their complete disappearance is it possible to say that this species was certainly not "better", i.e., more successful than any other species that still exists. For quite a long time many geneticists adhered to the view that apart from otherwise inevitable incest problems, a *minimum number* of approximately 600 pairs, which can be deduced by purely theoretical considerations, would be necessary for the survival of any population. In the meantime, some more recent data taken from real populations in the field have shown convincingly that the critical number nevertheless depends very strongly on the species under consideration and sometimes the number can be much lower than these 600 pairs (Remmert 1994; e.g., the wisent was able to be saved although only a few dozens have survived being hunted by humans).

In other words, there is no life on earth better or worse than any other, the only important thing is: to be or not to be. Accordingly, there are no better and less adapted organisms but only (at that particular moment) existing ones and, for any reasons, those no longer existing. At the same time this also means that during evolution there must not necessarily be any directed development towards any higher degree of complexity. Such a conclusion is evident if one takes into account that the property of possessing a complex structure for itself does not automatically have to give an evolutionary advantage. On the contrary, already at a very basic level it is not a trivial problem („Eigen paradox") to explain for example how life shortly after its first origin and under perhaps still favorable environmental conditions (surplus of certain nucleotides, amino acids and other monomers; Ferris 1987) managed to avoid an adaptive regression to much simpler, i.e., more and more shorter RNA-stages which from then on should have remained much more successful than any other possibly more complex forms (Eigen 1971, Eigen and Schuster 1977, Maynard Smith 1983, Szathmary 1989). In general, the fact that organisms sometimes nevertheless develop toward higher complexity is usually connected for the simple reason that most of the "cheaper" places and niches of evolution, which are associated with less expenditure of energy or resources, are already occupied, hence a development toward more refinement and increasing complexity has often been selectively enforced over long periods (Saunders, Work and Nikolaeva 1999). In the case of the previously mentioned RNA-stage, this was possibly imposed by a persistent deterioration in the environmental conditions favoring the formation of membranes and compartments as a basis for the more stable hypercycle (Eigen and Schuster 1977).

At the same time, once some specific complications in an organism's structure have been achieved during evolution, they are effective in such a way that a loss of such characters, which are already interrelated in an advantageous way, very rarely (exception: secondary parasites, cave animals) have a selective advantage. Thus it is practically impossible for us humans to lose again all our genetically fixed mammalian characteristics to regress again back perhaps to reptiles or even—something which one could think about when one considers the advantages of our modern whales—back to fishes. Exactly the same of course holds analogously for all other large groups of vertebrates, for bony and cartilaginous fishes likewise as for birds and reptiles. The evolutionary theorist Rupert Riedl characterized this system-dependent intertwining of complex morphologic traits in multi-cellular organisms very persuasively as the result of a historically developed "order in living organisms" (1975), as the driving force of which must be understood as a special kind of internal selection. However, such an internal selection by no means represents a contrast or even, as sometimes was argued, a direct contradiction to classical Darwinian selection, which usually is thought to work only from outside the organism, hence the specification "natural" selection (for a clarification of this misinterpretation, see Dawkins 1986, chap. 11). On the contrary, as a complementary part of it, it is primarily concerned with all those numerous directing components of natural selection which, in many accounts of evolutionary theory, simply have been omitted for the sake of simplicity. In the words of Riedl:

We are dealing here with a selection that goes much further than the harshness that the environment alone would dictate. Nevertheless, it is not necessary to differentiate between an "internal" or "external" selection since every test result [= survival or death] arises from the confrontation of the subject of the test [= the organism] with the test instruction [= genetic blueprint]. ... This "internal" selection has the same relationship to the "external" as has a selection within a company to one within the market. It is achieved due to the demands of the market [= ecological niche] but also through the functional conditions of the product [= adaptedness of the morphological character within the body] and the organisation of the company [= the complete organism]. This finally leads to a canalizing autonomy of the rules of testing and tolerance [= physiological limits of the metazoan], yet even to a proper system-bound lawfulness [= network character of the genetic coding system] ([...] = my own remarks) (translated from Rupert Riedl 1975, p. 217/298)

However, in a similar way, this was already understood by August Weismann who, while speaking about a special kind of "intra-selection", wrote that "in the course of generations, by constant selection of those germs the primary constituents of which are best suited to one another, the greatest possible degree of harmony may be reached" (cited in Ridley 1982). In this way, internal systems as well as competition-based external influences reduce equally the potential possibilities and chances of a successful regression back to ancient evolutionary stages. Concerning humans it is this intricate internal "harmony" which explains why there is a relatively small probability that we will develop back toward one of our more harmless ancestors as for example rodent-like ones. Perhaps this may be regretted by some ecologically thinking people, but at the same time this also means that our intricate phylogenetic past has now imposed on us the burden of staying a primate and especially a human for the rest of our evolution.

The striking frequency with which, from time to time, specific evaluation criteria are erroneously associated with the theory of evolution has certainly to do with Darwin's somewhat unfortunate choice of the term "natural selection". This was also the reason why Darwin himself, on the advice of Alfred Russel Wallace, one of his closest supporters (see below), adopted a term coined by Herbert Spencer in 1866, survival of the fittest which was already very close to today's commonly used differential reproduction. Initially, Darwin proceeded deliberately from the perspective of a human breeder, who subjects different available domestic animal and plant races to a directed selection based upon some well-defined evaluation criteria. During the breeding, individuals judged as being less apt are consciously excluded from reproduction whereas "better" individuals are consciously promoted. If one transfers this undoubtedly very illustrative and thus popular picture to evolution in general, the result is a view where a higher authority in the style of mother nature in a similar way to the breeder tries to separate the better from the less well-adapted. No comparison could be more misleading than this one when it is used to describe the actual processes and mechanisms which take place during evolution, because what applies here alone are the physical and ecological framework conditions, which in the end either allow a certain form of life to exist or not. This was already understood by Alfred Russel Wallace, Darwin's somewhat forgotten co- or even better: pre-discoverer (cf. Wallace 1855) of evolution:

My Dear Darwin, I have been so repeatedly struck by the utter inability of numbers of intelligent persons to see clearly, or at all, the self-acting and necessary effects of Natural selection, that I am led to conclude that the term itself, and your mode of illustrating it, however clear and beautiful to many of us, are yet not the best adapted to impress it on the general naturalist public ... "Natural selection" is a metaphorical expression ... and to a certain degree *indirect* and *incorrect*, since, even personifying Nature, she does not so much select special variations as exterminate the most unfavourable ones (Alfred R. Wallace 1866).

There is no place here for the common belief that sometimes a certain trait or ability of an animal could be realized or executed in a much better way than it is actually the case or that even the same special feature theoretically could be optimized. The sole criterion in evolution is mere existence or survival through successful reproduction and hence it is indeed somewhat tautological even if it is in a rather positive sense if, for example, behavioral ecologists succeed in showing that any arbitrarily chosen organism is in fact not a perfect one, but, under the respective environmental conditions existing at the time, the best one of its kind (Dupré 1987). Of course, it can sometimes occur that, under specific circumstances, one animal species reveals itself to be not so well-adapted in comparison to another one and finally becomes extinct if for example both species are placed by chance in an environment where they have to compete for the same natural resources. However, such scenarios could be constructed all the time for any given species in such a way that so-called "better" species could be imagined over and over again. Additionally, theoretical optimizations which would make all existing organisms appear to be completely underdeveloped and non-adaptive are quite easy to invent. Be that as it may, viewed from the background of the only relevant criterion, that is surviving by reproducing, this is of secondary importance.

Compared with the conceptual closing of the *Neo-Darwinian synthesis* during the first half of the 20th century by people like R. A. Fisher, J. B. S. Haldane, S. Wright, H. Muller, T. Dobzhansky and E. Mayr it was only recently that such a purely quantitative-abstract account of evolution has been impressively confirmed by the aspect of adaptive *neutrality* of many biological modifications.

If we follow the Darwinian chief ideologist Spencer we would then have to say, what is relevant, is "the survival of the at least equally fit". This important conceptual change was mainly due to the Japanese population geneticist Motoo Kimura (1982) who first succeeded in showing convincingly the neutral selective value of a whole variety of genetic mutations. Hence first of all it makes no difference at all if a newly acquired trait is neutral, optimal or only half as perfect as it could be in theory. What counts here is only that no real disadvantage compared with the reproductive success of the remaining competitors emerges, be it as small as may be. Secondly, however, and this has only very recently been demonstrated in a quantitative manner by Peter Schuster and Walter Fontana (1998), neutral mutations are of great evolutionary importance because they alone allow organisms to shift away from otherwise possibly selective dead ends into new pathways of continuing adaptation. The example they used to prove this fact is very illustrative since it proceeded from the simplest relationship one can imagine that exists between genotype and phenotype. Under controlled frame conditions they calculated fitness ranks for a variety of tRNA-sequences (= genotype) which, by simultaneously forming different 2-dimensional molecules („shape" = phenotype), had been selected with respect to their similarity to a predefined target shape. The main results of the subsequent computer simulations showed that 1) continuous phenotypic transitions into one direction can be accompanied by highly improbable, i.e., discontinuous transitions in the opposite direction and 2) this explains why discontinuous transitions are often preceded by extended periods of neutral drift. As a consequence so-called "neutral nets" (Bornberg-Bauer and Chan 1999) of genetic sequences play an important role in evolution as to repeatedly provide open access to increasingly more improbable structures (cf. the nice metaphor "Climbing Mount Improbable" by R. Dawkins 1996).

Hence to be the "fittest" must not automatically mean to be simply the strongest, most brutal, most merciless, swiftest, biggest, heaviest etc., but can also stand for most efficiently moving, avoiding conflict, cooperative, reactive, farsighted, sensible etc.. What is permanently being tested during evolution—and this is also very often overlooked—is the aggregate system of the *whole* individual organism (*individual selection*; cf. chap. 5) and not some isolated components which perhaps, if taken by themselves, could be viewed as being optimal. Thus, it is important to make a clear-cut distinction between, on the one hand, the *ultimate* principle in evolution, that is differential reproduction, and, on the other hand, its concrete *proximate* realizations that are different physiological and behavioral mechanisms. As an example, let us take the case of any species of animal, be it a bird, reptile or mammal, where in a very general sense all its members which, in addition, usually live in groups have to compete for limited resources of very heterogeneously distributed food. For the purpose of our present considerations, it is sufficient to assume two basically different strategies of withstanding this severe survival and reproduction test. The first strategy would consist of simply concentrating on the process of food intake and its subsequent digestion, that is, essentially feeding more rapidly and efficiently. In contrast, the second strategy would consist of forgetting about optimizing the food intake rate for the moment , but physically attacking potential competitors instead in order to provide oneself with better and longer-lasting access to the resource in question. Interestingly, many people tend to consider the second strategy as being the real Darwinian one, instead of the first one, and this first one is sometimes even held to be non-competitive. And this simply because of the quite erroneous and superficial impression that indeed, as everybody can easily see, the animals do not quarrel about anything. However, already Darwin made clear that he used the term "struggle for existence" in a much more abstract manner:

I should premise that I use the term Struggle for Existence in a large and metaphorical sense, including dependence of one being on another, and including (which is more important) not only the life of the individual, but success in leaving progeny. Two canine animals in a time of dearth, may be truly said to struggle with each other which shall get food and live. But a plant on the edge of a desert is said to struggle for life against the drought, though more properly it should be said to be dependent on the moisture. A plant which annually produces a thousand seeds, of which on an average only one comes to maturity, may be more truly said to struggle with the plants of the same and other kinds which already clothe the ground (Charles Darwin 1859, chap. 3).

Obviously, this very popular association of physical aggression with Darwinism has to do with the widespread misunderstanding of biological evolution as only working with a combination of mere brutality and open personal selfishness. But that is way off the mark: the true interpretation says that, for example in the above case of our animals competing for food, both—proximate—strategies by being genetically transmitted from one generation to the next cannot avoid being subjected to natural selection and hence themselves compete for their continued existence in the course of evolution, because they will both have an impact on the—ultimate—outcome of their application, that is survival and reproduction. In reality, however, an often quite complicated mixture of different behavioral patterns called the *evolutionarily stable strategy* (Maynard Smith and Price 1973) decides upon the definitive use of a concrete behavior category under certain internal and external conditions. Evolutionarily stable here means that, at least for the current situation for the species, no other alternative strategy is able to spread through the population. Accordingly, concerning the real course of competition for resources in animals, a whole set of more or less independent factors such as spatio-temporal distribution, abundance, nutritional value, urgency, kind of food, size, modes of handling, possibilities of monopolizations etc. determine the selectively advantageous outcome in terms of applied mechanisms (for details see Archer 1988; Maynard Smith 1991; Milinski and Parker 1991).

For example, a very simple rule says that physical aggression can only pay in cases where its costs in energy and danger of being injured are at least lower than the benefits obtained from the resource (Brown 1964). One can very easily test this prediction by varying the amount of food given to tame water fowl in anyone of the numerous city parks around the world, from Central Park in New York City to the "Stadtpark" in Vienna. If you give very little to a group of already impatiently waiting animals, you will provoke some often quite severe conflicts, if, however, you practice playing the big spender, most animals will be too busy enjoying their good fortune to indulge in aggression at all (by the way, the same is exactly true for human siblings of a similar age: compare the relationship 1 toy for 2 siblings with 2 toys for 2 siblings; not to mention in politics).

For behavioral research, the credit for having both inspired and promoted the introduction of this kind of basic theoretical analysis should go to Niko Tinbergen (1963). Since then it has helped us very much to decide which of the ecological factors are indeed responsible for the evolutionary adaptedness of a continually growing number of behavioral patterns from all kinds of domains (Krebs and Davies 1991). In particular, it now also regularly prevents most inexperienced researchers from making a mistake, which was still very common among many classical ethologists, namely to very rapidly and in most cases erroneously conclude from an apparently "non-perfect" behavior its non-adaptedness. Still during Lorenz's time, such considerations had the power to launch a quite extensive, but in the end absolutely fruitless discussion about the serious problem of "luxuration", that is the formation of expensive traits for which no immediate biological function could be specified (within this trend, even the expansion of the human brain was sometimes seen as a kind of unnecessary by-product of other morphological or physiological changes). In a somewhat comparable, but more moderate form, the discussions around "exaptations", a term coined by Gould and Vrba in 1982 to designate structures which fulfill a concrete biological function without having been explicitly selected for that purpose (example: turtle flippers used for sand-digging during egg laying) continue such a rather anti-adaptionist tradition up to this day (for the case of the brain, see Skoyles 1997). Today, however, it is much more Tinbergen's quite prophetic prediction which increasingly proves to be the correct one for the behavioral sciences:

„As one becomes better acquainted with a species, one notices more and more aspects with a possible survival value." (Niko Tinbergen 1963, p. 422)

Or, to finish again with the master himself Shakespeare: To be or not to be, which by itself is the decisively important question.

4 Chance as Necessity

Does evolution depend on random
search?

C. H. Waddington

Some admirable intellects, even today, appear still not to have accepted
or even understood that selection
alone has produced the whole con-
cert of living nature. Selection operates namely *on* the products of
random chance since it cannot feed
itself from any other source.

Jacques Monod

How does evolution actually function? Charles Darwin was the first to recognize that natural selection is the crucial factor in the environment: it is responsible for the fact that the variously endowed organisms, which stand in direct competition with one another, have correspondingly different probabilities of reproductive success. The logical consequence of this is that even relatively minor differences in the biological fitness of organisms lead to numerical changes in the composition of animal populations; these changes can ultimately even lead to the development of new species. The total of all physical
and ecological environmental conditions—in the form of natural selection—
therefore decides the evolutionary fate of an organism.

As indicated in the previous chapter, every animal that actually exists here and
now is at least theoretically endowed with the exactly same potential, even
though certain negative or positive trends can often be anticipated. Should an
entirely new species or merely a new type within the species assert itself, then,
after the fact, "better adapted" means little more than the mere observation
that the novel individuals have increased their numbers to the detriment of
other individuals that turn out to be less well adapted. This is the only acceptable formulation that avoids an anthropomorphic valuation. It was the discovery of this underlying correlation that encouraged Darwin to speak of a
theory of the origin of species and, concurrently, with one fell swoop to challenge all existing religious creationist teachings as well as other scientific crea-

tionist theories. For example, the catastrophe theory of the French zoologist and paleontologist Georges Cuvier (1769—1832) still held that natural catastrophes wiped out the entire fauna and flora at the end of each geological period, so that new creation had to take place repeatedly. Although the concept of repeated new creations is fundamentally incorrect, the idea of occasional perturbations of much of the world of living things—triggered by dramatic changes in prevailing selective pressures—already comes quite close to the more modern hypotheses of Gould and Eldredge involving the so-called punctuated equilibrium in evolution (Gould and Eldridge 1977, 1993; Elena, Cooper and Lenski 1996). Thus, Darwin's selection-based considerations have retained their validity to this very day, and the degree to which he predicted a fundamental element of modern evolutionary theory is astonishing indeed. While natural selection is a cornerstone for the evolution of life, it is based on prerequisites that it cannot explain, namely the variability of living organisms. In Darwin's time, this variability of life was considered to be a given fact—a phenomenon plainly visible in nature. The problem therefore seemed to be solved. The material on which natural selection acted was present in abundance. Thus the entire academic effort was directed at what might be referred to as a secondary issue, namely on the struggle for survival as an ensuing process. Today we know that the variety of living organisms has a single, common root, namely the minute changes in their genetic material, which can lead to a corresponding changes in form, metabolism or behavior. The chemical uniformity of the genetic material, specifically of the nucleic acids, which are relatively large and complexly structured biomolecules in the nuclei of organisms, has provided the evidence that all existing forms of life most probably originated from a single or from very few primeval systems. Darwin was among the first to have surmised as much, yet without ever having been able to present concrete evidence. The precise nature of the change of these nucleic acids, in which once more a largely uniform principle (the genetic code: three nucleotides known as a codon function as a code for a specific amino acid) constructs the cells and bodies of organisms, remained in the dark for generations and was first deciphered with the advent of modern molecular biological methods. Prior to this, speculation reigned, whereby two major groups of theorists long held irreconcilable positions. One school saw a more or less oriented tendency in the change, whereas the other tended to view these changes as the result of chance influences on the organism. The former came to be associated with the name Lamarck who died when Darwin was only twenty, and he is often portrayed as Darwin's direct antagonist. This is only partially true because Darwin himself was equally taken, as Lamarck was, with the notion that the evolution of organisms was driven by some type of as yet unexplained inheritance of acquired characters. Darwin's long defunct explanatory attempt, which involved so-called gemmules (minute particles of somatic cells), held that the acquired information was somehow transferred into the germ cells by a mechanism termed *pangenesis*. These considerations are actually less eccentric than they appear at first glance: after all, they are in principle based on the notion of particulate, i.e., molecular units of heredity—long before Weismann. The idea itself, however, was as simple as it was incorrect. As we can simply observe, animals adapt to a range of different and often hostile environmental conditions over the course of their lives. Darwin and his contemporaries only logically concluded that this ability or knowledge about the environment could in some form be passed on to the offspring. This was manifestly demonstrated by the fact that many organisms had clearly become ever better adapted to the respective conditions over the course of evolution. This concept was closely aligned with the notion of directed evolution, which soon proved to be diametrically opposed to later interpretations. The second group of scientists, who took a skeptical stance toward the idea of any directed, much less an overriding principle, first developed with the advent of modern analytical methods in biology. The insoluble problem that had divided scientists for generations was first resolved with the discovery of genetic mutation. Thus, all the changes that Darwin still had to accept as somehow given in nature, turned out to be attributable to haphazard changes in the underlying biochemical structures (Luria and Delbrück 1943). The geneticist and Nobel Prize winner Jacob Monod very poignantly described this act of disillusionment in a book titled "Chance and Necessity". With the typical French esprit of a radical existentialist, he bluntly drew the logical conclusion, namely that humankind as well as the phenomenon of life itself were cast into this inhospitable world by an act of mere chance. Vehement negative reactions from all philosophical corners showed that these new insights from the natural sciences did not meet with enthusiasm. In the present context, the political-ideological and the philosophical discussion is of interest only insofar as it shows that something akin to a universal societal taboo had been breached. The notion that life, in particular human life, can only change or be changed by pure coincidence is clearly discomforting to many people. Evolution through natural selection is a process that one might be willing to accept, but haphazard and unpredictable change as a foundation for our own development? This subjective fear of "senseless" genetic mutations shakes the foundations of how we see ourselves. Ultimately, a future type of human being would have to be viewed as the result of precisely such a senseless process. Not even the pre-Darwinian "evolutionary thinker" Herschel (1861), who actively promoted Darwin's ideas, was ready to accept this underlying haphazard character of biological evolution. In this sense, he failed to—or chose not to—understand the very essence of the theory:

We cannot accept the principle of capricious and random variation and natural selection *as such* as an adequate explanation for the present and past organic world, just as we cannot accept the Laputan method (note: random method in *Gulliver's Travels*) of writing books (to take things to extremes) as an adequate explanation of the coming into being of Shakespeare and the *Principia*. In both cases equally, there must be a purposeful *intelligence* involved in order to give direction to the changes—to regulate their size, to limit their size and to guide them further along a certain path. We do not believe that Darwin rejects the necessity of such intelligent guidance. However, as far as we can see, this does not enter into the formula for this law, and without it, we cannot imagine how the law led to such a result. ... Accepting this and with some reservations about the origin of mankind, we are a long way away from refuting the view of this mysterious object expressed by Darwin in his book (John F. W. Herschel 1861, p. 12; retranslated from Engels 1995, p. 88/89).

The fundamental and highly emotional antipathy to chance as a creative principle in the evolution of life is far from being overcome today. This is manifested in a range of efforts, many stemming from the natural sciences themselves, for example from biology. These mostly involve the claim to have discovered a new basic principle in nature—one that is not so entirely based on mere chance. Alternatively, others claim to have developed some type of compromise between Monod's chance and an overwhelming biological directionality of evolution. Nonetheless, such recurrent, fundamental changes or putative refutations of the evolutionary theory in its current form generally disappear as fast as they appear. What remains once more is mutations as uncannily persistent, haphazard changes in the genetic material. Clearly, quite a number of people continue to have great difficulty with the notion of evolution based purely on chance. One explanation is the fact that other cogwheels in nature that interest us make such an orderly impression. The mere idea that pure chance plays a crucial role is reminiscent of a technical engineer who, in attempting to analyze and, above all, understand an unfamiliar machine, concentrates on entirely peripheral noises. The analogy is not perfect because precisely this (among many other factors) marks a key difference between living and artificial systems. Life evolves and changes itself in an entirely unplanned manner („the living machine differs fundamentally from other machines in that it builds itself" A. Weismann 1904, Vol. 1, p. 329), whereas machines constructed by humans are entirely dependent on their creator to change them in a minutely planned manner.

This peculiar discrepancy between the so clearly visible harmony and order of living nature and the notion of haphazardly driven change thereof can be easily rectified by more closely examining the level at which the demon chance in fact rules. Many people who hold creationist views envision that an organism is entirely dismantled and, much like in a lottery, its individual pieces then tumbled and mixed haphazardly in an attempt to produce a new, improved version. This is simply incorrect. Clearly, the chances of such a procedure meeting with success would be minimal. In fact, it would never even have given rise to life in its simplest forms, for example to self-reproducing molecules. Most scientists have reached agreement here, and repeated objections of this type have become somewhat tiring because all are based on entirely unrealistic assumptions, much like Herschel's purportedly crushing "Laputan blow" against Darwin (see above). The key point in the modern interpretation of the evolutionary theory is that, although nature is characterized by sheer immeasurable structure and order, its evolutionary change can only proceed through a diametrically opposed force, namely through pure chance. While chance may be constricted in many respects, it ultimately holds sway because it determines which changes occur and which do not.

Finally, a simple theoretical consideration can be used to demonstrate the compelling necessity of chance in the evolution of living organisms. We merely need to turn the earlier argument on its head. For the sake of simplicity, let us assume that both plants and animals can take fate into their own hands and thus, in a very targeted and deliberate manner, change their body plan, their body functions and their behavior. Accordingly, mutations of the genetic material could be directed to the probable advantage of the respective organism, not by pure chance as was once the case, but, logically, such advantages could be achieved at every individual step. The consequences of such a possibility would be as improbable as they are fantastic. On one hand, every organism could purposefully change itself to take ever better advantage of its environment. At the same time, this would trigger an enormous contest toward a life form that is absolute in its perfection and de facto immortal. Accordingly, any organism that ever succeeded in actually implementing such a fundamentally new, previously unheard-of method of evolutionary progress should would become absolutely immortal, angel-like beings immune to any of the potential problems of this world. At the same time, evolution in its true sense, i.e., via uncertain changes and therefore potential adaptation to ever new environments, would also be a thing of the past. It is reassuring to note that no such wondrous being has yet been identified. Moreover, human beings, who like to view themselves as unique on the evolutionary playing field, have also failed to make this miraculous step—even if some of us would like to believe so.

5 The Indivisible Individual

No one supposes that all the individuals of the same species are cast in
the same actual mould. These individual differences are of the highest
importance for us, for they are often
inherited, as must be familiar to
every one; and they thus afford materials for natural selection to act on
and accumulate, in the same manner
as man accumulates in any given direction individual differences in his
domesticated productions.

Charles Darwin

Living organisms change in the course of many generations due to mutation
and selection. To portray this as the "mechanism" of evolution as has been
done in the past and how it usually appears in many contemporary biology text
books is, strictly speaking, confusing. The term "mechanism", borrowed from
the language of human technology, implies that a product is made from certain materials in the same way as a machine. The results of this process are then
called living organisms and represent the provisional products of the manufacturer who, in this case, is evolution. No matter how lucid this metaphorical
representation may be, when used uncritically it has proved itself to be just as
confusing as most other expressions borrowed from human affairs. Animals
are not subject to a superordinate or completely methodical controlling mechanism which allows them to become increasingly better adapted to living on
this planet. Animals live, reproduce, change, and, while doing this, experience
various destinies some successful, others less so—and that is the whole
story, that is all there is to it. The result is called biological evolution without
any special principle or natural law having to be associated with it. Again, cau-
tion is advised with regard to old entrenched terms and concepts.

That the genetic make up of an organism must have a very special significance for its evolution has been known since the discovery and the first exact
description of the corresponding hereditary molecule by James Watson,
Francis Crick, and Maurice Wilkins. For this, they were awarded the Nobel

Prize in 1962. From that time on, genetics—the investigation of the genetic make up of living organisms—has played a primary role in fundamental research in biology. How important an organism's genes are to evolution becomes clear when one attempts to imagine it without these basic control units of life. It would quite simply not function since these genetic elements are inseparably interwoven with practically all the functions of life. Genes are responsible for the production of the body proteins, for almost all the subsequently synthesized molecules in the body, for the coordinated interplay of all these elements—and consequently for all the levels of complexity up to the harmonized reactions of the whole of the animal's body. Now, one could deduce that the gene alone is the material on which evolution takes place and the rest of the various molecules and structures found in the organism are there only for the purpose of guaranteeing the optimum conditions of life for their creators, namely the genes. Richard Dawkins carried this reductionist viewpoint to its ultimate conclusion when selfish genes use, or one could even say misuse, the body they have constructed simply as a "vehicle" for the purpose of their own multiplication. If one observes the spread and multiplication of certain genes in isolation, e.g., for hair length in mammals or their body size, etc., over many generations, such a picture does indeed emerge. The individuals themselves perish but the genes remain (potentially) immortal. The picture changes drastically, however, when we put the example to the test and then observe what the genes are capable of doing at all when they are placed on their own. Such a marvelously conjectured emancipation then founders without any kind of rescue being in sight.

Let us imagine the decisive experiment, a so-called experimentum cruci, in the following way: we remove from the nucleus of the cells of a living organism all the DNA (deoxyribonucleic acid) present, i.e., all the nucleic acids which alone are responsible for the genetic characteristics of the organism. The result of such an experiment will naturally be disappointing since the whole of the valuable hereditary material will not, as would otherwise be the case, manifest itself in a creative self-organizing way as a living organism but will drift along towards its complete dissolution. In other words, the extremely complex information contained in the genes emerges, or has meaning or significance only when it is integrated in an existing organism. Especially, it is the proteins and other important molecules (carbohydrates, fats, cofactors, etc.) that, in cooperation with the genes, produce what we call life characterized by its intransigence towards exigencies of the environment. Genes on their own have no chance in this conflict and this is especially true of a single isolated gene. That the gene selection theory is an disallowed reduction of complex biological phenomena to an apparently simple chemical doubling process, like the so-called autocatalytic process which occasionally occurs in the inanimate world, is demonstrated more by the fact that even the definition of the gene with all its effects in an organism is completely dependent on the exact structure of the whole genome and is therefore inseparable from the dynamics of all the participating structures and processes. Proteins do not fare any better in this respect for they are also dependent on this structure, coded by the corresponding genes, they are produced as a consequence of many independent steps. One could now, in exactly the same way as for the gene, consider the reproduction of the proteins, which double in a manner which is mirror image of that of the genes, as the major principle of evolution. In practice, this was the very first thing to happen, since it is logically possible, using a family tree of the proteins, to construct an almost perfect mirror-image representation of the genetic relationship. This is true despite a certain amount of blurring due to the presence of 64 possible codons (nucleotide triplets; 3 of them, UAA, UAG, and UGA, are so-called stop-codons responsible for stopping protein biosynthesis) for coding only 23 naturally occurring amino acids (e.g., leucine, arginine, serine with 6-times redundancy). In other words, it is basically possible to use the phenotype to explain the genetic correlations and vice versa, to use the genotype to explain at least some of the molecular phenotypic correlations (Doolittle 1999; Holm and Sander 1996; Tatusov, Koonin and Lipman 1997). In this way, the basic methodological equivalence of the phenotypical and genotypic approaches for producing family trees has been recognized for some time by most systematists (Dayhoff 1969). This is even admitted, at least in part, by some earnest skeptics of the methods used by geneticists, even when there remains a certain unmistakable bias:

Phenotypic correlations between characteristics can be determined with greater accuracy and they are much easier to obtain than their genetic equivalents. If genetic correlations are well-determined, they tend not to differ significantly—neither in their order of magnitude nor in their structure from their phenotypic counterparts (J M. Cheverud 1988, p. 966).

In reality, it is the genetic analyses which are simply many times more precise than the purely anatomical ones which naturally, just from the statistics, yield far more reliable results (for an impressive attempt of synthesis of both approaches in the case of placental mammals, see Liu et al. 2001; for still existing methodological problems of the purely molecular approach, see Naylor and Brown 1997). An elegant example of this is the recent confirmation of the maternal-mitochondrial (mitochondria, important providers of energy of a cell, are passed on to the next generation only by the egg cells) "Out of Africa" hypothesis (Cann et al. 1987, Lewin 1987, Nei and Livshits 1989) by the reconstruction of Adam's African Y-chromosome roots (overview in Gibbons1997a, b; Wood 1997; Hedges 2000; for the still open alternative of an additional multiple origin, see Wolpoff et al. 2001). A family tree analysis could hardly be more accurate than this and this is over a period of as long as 200,000 years. It is therefore no wonder that lately, in contrast to Cheverud's opinion, we can determine a departure of molecular taxonomy from the good old family tree. An up-to-date example for this is the separation of the arthropods, (e.g., insect), from the annelids, (e.g., earthworm). Up till then, they were always placed in one group (clade). Or then the redemption of the world-famous, fleshy-finned coelacanth fish, *Latimeria chalumnae*, only discovered in 1938, from its position as a "living fossil". This position has now been taken over by the lung fishes—the group of fish credited with growing legs and coming ashore—the transition from fishes to the first land vertebrates (for an overview, see Roush 1998). Every such new classification naturally brings inevitable and important consequences for all phylogenetic research by which the whole area of study automatically profits (Graham 2000; Klicka and Zink 1998; Rosenzweig 1998).

What is now the most important—the "selfish" genes or the "subjugated" proteins? Both are indissolubly bound together in their structure as well as in their function (cf. PROTEOME. http: //link/service/journals/10216/index.htm), which finally can only be defined by the wholeness of the organism. However, one significant, apparently immutable difference has established itself during the course of evolution since only the genes are subject to a certain mutability over the generations upon which the changeability of protein structure depends completely. In principle, it could have turned out completely differently in the early days of the evolution of living systems, perhaps just the opposite. So, for example, the proteins-first concept, in contrast to the nucleic acids-first concept advocates that proteins and not nucleic acids were the first independent replicating molecules (Dose 1982). From time to time, this thesis has become of interest, for instance, through new results of investigations into, at least partially, self-replication of short protein segments called peptides (Lee 1996, Saghatelian et al. 2001) or, currently more relevant, the infection-like transmission of a changed 3-dimensional configuration of complete proteins—as given in the dangerous "prions" (= "proteinaceous infectious particle") epidemic (see Kretzschmar 1997; discovery by Prusiner et al. 1992) whose particular mechanism is still disputed (Aguzzi and Weissmann 1997; for the possibly important capacity of temporarily "hiding" deleterious genetic variation, see True and Lindquist 2000).

Nevertheless, the two different kinds of nucleotides, i.e., RNA which very probably arose first during biogenesis and DNA which is a later, through a quite intricate process of chemical reduction, substantially more stable form of nucleic acid, have managed to retain the role of being in charge of driving evolution forward. By contrast, proteins as the much more versatile chemical catalysts—compare the set of 23 natural amino acids with the more limited "alphabet" of only 4 nucleotides (cf. Szathmary 1999)—seem to have always been responsible for the whole variety of diverse physiological "homework" to be done within the system and at the same time as comparably unreliable replicators remained dependent on the presence of nucleic acids taking care of, that means "coding" for their particular molecular structure through the definition of the sequence of amino acids (for a detailed discussion of this fascinating topic, see Freeland, Knight and Landweber 1999). In this special sense it is permissible to speak of both types of nucleic acids as being the important carriers of information in the biological realm. However, it is still the presence of a complete functioning living system which alone legitimizes the use of such a privileged designation.

Nowadays, when a biologist speaks of the primary importance of the organism for the evolution of living forms, he is suspected of wanting to revive an antiquated concept which has already been replaced by modern expressions from molecular biology and biochemistry and has become redundant. This picture changes the moment when we realize that the term "organism" means nothing more than the physicochemical definable expression "living system" with which we include a multitude of very different living forms. Naked viruses, the simplest bacteria without a nucleus, astonishingly complicated protozoans and also complexly structured metazoans are those individual systems which are involved in the life and death struggle of evolution. The single individual as a physically distinguished unit of living phenomena undoubtedly plays the most important role. "Individual" here can be understood literally as that "indivisible thing" on which the harmony and wholeness of life can suddenly be destroyed. In principle, this completeness, in the majority of scientific approaches to investigations and explanations of the phenomenon of life is a precondition even when it is not often explicitly stated. A remark such as the one given by the experienced chemist, when asked for his opinion, summarized his research with the pregnant words, "All life is chemistry", is very likely to initiate a certain emotional resistance. Nevertheless, it is very improbable that only one of the many biochemists who are seriously involved with the coming into existence and functioning of living systems would go so far as to say that simple chemical reactions independent of the whole system, or organism, could constitute the essence of life. The complete system is mostly not expressed but is automatically taken into account because it is so self-evident. So, for example, the gene selection theory, which says that it is sufficient with just individual selfish genes to explain evolution, has as a prerequisite that a complete cellular replication apparatus is available, without which not a single gene can reproduce itself. Even Dawkins, the most well-known protagonist of this view, recently conceded that the concentration of his thinking on one single gene only represents another perspective on the complete individual and in the long run, both views are different ways of saying the same thing (Dawkins 1994).

From a completely different view of things with reference to the unit of selection, we however must distance ourselves as rapidly as possible. The concept of group selection, as was first proposed by Wynne-Edwards (1962) and today is again advanced by people like Wilson and Sober (1994), states that natural selection works not so much on individuals but above all on groups for their fitness for survival. It is thus claimed that "group selection" is more effective in bringing about changes than individual selection, hence the biological fitness of the single individual may suffer since it is of secondary importance:

Altruism can evolve by group selection even if it is hindered by individual selection (David Wilson and Elliot Sober 1994, p. 640).

That such a view already contains an obvious logical inconsistency on a purely theoretical level can be demonstrated by a simple evolutionary scenario. We assume that each of us belongs to such a group which meets its fate decided by a real group selection. Simultaneously, we presume that the readiness to take part in an altruistic, co-operative manner in group formation has a genetic basis. Consequently, we would have to reckon with at least a slight reduction in our personal fitness and a concomitant long-term but noticeable drop in our personal reproductive rate. This would inevitably mean that, in time, exactly those most important to the group, namely those motivated for building the group, and above all, individuals capable of building a group as a part of the population would disappear. In the long run, this is identical with a disappearance of the group. In addition, we can now see very well a problem which is not at all easy to solve, namely that of defining the identity of a group with comparable precision to that with which we define multicellular individuals: a permanent and compulsory union of genetically identical cells. Are two people sufficient or must it consist of at least ten, or does it have to be a unit of related individuals, e.g., a family or clan, and are 10 million, e.g., a small country, also a group? If the group to which I belong breaks up, has it "died" and does it form a new one when I simply transfer to another? A whole barrage of never-ending foolish questions, as one can see.

The formation of social groups can only be understood, if at all, only through the effects of individual selection, since it needs decidedly evolutionary stable types of individual who are prepared to submit to the various conditions in order to enjoy the selective advantages of group living (the attempt at a conciliatory, but nevertheless empirically untenable "compromise solution" to this controversy can be found in Dugatkin and Reeve 1994). Then groups can arise without the need of mysterious group selection. These arguments are also valid for all those amusing speculations about a possible "species selection" which, during the course of evolution, has supposedly safeguarded the welfare of individual species. Without stable populations, there are no species, without stable groups there are no populations. However, without stable individuals, there can be neither groups nor populations, not to mention species. Therefore, below the level of the individual organism, the relationships are turned round as if on an axis of symmetry to give its exact opposite: without a stable organism, there is, in contrast to the individual who leaves the group and continues to live, no individual continuation of life for the single cells of the body and certainly no further existence of an individual gene (Heschl 1994a). Hence, in the end it is all a question of pure physics, nothing more.

However, that does not mean that, by the integration of previously independent individuals, new types could not arise. Just the opposite, a complete row of important evolutionary transitions are characterized by such an integration, so, for example, the coming into existence of the modern eukaryotic cells, which, with great probability, were created by a stable fusion of prokaryotes (cf. Margulis 1970: Mitochondria, Chloroplasts and Cilia as Endosymbionts). In addition, that peculiar group, the lichens, i.e., fungi which formed an association with certain algae, and, naturally, also the multicellular bodies which emerged from such a mutually beneficial union. The possibility, that in principle, a "composite animal" can arise was already recognized by the unappreciated termite researcher, Eugène Marais (1976, p. 66) who, at the same time, understood very well that the fine difference between a termite colony and a metazoan must be sought in a special still-to-be-reached characteristic of the latter: "Should natural selection advance further, we could experience that the termite mound sets itself in motion and wanders slowly over the veldt" (p. 89). The siphonophores took this evolutionary advice seriously and, sometime in the past, they released themselves from the seabed and now, in the meantime, even swim in a coordinated way through our seas. Potentially, however, each individual animal can still, when artificially isolated, independently grow into a new colony. This means that a 100% dependence on the whole system—as it is in the metazoan—does not yet exist. The most decisive factor is the narrow physical and the concomitant causal connection of the system components. These are, at first, a loose assemblage of previously independent units which then develop into a new kind of individual. This is naturally not the case with termite nests or colonies of ants. Wynne-Edwards and his followers, in this respect have overshot the mark, while Dawkins, with his exaggeratedly reductionist analysis, has broken everything down in order to finally let the mysterious memes, like resurrected mental genes, rise from the evolutionary ashes (Dawkins 1976, Chap. 11). The living organism alone has all what is necessary—a clear symbol of its evolutionarily proven robustness—to survive completely undamaged.

Evolution is therefore always related to individual, distinguishable and consequently physically separated units which, according to personal preference, can be called organisms, as has been the case up till now, or, in modern parlance, designated as living systems. These systems are characterized, when considered physico-thermodynamically, by the extraordinary stability and structural robustness of all their underlying chemical processes. Or, as the systems theorist Paul Weiss would put it:

A system could, therefore, be defined as a complex unit in space and time so constituted that its component subunits, by "systematic" cooperation, preserve its integral configuration of structure and behavior and tend to restore it after non-destructive disturbances (Paul Weiss 1971, p. 14).

This robustness is demonstrated by the astonishing ability of living organisms to overcome disadvantageous or even dangerous environmental influences and therefore to live as long as possible, and, what is even more important as far as evolution is concerned, the ability to reproduce physically and therefore to achieve a kind of potential immortality. Consequently, living systems can justifiably be regarded as the first real systems on this planet which—in contrast to non-living material which is subordinate to the thermodynamic equilibrium with the environment—are equipped with the extraordinary uniqueness of retaining their individual identity in space and time, and therefore of having the first reconstructible life history in the form of biological evolution.

6 L = C*

We now find ourselves at the fundamental level which leads from chemistry to biology: here a completely
new quality emerges which does not
exist in the terminological world of
physicochemistry, where one speaks
of material interaction, of atoms,
molecules or crystals, of forms of
energy and their transformation. The
quality is information.

Manfred Eigen

The Shannon information concept
does not say whether a message is
meaningful or meaningless, valuable
or valueless, i.e., it is not concerned
with meaning at all, in other words,
the semantics are lacking. It is precisely in the field of biology that this
lack is a significant shortcoming.

Hermann Haken and Maria Haken-Krell

The biological sciences, as those disciplines most intensely involved in the investigation of living systems, have, since the middle of the 20th century, made
great strides comparable with the great discoveries and development achieved
in physics in the first half of the 20th century. Nowadays, modern genetics,
molecular biochemistry and physiology are developing at such breakneck
speed that it has become impossible to keep an overview of what the detailed
state of knowledge is even in very narrowly defined fields. Nevertheless—and
this fact must amaze every person interested in science—there are not a few
basic questions of biology that still remain unanswered to a high degree. No
more so than in the difficulties faced in the investigations into the origin of life
itself (*biogenesis*). Although innumerable elemental processes of the phenomenon of life have been explained in the greatest detail, e.g., the basic biochemical relationships and interactions between the nucleic acids which make up
the genetic material and the proteins which make up the cell bodies, up till now

---

* (L ... Life, C ... Cognition)

nobody has managed to work out a completely convincing model of the origin of life. This is expressed by the fact that no-one has not yet succeeded in producing a man-made living system in the laboratory. A somewhat different, but by its close proximity to natural systems perhaps more successful experiment has been recently set up by Peter Schultz at the Scripps Research Institute in La Jolla. He is trying to produce living systems artificially with more than 4 nucleotides and derive proteins from them with more than the 23 amino acids normally used in living organisms (report in Service 2000). Ignoring the fact that there are not a few promising theoretical and also empirical attempts to explain the phenomenon of life, producing such systems artificially from non-living components remains an unsolved problem. Is life therefore an insoluble puzzle of Mother Nature? If we use as our basis, and as biologists we have good reason to, that, at one time, conditions must have existed without any kind of life, we would be ill-advised to accept the origin of life as indeed a wonder of nature and to refrain from setting up any more experiments. That for a long time there were no living creatures on the Earth and then at one time living organisms become noticeable makes it inevitable that there must have been a transition from the inanimate to the animate and that is an unshakable fact. The acceptance of a supernatural creation only begs the question and leads further to the problem of the origin of the Creator and make things even less understandable instead of deliberately trying to bring the required solution within reach. The actual difficulties of investigating the origin of life lie in the novel qualities which, with this phenomenon, were able to manifest themselves on the earth. In particular, the new and extraordinary ability of autonomic self-organization, an ability which could only be ascribed to the inanimate "not-yet systems" to a very small extent. The special feature of the life processes can be found in the unusual robustness and stability of the integrated completeness of all the processes involved—which is a possible explanation of the extreme difficulties we face in trying to simulate them experimentally. Obviously, life must have originated in a critical phase practically from alone, and every additional interference from outside has the opposite of the desired effect, namely it does not allow the system to begin living by itself. That the origin of living systems could have been an event with a critical threshold was suggested by the biochemist Klaus Dose:

According to the views developed here, the first appearance of a primitive system which could be considered as living must be regarded as a sudden step (Hans Kuhn and Jürg Waser 1982, p. 901).

We must therefore simulate very exactly the complex framework of conditions which must have existed over a very long period at the time of biogenesis if we are to see the wonder of the origin of life in the laboratory. The time factor will play a very important role since it is not possible to exclude that first a certain sequence of complex chemical processes was the signal for the countdown to the emergence of life. The model of the *hypercycle* developed by Manfred Eigen and Peter Schuster which postulates the spontaneous and, above all, cooperative integration of previously independent replication units consisting of nucleic acids and their accompanying proteins points exactly in this direction. A *hypercycle* comprises a permanently stable sequence of a definite number of macromolecules which give mutual support to each other during their reproduction [e.g., $N_1 \Rightarrow P_1 \Rightarrow N_2 \Rightarrow P_2 \Rightarrow N_3 \Rightarrow P_3 \Rightarrow N_1 \Rightarrow ...; N_i =$ nucleic acid, $P_i =$ protein], through which, a random linearly causative process in the realm of the still inanimate becomes a permanently cyclic self-organizing living process (Eigen and Schuster 1977). The theoretical chemist Walter Fontana, in the meantime, has attempted to describe this obviously difficult to control emergence of a permanent and stable dynamic system from unstable preceding phases by purely mathematical means. It appears that we will indeed soon be in a position to be able to explain what in scientific jargon is almost informally and trivially designated as a "biological structure" by means of a special kind of self-organizing origin of life formula which will only need confirmation by molecular chemistry (Fontana 1994). It will be a long time, however, before we will have "live" experience, so to speak, of the sublime moment of the origin of life, not in the usual form of human parenthood but in the real meaning of the word as the spontaneous emergence from inanimate materials.

The unexpected problems with which we are confronted with the moment of creating life in the laboratory leads us to another problem. To the present day, there is in fact no common universally accepted definition of life. There are a number of lists of obviously essential prerequisites and characteristics that are associated with life but, nevertheless, the quintessence of the phenomena still seems difficult to grasp. There are even some researchers, because of these considerable theoretical as well as experimental difficulties, who hold the view that, in principle, it cannot be defined exactly with scientific concepts. For example, the biochemist Klaus Dose comes to the conclusion:

At the moment, we as scientists cannot give a comprehensive definition of a living system. We are only in a position to ascribe certain characteristics to living systems. Many characteristics of living organisms, however, are not sufficiently researched and cannot be specified exactly. For this reason, a listing of characteristic properties of living systems still remains incomplete today. ... It will remain a matter of definition or opinion as to whether such a system is considered "living". We recognize that a clear-cut boundary between "living" and "non-living" does not exist at all (translated from Klaus Dose 1982, p. 948/949).

Which exceptional characteristics of living systems can be listed? Here follow perhaps the most important ones which are continually repeated when life is described from a scientific perspective:

1. Metabolism of substances from the environment for the maintenance of the internal equilibrium of substances and energy (homeostasis)
2. A demarcation between the system and the environment in the form of a membrane
3. Ability to reproduce identically
4. Passing on genetic material to the offspring
5. Possibility of evolution through random changes of the genetic material

Essentially, these properties paraphrase the various aspects of the same special ability consisting of maintaining their purely physical existence and their own structural identity. Beyond this, associated with the ability to reproduce is an aspect of immortality which serves as a basis for the possibility of Darwinian evolution for systems awoken to life. The question now arises as to whether life can be characterized by the sum of a row of special singular properties or after all by some fundamentally new property, not previously encountered in the initial stages, which manifested itself with the origin of life. On closer examination, this question—against all reductionist skepsis—can only be answered with a very clear yes.

The maintenance as well as the passing on of the physical structure—here we mean the sum of all the life processes of a living organism—presupposes a characteristic which, indeed, proves to be novel in the history of the cosmos: the coming into being of information, or in other words, the active knowledge of the how of maintaining one's own existence. With good reasons, this characteristic on its own can be declared as that very special novelty of the phenomenon of life and all the other clearly subordinate characteristics are an integral part of it. This means that for life something like a material knowledge of its own existence and its relationship to the newly emerged environment of the organism is the deciding factor. With the first appearance of a living system as the first aware "subject", the "object" appears simultaneously, namely a specific environment and with it, probably one of the most astonishing relationships within the whole of cosmic development. Before, where only a completely passive submittal to adjustment processes was possible, e.g., a rock rolling down a hillside, now we suddenly have a completely novel ability of actively existing with the background of an outside world which, just as suddenly, has been miraculously created from nothing.

It would be presumptuous to claim that this novel quality of the phenomenon life had not yet been recognized by scientists. Just the opposite is true—most concepts already implicitly include this new quality in some way or another but without fully appreciating its fundamental significance, instead they consider it to be a subsidiary property. As a rule, the importance of passing on the biological information from one generation to the next is talked about as though it were primarily just to ensure the identical reproduction of certain material particles—the genes. In reality, the result of a successful replication or doubling of the genetic material only represents a prerequisite and not the realization itself of this new quality:

How should one define life? With one exception, it seems arbitrary to consider one or other of the steps previously described as the boundary between the animate and inanimate and this applies to all the steps still to be described. The exception is the very first step of all— the first appearance of systems which can reproduce themselves and mutate and through this are subject to selection. This step is the only one which leads to the creation of a novel property of matter, whereas all the subsequent steps are gradual improvements of this new property. This property, whose presence was previously not even hinted at, manifests itself with the existence of systems, ..., that one can describe as *capable of learning*, systems that begin to carry information, i.e., a meaningful message whose content grows in size as the learning process progresses. ... Although this definition of primitive life may seem grotesque, it has to be conceded that it is the only one which is completely logically unambiguous. Definitions related to later steps appear groundless (translated from H. Kuhn and J. Waser 1982, p. 866/867).

It should be emphasized here that, also for biologists, this definition represents the sole meaningful criterion for deciding if something is animate and not inanimate. Life can be best understood as the creation, maintenance, and new acquisition of biological information. To avoid misunderstanding, it is still important, however, to clarify two completely different interpretations of the information concept. On one hand, one speaks generally of information when certain facts are meant. In such cases we speak of semantic information or information that carries a meaning. On the other hand, there is a purely physical information term related to the purely theoretically possible number of potentially achievable states and can be given numerically as their dual logarithm ld in so-called *bits* (or *bytes*). The game of roulette can help us understand the fundamental difference between these diametrically opposed concepts. In the first case, our knowledge, i.e., our exact information about the fate of the ball after being thrown into play by the croupier, is zero since, normally, we do not have an exact insight into the process. In the second case, each individual compartment of the wheel, whether we can predict the end result or not, holds the exact physical information of ld 37 (= 5.21). Eighteen black and exactly so many red numbers plus the zero on which the Casino makes its main profit gives 37 equally probable possibilities. The binary logarithm (ld) describes simply the number of necessary decisions (e.g., right/left branches in a decision tree) till a single event is reached. Viewed physically, we have almost 200 (37x5.21) bits of information in front of us, looked at purely cognitively, however, we have no information at all about where the ball will fall apart from the fact that it will come to rest in a final position somewhere on the wheel.

Consequently, in the case of the origin of life, nothing else happened except the realization of a first-time and up till now unique—for reasons of competition, a new life would have had no chance—transition from purely physical information which only describes the possibilities of which state the substance can be in, to the origin of real semantic information which, for the first time, contains knowledge of the outside world as significance for a specific subject. "Life as a cognitive process" was a pivotal chapter in perhaps the most important book written by one of the founders of comparative ethology—*Behind the Mirror* by Konrad Lorenz (1973). Now, if one takes the results obtained by modern biogenesis research seriously, one comes to the conclusion that this fascinating metaphor can be understood entirely literally. We can put forward the claim that living systems are nothing more than cognitive systems that, for this reason, carry on living. Therefore the formula L = C is valid, life *is* cognition. In the meantime, even molecular biologists no longer shrink from using such apparently metaphysical expressions like "cognition", "knowledge" or related terms:

The knowledge K, after reaching a certain evolutionary step, is that information (concerned with the overcoming of each new barrier that preceded the step under consideration) which has to be completely thrown away (translated from H. Kuhn and J. Waser 1982, p. 901).

The philosopher and game theorist Werner Leinfellner considered this relationship to be more than just a cognitivist analogy:

Space and time are, according to Lorenz, Vollmer, and Wuketits, "preprogrammed" instructions fixed in the genes, ... The author agrees with the main thesis of evolutionary epistemology but replaces the "selection paradigm" by a structural proof that natural evolution and the evolution of intelligence are, indeed, one and the same process (translated from Werner Leinfellner 1983, p. 246).

With this, the term "knowledge", due to the reference to the active process of cognition, is significantly more adequate than the undoubtedly more scientifically sounding term "information" (see the citation from Manfred Eigen at the beginning of the chapter) since the latter is unfortunately—as previously mentioned—used in completely different and therefore misleading ways. Knowledge can, in this connection, be understood generally as *assigning of meaning* to some object or event, and thereby includes, because of this very fundamental definition, reactions and behavior of living organisms of all stages of complexity, from primitive protozoans up to *Homo sapiens*, the thinking man. With reference to this new quality, which was not established until the emergence of life on earth, the physical information concept ultimately proves itself to have a subordinate meaning since it is only through the act of cognition that the so-called physical "facts" are transformed into meaningful information and therewith, at the same time, into real, i.e., "living" knowledge. Or, as Maynard Smith has recently put it for the domain of ontogeny: "development is going to require both a dynamical and an informational approach" (1998). This clarification is, therefore, of special importance because it directs us to a further fundamental connection. Semantic or biological information, very general cognition and knowledge about something, can only come about with the interaction of a living system with its environment although this relationship is primarily an asymmetric one. The system itself can alone possess knowledge completely independent of the abundance of physical information that may inundate each system. Only after adaptively overcoming these disturbing physical influences from outside, is the semantic information shown to be present in the living system.

One of the best examples of this are books written as an artificial store of human knowledge. Such a written record can hardly lead to confusion. Books are stuffed full with huge amounts of physical information, mostly black on white, but it is only after a completely specific interaction with a living organism, the human reader for example, that from this—under certain circumstances—something like cognition can occur, which means only in this organism and nowhere else. How dependent the origin of knowledge is on the activity or, more generally, the organisms' being alive and, being more "specific", the actual species of living organism we have before us, is shown by all the many cases in which there is information which we can access and which is capable of being interpreted by us as humans but in the case of other species "makes no sense" in the truest meaning of the word. One only has to think about mathematical theories, musical compositions, and similar complicated matters which very quickly rise over the mental level of fruit flies and wood lice. Naturally, the opposite is also true in those cases where humans are not in a position to open the door into the world of other animals, e.g., the perception of ultra sound, or infra red, or the resolution of very rapid movements. This inaccessibility is, as a rule, a principle and simultaneously cannot be circumvented even by our fantasy since we can only experience through one of our actual senses (e.g., by reading optical recording instruments) that we cannot hear ultra sound. The actual access itself to the transcendental must, however, remain closed.

Let us return to our starting point. The transition from inanimate substances to the first real systems which we may construe as living can only be adequately described and indeed understood through scientific terms when we are prepared to adopt a concept that is apparently from a completely different field and to use it for the phenomenon of life. What we mean here is, at first sight, a seemingly very esoteric term, namely cognition. which, in its narrow meaning, stems from philosophical-ideological traditions but, when examined closely, corresponds exactly to that quality which did not come into the world until and as a direct result of the emergence of living systems. Cognition is always necessary when a physically independent subject is confronted with the whole of the nonsubjective, i.e., the so-called objective world. Only then is it legitimate and, consequently unavoidable, to introduce cognitive concepts into the natural sciences. Concepts which, intuitively, would only seem to have meaning for the higher levels of typically human intellectual existence. This is way off the mark since this setting up of barriers is absolutely arbitrary. This can be concluded just from the fact of the biological development of modern man that it is not possible to draw a reliable boundary after considering that, at this or that point, the creation of the human intellect must have taken place. Before this, there could have been no cognitive ability, and not at all in animals. The difficulty with Darwin's theory is that one can always question what the actual pre-stage of each extraordinary acquisition looked like, and there is no conceivable escape from this constantly posed crucial question.

Let us assume something for which there is a lot of evidence, namely that man is directly and physically related to those first living systems that came into being about four thousand million years ago. Now we can be provocative and put forward the thesis that, even at this early point in time, something like cognition must have existed. This conclusion may seem highly exaggerated but only a few supplementary considerations will allow things to appear in a completely different light. On one hand cognition can, without difficulty, be described generally as behavior and this behavior can now be much more easily pictured as an evolutionary series of intermediate forms that must ultimately have had their beginning at the stage represented by the very first system. Now, a typical characteristic of living systems is that they can independently change themselves, move and consequently behave purposely. We may even assume that the most primitive life forms already showed a certain set pattern of behavior even if it was only in the form of comparatively simple chemical exchange processes with the environment. This means nothing other than that the primitive life form already "knew" how to behave in a meaningful manner in respect to the environment and in this way to survive. The wonderful connection of the whole evolution of the intellect is reflected more distinctly and directly in the actual development of every human individual. This begins with the inconspicuous fertilized egg cell which develops into a cooperative community of unbelievable complexity in which millions of individual cells take part. We have to ask where does this phenomenon cognition begin? Right at the beginning, not until the third month, or even later? One is inclined to grant this to a human adult, but to a newly fertilized zygote? If one follows the stages all the way through there is no break, no single place where one can justifiably claim that here something very exceptional has appeared in the plan, something such as a sudden and unexpected introduction of a reaction capability and with it cognitional ability.

The logical conclusion from such considerations is obvious and only confirms that what began above as an attempt to give a scientific definition of life has become more exactly defined. As already mentioned, most biologists and evolutionary theorists have naturally recognized the necessity of introducing cognitive concepts into their science, and this has also been realized in the form of the meanwhile wide-spread term of biological information. However, it is now clear that the cognition concept is much superior and furthermore theoretically more convincing than information in its usual form as a description of static conditions insomuch that cognition is more better adapted to the process of applying semantic information, i.e., to the life process itself (that the DNA, the so-called "hereditary molecule", is far from being a static store of physical information, is nicely demonstrated by the fact that around 10,000 hydrolyses of a purine base occur every day in a living human cell; cf. Strauss 1993). In this way, it is only through the continuous act of cognition itself, life "hates" interruptions, that what we understand as information becomes indeed meaningful knowledge of ones own identity and the world around us. It may sound trivial but only the activation of the information stored in a living cell allows us to speak justifiably about the authenticity of this information what otherwise—that means without the concrete reference to the whole phenomenon of a self-maintaining dynamic system called a life-form—would give no meaning at all. When scientists talk about the creation and passing on of biological information, they do not mean a static pattern of some biochemical states in which the complete knowledge of living creatures is thought to be written down as if in a book—this erroneous comparison is unfortunately used far too often. What is meant is their functional actualization within a background of the permanent dynamics of the living process. In other words, the process of life can only be correctly understood as being identical with the process of cognition or to be brief and to the point: L = C (Heschl 1990).

7 The Central Dogma Reformulated

If it is true that the human intellect has reached its present height by the same slow selection process by which all developments have been guided and raised to their adaptational heights, we must see therein the strong indication that even the highest intellect among us cannot see beyond the relationships governing our ability to survive.

August Weismann

In this chapter we will occupy ourselves with the central statement of this book, namely the claim that the only thing we can know about the possibilities of advances in cognition must be that, in principle, we cannot know anything about it. It is only logical, some would agree with a saying that sounds as if it came from Socrates (469—399 B.C.): "I know that I know nothing" which they may have heard sometime. However, Socrates' wise saying misses the mark somewhat since, firstly, he claims something paradoxical and, secondly, he does not take the fundamental question of cognitive acquisition into consideration. With regard to the latter, it makes no difference if one maintains that one knows something or knows nothing. Both claims reflect at least some knowledge. This is generally valid for all ontological puzzles of philosophy that are based on the well-tried patterns "there is a reality of something (e.g., soul, intellect, thing per se, etc.)" versus "there is no reality of anything (e.g., soul, intellect, thing per se, etc.)".

What is surprising is that the qualified experts in this subject, the philosophers, have concerned themselves only a little, or not at all, with the actual consequences of such claims. Therefore, we can fall back on the well-tried process of putting the whole of philosophy, and as we will see later (cf. chap. 15) the complete philosophy of science into one basket namely the voluminous basket of method. Firstly, however, we should give an accurate definition of this all-powerful word. For this we can quote Merriam-Webster's Collegiate Dictionary (1995):

1: a procedure or process for attaining an object: as a (1): a systematic procedure, technique, or mode of inquiry employed by or proper to a particular discipline or art (2): a sys-

tematic plan followed in presenting material for instruction b (1) : a way, technique, or process of or for doing something (2): a body of skills or techniques
2: a discipline that deals with the principles and techniques of scientific inquiry

Philosophers have added, to this all too dispassionately applied definition, a further considerably more elevated one: "by research guided by reason, way of truth" (Ricken 1984). It is a fact that almost all the philosophers and therefore most of the cognitive scientists up till now are the only ones to have occupied themselves with what is obviously considered to be the crucial problem of the correct method with the conviction, sometimes open and sometimes hidden, that such a thing must exist. We will show that it is for exactly this reason that it is impossible if one wishes to attempt to find an answer to the nature of progress in cognition. It gets even worse—all the countless discussions and debates about some supposedly correct or incorrect methods of gaining new knowledge show us exactly why it is completely impossible to have knowledge of the attainment of really new knowledge. Such knowledge can in no way exist, neither in advance, which is obvious, nor in retrospect. which is less obvious. It has nothing to do with considering how we, in the opinion of Popper, arrive at relatively certain (e.g., empirical) knowledge or how we, in the opinion of Kant, can advance to a completely objective knowledge of the world, the legendary "things themselves". It is rather another somewhat different facet of the usual philosophical debate about the intellectual advancement of the human race. All this can be safely ignored since it shows only what cognition acquisition *cannot* be, namely, any human acquisition- or even comprehensible method of progress.

It is strange but it is exactly this central point in the question about the nature of human cognition, which would give us the possibility of making real progress, that we have never really queried. We are so self sure and have remained convinced representatives of the species *Homo sapiens* that we have never think even for a moment that we have not been called to a higher rank of cognition. Our species would be more suitably named *Homo arrogans* for our conceit. The great illusion that lies behind such a conviction can be described best with the help of a metaphor. For this purpose let us imagine that we find ourselves on the actual border of our cognitive ability but sense within us the burning desire to take at least a small step into a genuine cognitively unexplored land. Behind us we can see the broad field of our previously existing knowledge which gives us hope for the success for our new undertaking. What can we say about the first step? Can we tell what direction we should go? Can we conjecture at any relationship to our existing knowledge? Can we finally, after we have somehow managed the first step, on looking back, see from where we have come to reach this point, be proud and enjoy the gaining of new knowledge. e.g., a discovery (of a continent) or even an invention (a machine or a new theory). The only possible answer to such questions can only be no and no again since as soon as we can give a direction or state any kind of relationship to our present knowledge, or even in retrospect after a deed, be able to understand what we have apparently obtained, we must have always possessed this knowledge even if it was not consciously. Arguing in reverse, that means that we can exclude with certainty the existence of cognitive acquisition when we produce an intrinsic relationship to our present state of knowledge, be it by some scientific or other method, by subjective intuition, or by whatever tricks. What is then the real nature of cognitive acquisition? It is exactly what for us must be, in principle, classed as absolutely incomprehensible, namely, acquiring completely new knowledge. This cannot be understood in any way before nor after the event and therefore must always elude every methodical or systematic approach. To summarize our short excursion to the fictitious border of our knowledge, we can now make the following apparently purely negative statement: The acquisition of knowledge can never occur in any directed form. If we want to make a positive statement, we arrive at the essentially important axiom about the nature of acquiring knowledge: a real growth of knowledge can only be realized by the roundabout route of chance processes.

If this investigation were not about the close relationship between evolution and cognition, one could confidently call a stop and add a further variant of a philosophical epistemology to the many already existing in the hope of obtaining as many supporters as possible for this new theory. One could think up an attractive sounding label for this thesis—the chaotic theory of cognition would be an attractive example, a very fashionable one if also confusing (cf. deterministic chaos)—in such a way to attempt by exaggeration and polemicizing to promote the awareness and popularity of a new philosophical trend, e.g., of indeterminism or anarchism. On one hand, the term "anarchism" has, unfortunately, already been appropriated by the philosopher of science Paul Feyerabend from Vienna with the completely different meaning of a desirable social freedom in the choice of methods (Feyerabend 1976). On the other, this so-called "anarchism of theories" has nothing to do with the claimed basic limitation of every method when it involves achieving cognitive progress. What has been worked out here by some short epistemological considerations can be set in opposition to all previous philosophical concepts. It is nothing more than the unexpected confirmation of probably the most important axiom in evolutionary theory, i.e., that about the basically random character of mutations as the raw material on which natural selection operates. Genetic variation in the form of actual random changes in the structure of living systems is exactly, for these reasons, the indispensable precondition for the possibility of biological evolution. No specified or understandable method here, random undirected structural changes there, finally absolutely no predetermination that equates the cognitive mode of advancement in knowledge with the material mechanism of advancement in evolution—all this corroborates and simultaneously takes us even further with what was established at the beginning of this investigation: life is identical with the process of cognition and as it has consequently become apparent, the nature of cognitive advancement is identical with the manner of life advancement. If we claim here that progress in cognition, if it happens at all, generally can only take place through complete "unawareness" of the subject concerned, such an epistemological thesis represents nothing more than an indirect confirmation of a statement that can apply in exactly the same way to the development of living systems. In the words of Richard Dawkins, it sounds as follows:

There is randomness and randomness, and many people confuse different meanings of the word. There are, in truth, many respects in which mutation is not random. All I would insist on is that these respects do *not* include anything equivalent to anticipation of what would make life better for the animal. ... It is only if you define "random" as meaning "no general bias towards bodily improvement" that mutation is truly random. ... The Darwinian says that variation is random in the sense that it is not directed towards improvement, and that the tendency towards improvement in evolution comes from selection (Richard Dawkins 2000, pp. 377—379).

We have to give him our full agreement since what would inevitably happen if the opposite were possible? As soon as an organism could indeed predict even the smallest fraction of a selective advantage of a morphological or physiological change, it would be then no longer understandable why it did not utilize this extraordinary capability and by doing this outstrip and hence supplant the still blind reacting competitors in a very short time. All this means nothing less than that the living organism must be completely helpless in that point that is the most important of all, namely the adaptation to the constantly changing conditions in the environment by a corresponding improvement in its internal equipment. There is no possibility of a directed anticipation of an advantageous adaptation, as repeatedly postulated in some form or another in Lamarckism. It is therefore no wonder that in spite of numerous attempts by some prominent researchers, so far, not the slightest evidence has been found for this theory. The embryologist Charles Waddington tried in vain to demonstrate a kind of genetic assimilation of phenotypic modifications, the biologist and developmental psychologist Jean Piaget attempted, in a similar way to document the inheritance of acquired characteristics in snails (1929), and the unfortunate zoologist Paul Kammerer became famous before this—he committed suicide in despair because his work was repudiated—for a similar experiment on salamanders. One could now simply claim that the conditions for successful functioning of Lamarckian evolution were so specific and for exactly this reason the empirical proof of this kind of evolution must be correspondingly difficult. Is it possible to oppose this objection with a principle, a theoretical consideration perhaps, which can point out the incoherence of the Lamarckian approach from the basic rules of evolution?

Our epistemological considerations from a little while ago provide exactly what was seriously lacking from Darwin's theory of evolution, namely, a purely theoretical reason for the fact that the evolutionary changes in living things can only occur from randomly acting processes, genetic mutations which in the meantime are known exactly through their own separate research field (Auerbach 1976: Mutation Research). Cognitive acquisition, which means nothing other than successful adaptation of an organism to its environment, can never be directed since this automatically presupposes the possession of the knowledge that was, in fact, to be sought for. Directed acquisition of new knowledge and directed adaptation to changing environmental conditions are identical assertions and therefore equally impossible and this just from the theory on its own. A consistent, well-thought out epistemology, no matter how fantastic and far-fetched it may seem at first, now provides that convincing argument that was missing from evolutionary theory and whose lack caused it to appear controversial again and again.

The purely epistemological considerations are so informative that it is worthwhile following them up further in order to see the correlation with evolutionary theory even more clearly. Let us assume, in a similar manner to the way we did for the realm of organic evolution, that it is possible to get near to the achievement of cognitive gain at least in very small but, for all that, directed steps. How can we exclude this possibility conclusively? We can ask ourselves again, at that moment when the cognitive subject is aware of the possibility of taking one of these small directed steps into virgin country, can there be any let or hindrance to making further directed steps be they large or small? The final conclusion would again be an unnatural organism, since a being so endowed must, in the short or long term, inevitably achieve perfect omniscience. At least, there would be no real reason that could prevent the organism from taking the next step, if again small, in the direction of cognitive growth, then the next one, and then the one after that etc. In principle, it could also, with one single directed step from its particular starting point, irrespective of whether it was an ameba or Einstein, arrive at the condition of perfect omniscience, in short, immortal and universal adaptation.

A purely formal confirmation of the thesis that cognitive acquisition can only be reached by a deductively derivable and therefore random "infringement of the rules" can be found in Gödel's famous theorem of incompleteness published in 1931 under the title "On indeterminable propositions of the Principia Mathematica and related systems". In this work, he showed that every logical system—and living systems are in a wider sense also simultaneously (bio-)logical systems—must remain incomplete since it is not possible, by the means of the logic selected at a given time, to prove freedom from contradictions, in our case, the functionality of the system itself. For example, it is out of the question to prove by argumentation the freedom from those contradictions in arithmetic that are represented in formal arithmetical calculations. Only after an extension by supplementary axioms, can this be remedied, but this is only temporary since then their own freedom from contradictions becomes a problem. With this, Gödel demonstrated elegantly that logical systems from their nature are ultimately tautologically closed entities, whose change, from itself, cannot be derived logically. From an evolutionary perspective, this insight is complemented by the empirical finding that living systems are characterized by the unusually robust dynamic stability of their self-organizational structures (cf. circular causality realized by the Hakenian "slaving principle", Haken 1989, p. 13—29; Paul Weiss speaks of a "'control' of the components by their collective state", Weiss 1971, p. 13; for a nice artistic realization of circularity see M. C. Escher's famous "Drawing Hands" from 1948 which depicts a hand on a sheet of paper drawing a second hand that draws the first one and so on), that on the one side approaches the criterion of a true physical perpetuum mobile (= the living process itself) but on the other can only be altered structurally by ever potentially hazardous disruptions of a random nature called mutations. It is exactly this inadmissibility of Lamarckian directed instructions, whether they are from the adaptation of "intentional" behavior (example: the giraffe stretching its neck) or imposed or imparted by the environment (example: "information transfer"), which determines the nature and above all the limitations of cognitive and therefore living systems. So, paradoxically, it turns out that Gödel's incompleteness theorem—but certainly not his somewhat inconsiderate Platonistic realism (Gödel 1944)—is able to complete Darwin's theory of evolution in that, for both cases, it shows and establishes what distinguishes the evolution of cognitive and therefore living systems: the indisputable necessity of chance of going beyond the limits at a particular time and simultaneously the principle of the impossibility of an individual having insight into this process, in short, blind mutation that is subjected to natural selection (Heschl 2001).

A concrete example from biology may serve as an illustration. We proceed from the fact that a whole cascade of chemical reactions keep an actual living system "alive" so to speak. We assume that a DNA sequence produces, via messenger- and transfer-RNA, a so-called initiator protein which, by contact with another specific DNA region, in turn, activates other gene sequences and the production of further proteins ultimately producing one or more complex protein molecules. These, together in cooperation, from time to time, bring about the duplication of the whole strand of DNA of a living organism (DNA polymerases + exonucleases, various auxiliary proteins: primases, DNA helicases, DNA topoisomerases, DNA ligases, additional binding proteins: SSB, RPA, regulation by proteinkinases; details in Knippers 1997). Now the possibility arises that, with the help of a further protein molecule, this replication of the original strand could be accelerated, stabilized or simply made more reliable whereby this new molecule can be directed or produced by a slight variant of an already existing DNA sequence.

Now comes the relevant Gödelian question: How is it possible to get the desired complementary information which would "explain" the organism better, i.e., could allow it to function better, into the system? Gödel says that for the field of formal logic, this cannot occur by a simple proof that could have been derived from the available logical system. Toward the end of his famous dissertation, he arrives at the conclusion that, if arithmetic is free of contradictions, it must be incomplete. This can also be expressed by the following *provable* equation:

$$(\exists y)(x) \sim \text{Dem } (x,y)$$

which means, if one follows Nagel, Newman and Hofstadter's (2001) metamathematical interpretation, that—for the case of arithmetic—there exists at least *one* arithmetical formula, for which *no* series of formulas can form a proof. Today, modern molecular biology shows that going through all the possible derivations, or chemical cascades of the system in question, does not lead to any fundamental change or even an extension of its possibilities until the moment when an additional "input", an infringement of the rules of chemistry, so to speak, in the form of an unforeseen random change in the "axioms", i.e., the basic constituents of the system (e.g., nucleotide sequences, amino acid sequences) takes place.

At this point, logic and bio-logic come together in an absolutely nontrivial sense, which means that the connection is not just metaphorical but rather should be understood as a connection of an ontological nature. This fact becomes even clearer if one looks more closely at the deeper connection of Gödel's incompleteness theory with the perhaps most important empirical finding in modern evolutionary research—the central dogma of molecular biology. This "dogma" only recapitulates what up to now seems to be a generally valid fact, namely that a transfer of information is only possible from the genotype (DNA, RNA) to the phenotype (proteins). This means from the invisible "causators", the genes—in the meantime, the genes themselves have become more or less visible through sophisticated molecular tracing techniques (see "Gene chips offer first look at genetic behavior in brain", Salk Institute, Press Release Sept. 25, 2000)—to the phenotypically visible characters of an organism and never the other way round (Crick 1970):

$$\text{DNA} \rightarrow \text{RNA} \rightarrow \text{Protein}$$

The term "dogma" gives us the idea that a fundamental principle is meant although, to this day, its arguments can only be provided by empirical data. Now Gödel's theorem is, for the first time, in a position to provide us with a purely theoretical explanation of the universal validity of the central dogma. The point is to understand that it is basically irrelevant which structure or molecule randomly mutates in a living system, the only important consideration is that random variation is the *only* possible way to evolve. That means, however, that all the many other processes that simply obey certain general physico-chemical laws within the cell can have nothing to do with a growth of information. This also means that a postulated Lamarckian retransmission of information from proteins to the genes, i.e., from the body to possibly the next generation, has no purpose since this supposedly "new" information already exists in the genes. This equally applies to a transcription of an RNA sequence back to a DNA,

$$DNA \leftrightarrow RNA \ (\rightarrow Protein)$$

a possibility known only since the discovery of reverse transcriptase in viruses and the retroposons in eukaryotes. It is not so much a violation of the central dogma, as it is sometimes confusingly portrayed, but nothing other than an internal transmission of already existing information, very similar to the case of the famous "jumping genes" (McClintock 1951) that consist of mobile DNA elements called transponsons. In the meantime, modern genetic engineering has shown how mobile, even across species, at least certain genetic elements can be. However, that is not yet everything since the central dogma could even be rewritten in its hitherto "forbidden" form, namely

$$DNA \leftrightarrow RNA \leftrightarrow Protein$$

without it being a problem because ultimately its essential point is only to forbid any possibility of a directed acquisition of information. Consequently, the first protein which would be written back to RNA or DNA would not conflict at all with the dogma since, as long as it is itself the result of a transcription from DNA or RNA, this protein would not confer any new information. On the other hand, if this same protein would have meanwhile succeeded in changing its particular sequence of amino acids by any chance process, it is not to see why such a "creative" protein could not push real evolution. However, at least in our known world reality is different since proteins are unable to evolve in that manner. Hence it must well be for quite other reasons as for example reduced thermodynamic stability and highly complex 3-dimensional form which makes unfolding rather difficult that proteins are not transcribed to nucleic acids or any other type of molecules and in this way automatically have opted out during evolution as potential, i.e., reliable carriers of information.

Let us give a brief recapitulation of this chapter. The decisive and most important question in epistemology, as it has become evident, is associated with the nature of cognitive gain. Can this come about in some form or other in a directed manner or are there more basic limits to our thinking and understanding? The result states nothing less than that directed cognitive progress must be excluded on purely theoretical grounds—or in other words *by definition*. Gaining new knowledge can only result when there is absolutely no orientation or cognition involved. A modern biology, extended by a scientific epistemology, can here, on both the lowest and highest levels, produce its own theory of living things since, once a mutation has become established, i.e., is successful, it represents a real increase in information. In the meantime, most natural scientists, especially those biologists who are seriously involved in evolutionary epistemology, are in complete agreement with this view. However, how do things look on the other, namely higher levels of cognition where ultimately the epistemological significance of all our inspirations and creative ideas is dealt with?

At first sight, it looks as if the consequences for this still newly formulated interpretation of our learning capacity, our thinking—in general, our whole intellect are not serious. This is because of the strong positive impression at the moment of a personal discovery itself, that difficult-to-describe instant of the completely new inspiration that allows a fascinating novel idea to emerge as if from nothing. This totally strange impression seems to match exactly the conditions that we have just considered as absolutely essential for the existence of a gain in cognition. The acquisition of new knowledge appears to be arrived at in some incomprehensible way whose process cannot be reconstructed in any detail. We could very well accept our many small individual discoveries and new inspirations as real acquisitions. Hence it was not for nothing that Karl Popper refused throughout his life to say exactly how we arrive at new ideas and inspirations. They exist in the lives of every one of us and that is that. How they may actually come into existence cannot be as important for followers of Popper as the much more significant question as to whether they can reproduce the world outside in an appropriately empirical and correct manner. Popper also claimed that these inspirations and new ideas cannot be derived in a logical way from already existing knowledge. One remembers perhaps his very dogmatic prohibition of induction as a method of gaining new knowledge about the world and compare this with the axiom elaborated here of random undirected cognitive progress. Our intuitive feelings, in any case, make us admit that he must be correct since we appear to have really no idea of how we arrive at our resourceful flashes of inspiration. Have we then ultimately become an endowed species emancipated from further evolution who can, what otherwise in all other organisms has to be accomplished laboriously over many generations, after overcoming genetic mutations rise to even greater, but now purely intellectual levels and in addition much, much more swiftly. Has our brain, as it indeed appears, its own evolution within its power?

8 Learning: Appearances Are Deceptive

We are so apt to admire instincts as something very extraordinary by all the disquisitions of human understanding. But our wonder will perhaps cease, when we consider, that the experimental reasoning itself, which we possess in common with beasts, and on which the whole conduct of life depends, is nothing but a species of instinct or mechanical power, that acts in us unknown to ourselves.

David Hume

Likewise, my CS-US-FB sequence is a mnemonic device for recalling that some species, like us, often fabricate associative behavioral strategies with Tolman's lucid cognitions in the service of Lorenz's blind instinctive processes which have been embedded in the anatomical structure of the species by natural selection (CS ... conditioned stimulus, US ... unconditioned stimulus, FB ... instinctive feedback).

John Garcia

Let us begin our critical stock-taking straight away with that category of behavior that perhaps demonstrates best the embodiment of development and progress concerning cognition. If something is learned, whether by an animal or a person, this formulation alone seems to include a growth in cognition. No other behavior is so strongly associated with cognitive progress as everything that has the slightest thing to do with learning. Classical behavioral research itself has contributed a great deal to this point of view in that it has always emphasized the sharp contrast between innate instincts and learned reactions (Tinbergen 1951). In this way, the one was ultimately defined by exclusion of the other. If a certain behavior could not be incontrovertibly proved to be innate in the course of certain experiments, it was then thought logically to represent an individually acquired ability, i.e., something really new. Experiments with so-called experience deprivation were used to find out if the behavior of an animal was innate or it could only have been acquired by its own individual experience. When a very definite reaction (courting display to the opposite sex, flight behavior with regard to predators, feeding behavior towards the natural food of the species), on first contact with the object which triggers the response, was shown correctly and, above all completely, it was concluded that the animal already possessed the appropriate, probably innate knowledge that enabled it to react in such a way as to increase its chances of survival.

It was indeed impressive to see the apparent extent of such innate knowledge in many animals before they had had any kind of close contact with the situation in question. This early period of behavioral research bore the stamp of the first great discoveries of ones of their most important pioneers (Lorenz, Tinbergen, v. Frisch, Heinroth) and it was therefore more than understandable that great emphasis was laid on the significance of the innate elements of behavior. In addition, during the early years, what is now known as ethology—comparative systematic research into behavior—was generally still considerably underdeveloped and for the sciences advancing more and more into the realm of the living was for a long time considered taboo. Behavior was either "studied" by selected philosophers or, if, in its special form, was more concerned with humans, it became a topic for the rapidly developing study of medicine. In this sense, the British empiricists (Hume, Locke, Berkeley) and some of the Greek thinkers of antiquity (Aristotle, Epicurus, Sextus Empiricus) were not such bad behavioral researchers as one could rashly claim with the advantage of hindsight. Sigmund Freud (1856—1939) was, in this sense too, and remains so to the present day, an enormously influential behavioral researcher at the turn of the last century with the not insignificant reservation that he knew next to nothing about the principles of Darwinian evolution. Nevertheless, his concept of the subconscious id, that mysterious reservoir of all biological needs and drives, without doubt originates from inspirations of the advocates of early evolutionary theory. With his own discipline, the analysis and therapy of psychological disturbances, he had great success at first but today this has become less and less: in the meantime, he is even blamed for grave Freudian slips in the area of diagnosis which was Freud's own special field, the famous *Interpretation of Dreams* of 1900 (cf. Webster 1995).

In the meantime, the latest results on the dreaming in both humans (Maquet et al. 2000) and in rats (Skaggs and McNaughton 1996) have shown that, in sleep, exactly the same neuronal patterns are activated as those previously in operation during wakefulness when solving concrete problems. Dreaming, therefore, could fulfill the function of learning consolidation that separates the many inessential things from the essential ones that we (and rats) experience every day in that an activation threshold only leaves the peaks of patterns. This would, among other things, explain dreaming in babies who definitely do not possess that span of awareness that would be necessary for the content-rich basis for producing a complicated Freudian enigmatic dream. However, also in relation to the subconscious, Freud speculated somewhat incorrectly: a subconscious experience which by some mechanism of recollection (normally: memory, in psychological cases of emergency: psychoanalysis) becomes a conscious one is not a mystery but nothing more than something once consciously experienced that after a certain time, for some reason or other, is remembered. This is even valid for relatively simple motoric learning process (e.g., riding a bike, swimming, playing football; Shadmehr and Holcomb 1997).

The speculative character of all this changed dramatically during the 20th century when more and more biologists concerned themselves systematically with the behavior of animals. Only then, did the scientific approach to the theory of evolution become the highest maxim for a completely novel approach to already well-known phenomena and enthusiastic research was carried out into the so-called preservation of species value of types of behavior. Should a certain type of behavior have been able to have an evolutionary beneficial value for a species, it was first necessary, if not to prove its inheritability, nevertheless to make it probable. A type of behavior that was only learned, everyone was in complete agreement in being opposed to Lamarck's view, could never bestow any significant advantage since it is not genetically transmissible to the next generation.

It is therefore not surprising that the birth of classical European behavioral research was characterized by the discovery of innate behavior and the accompanying innate releasing mechanisms (IRM). The American approach to behavioral research in contrast can be found in the diametrically opposed dogma of Behaviorism which describes humans in particular but also living organisms in general as a blank sheet of paper to be written on by the environment. Sociologically, this is very understandable since what counted in the newly formed society of America was, above all, not your lineage (e.g., nobility) nor your place of origin (nationality) but your particular personal abilities and the corresponding will to work. This was said and is still said by the official credo. Back in Europe, these mechanisms were put forward as more or less fixed sensory filters which tested the incoming sensory data for their meaningfulness and only let very specific and highly specialized stimuli through to the animal for it to respond. In consequence, the proof could be found for a great number of these mechanisms, whereby it turned out that this kind of stimulus filter could be in very different parts of the nervous system and not just on the periphery of the sense organs. The first clear-cut proof for the existence of selectively reacting neurons was obtained by the neurophysiologists D. H. Hubel and T. Wiesel (1962) for which they were awarded the Nobel Prize in 1981. They were able to show that the individual nerve cells in the visual cortex of cats only responded to certain stimuli with an increase in their impulse rate whereas to other stimuli they did not react at all. As the experimental stimulus, they used beams of bright light that were moved through the field of view at various angles of inclination.

How is it that studying innate behavior can be of any importance if one wants to deal with, not this kind of behavior, but its exact opposite? Quite simply because one can only be defined by the exclusion of the other and strictly speaking cannot be investigated separately. The decisive criterion now, allowed by the consideration held up to the present day, namely, that an exact differentiation of them is that they are two fundamental categories, is ultimately an information- or epistemological one—only how it arises and the specific origin of a behavior can give us an explanation of the concrete case with which we are dealing. From this perspective, innate behavior is behavior biologically acquired in the course of ancestral history, which, for the individual, is genetically determined and therefore unchangeable, whereas learning opens up for the individual the fantastic possibility of being able to overcome more or less the biological limits imposed by his genetic makeup and therewith to make a new acquisition. Such a new acquisition means nothing more than that a previously unknown piece of knowledge about itself or the environment has been acquired represented by this new behavior. The only significant limitation imposed on this learning is that oddly enough it cannot be easily transmitted to the individual's offspring. This limitation on what otherwise would be in principle an unlimited flexible behavior will be immediately understandable when we consider that such a transmission would mean a confirmation of already refuted Lamarckism. Learning opens up for the fortunate individuals of species possessing this ability, denied to "more stupid" ones, the unique opportunity of gaining information about themselves and their environment and, furthermore also in most cases to have the not unimportant ability to revise the knowledge acquired from new again and again as required. What has been learned can be forgotten in time and can therefore, in principle as often as desired, make room for even newer and, under some circumstances, even better understanding of new kinds of problem posed by the environment.

This optimistic picture of learning emphasizes the positive contrast to the biologically preordained one, to the predetermined genetic inheritance of phylogeny, which we humans in some way are bound to believe, but which—and in this both ethologists and behaviorists are in complete agreement—through personal endeavor and one's own individual abilities may be outwitted. The ability to overcome one's biology through new acquisitions is not so emphasized by the behaviorists as much as one may prematurely believe. They have always put learning in a special position in behavior but have never seen any contradiction to supposedly inborn behavior. For them, the laws of conditioning were absolutely universal rules, which in every area shaped not only animal behavior but especially that of humans as well and in an almost arbitrary manner could modify and continually renew it. A contradiction to innate behavior was not taken into consideration since such a thing was thought of as simply nonexistent. So it was a special service performed by the classical ethologists to proceed from a dichotomy, a basic splitting of behavior into a phylogenetic innate part and an ontogenetically acquired part. Even the title of the dissertation published almost forty years ago by Konrad Lorenz (Phylogenetic Adaptation and Adaptive Modification of Behavior, Lorenz 1961) speaks in a paradigmatic way in favor of splitting behavior into two fundamentally different parts. In spite of strong criticism from some behavioral researchers, this is still considered as being important. One of the most important critics of this incorrect dichotomization was Daniel S. Lehrman, who unfortunately was completely misunderstood by Lorenz. Lehrman rejected neither instinct nor was he a dogmatic "milieu theorist" as commonly thought, but rather he showed how to "question the value of the dichotomy itself *without* accentuating one side of the argument or the other" (retranslated from Lehrman 1974, p. 77). Today, Lehrman's position in this question can be easily reconciled with ethological thinking (Heschl 1989).

In the meantime, however, radicalism has suffered a number of severe defeats since more and more often empirical investigations have shown that many kinds of behavior are very clearly a result of well-defined genetic instructions. A new scientific discipline has even been created—behavioral genetics—and has carried out research into this relationship with increasing success. The answer to the question about the existence of innate behavior can be considered as accomplished since there is no more doubt that such a genetically directed behavior does indeed exist. However, what remains is the much more difficult question of the evolutionary status of learning. The behaviorists very obviously made a mistake in thinking *all* behavior must have been learned. For this reason are the ethologists alone correct if they claim that learning is a behavior that could be defined as being not innate but represents rather a genuinely new acquisition? An evolutionary theory that is based on the idea that the development of a new life form is simultaneously a cognitive process should provide us with the means for deciding what a genuinely evolutionary new acquisition may be and may not. What follows is the application of the present view of evolutionary processes now extended by epistemology to an actual case.

In order to clarify what may be extraordinary about learning or to examine whether there is anything at all extraordinary to be found in learning, it is sufficient to take a fictitious learning process and, at least purely theoretically, to consider it broken down into its components. Learning is, viewed outwardly, nothing other than behavior changing with time that can be considered as consisting of numerous closely related single steps. Let us begin with the very first step in a learning sequence, the recognition of a learnable situation, and ask ourselves what must happen at this point so that learning can take place. Here, things are relatively simple since it is always the learning subject itself that has to know where it has to begin otherwise the result must be completely arbitrary. The next step is where an especially attention-arousing stimulus has already been discovered and now this arousing stimulus has, in some way, to be brought into association with another stimulus. How can this happen? Who in the world whispers to a learning animal or human and says what has to be done next? Again, the previous situation repeats itself. We either already know which very specific stimulus under which temporal or spatial conditions can be attached to the first one or we lose ourselves again in irrationality. For the reason that we can find no compulsive criteria or that an infinity of optional ones for a connection in the environment, we must conclude that in some way the corresponding knowledge of an adequate relationship between the two stimuli must have already been present within the learning system. This naturally contradicts in a very direct way the usual, so obvious opinion of a process of instruction of the learning subject through the outer world.

Learning, according to the current doctrine, is in principle nothing other than passive or, at best, active instruction—a direct kind of information intake through and from the environment. However, a closer examination of the individual steps involved immediately shows the absolute impossibility of such an opinion since a theoretically infinite number of conceivable relationships between any events in the environment is diametrically opposed to a permanently required need to decide what to do and, in particular, what to associate. Which from the millions of stimuli bombarding the system should be brought into association with another or, which makes things even more complicated, which ones may, on no account, be brought into association with each other because they can bring about a non-adaptive, and hence fitness-reducing reaction of the subject. The knowledge of the relationships—and this is the decisive point—must therefore, in every single case, have already been a prerequisite since otherwise the result would have been a completely randomly oriented behavior of the learning system. Here, one could argue that it is known that not a few learning strategies follow the pattern of trial and error, so that ultimately at least something like instruction from the environment must be present. The subject, by trying out the most varied ways of behaving until he is successful, apparently without any plan, gains new knowledge from his environment that he obviously could not have had before. Appearances are again deceptive, when we consider critically what prerequisites are essential for this supposedly random learning. In order to learn anything from success in reality, e.g., finding food, a resting place, or a mate, one has to know exactly which action led to this success—this means, in other words, the action itself cannot be random or carried out without any plan, because otherwise the existing information about it could neither be used nor stored. On the other hand, to be able to interpret the success of an action one must know exactly the required relationship between action and result pertaining to survival, otherwise arbitrary behavior would be expected. In fact, experimental research has shown that for example humans are unable to produce real random sequences intentionally (Guttmann 1966; Mittenecker 1958, 1960; Mittenecker and Raab 1973; Remschmidt 1970; Trappl 1971) which proves that the highly *systematic* method of trial and error cannot be uncritically equated with purely physical chance and selection processes. In addition, the distinction between real chance versus determinate processes has also been shown to follow some well-determined behavioral rules in both animals (Smith et al. 1997) and humans (Gigerenzer, Hoffrage and Kleinbölting 1991; Lalouschek, Lang and Deecke 1995). As an explanation of the latter phenomenon people are thought to act as given in the following description: "It seems as if the ideal picture of a random series would correspond to the urn model without putting back the balls, i.e., one ball after the next is taken out in a supposed random manner until the urn is empty and then the whole process is started again" (Mittenecker and Raab 1973, p. 240). This further supports the hypothesis that no doubt organisms are good, i.e., natural selection-adapted players, but bad or at least false gamblers only more or less good mimicking true randomness of behavior.

An imaginative experiment, carried out as long ago as 1932 by Krechevsky, showed that the movement of animals before they encounter a reinforcing stimulus—one that actuates learning (e.g., food)—can be anything but arbitrary. This experiment consisted of placing laboratory rats in a labyrinth that could not be solved. What was its purpose? Krechevsky's clever idea was to put the animals in a labyrinth in which it made no difference which route they decided on since they were rewarded no matter which of the end boxes they reached. This meant that there was nothing to learn and therefore considered insoluble. The rats were confronted with four junctions one after the other where they had to choose whether to turn right or left. At these junctions, there were optical stimuli that varied arbitrarily with regard to the spatial relationships (left/right). What was the result? Instead of the animals taking any way or other to reach their goal as expected, they preferred particular routes without any recognizable reason for doing so. The random variation of the external stimuli together with the universal 100%-rewarding of the animals should have produced a perfectly random pattern of movement. However, just the opposite happened. Each animal had his own special way of "solving" this labyrinth and one could surmise each animal to be convinced that his was the only correct one. Krechevsky (1932) chose the term "hypothesis" to characterize the behavior of his experimental animals and this for good reasons. Each animal had used his individual and precisely not by chance strategy and formed its own "hypothesis" although from the organization of the experiment it had not been necessary. One could hardly devise a more ingenious method of demonstrating the inherent autonomy of learning behavior.

A completely different species was used in an experiment to prove the presence of specific individual learning patterns by Viviani and Terzuolo (1980), namely that other classical "guinea pig" *Homo sapiens*. What these two physiologists, by using a simple method—they employed an electronic digital table (622 RP CALCOMP) to register complex hand movements with both high spatial (0.025 mm) and temporal (100 Hz) resolution—discovered was what Krechevsky many years before had investigated with his rats. During their observation of professional typists who had been asked to demonstrate their competence on a typewriter, they noticed that there was an extraordinarily stable temporal pattern of movement. They discovered, for example, that the motoric sequence for each word had an invariably specific structure, a kind of typing pattern, that was completely independent of the typing speed. This invariance in the pattern of movement was aptly designated as homothetic or "self resembling" behavior.

From these first observations, it was then possible to speculate that, behind this special case of invariant behavior, there was an underlying general principle for learning complex sequences of movement. The authors in subsequent experiments allowed the subjects of the experiments to move their hands freely instead of the movements in typing which, of course, are more limited. The subjects were asked to write letters, then words, and finally complete sentences by hand. The results were astonishing. The resulting spatiotemporal pattern indicated an almost perfect spatiotemporal invariance in the handwriting. This demonstrated that even the most complicated learning processes in humans are based on an individually characterized program of movement that in no way could be described as random. Every single letter, word, and sentence was typical and unmistakably the product of a single person, no matter under what conditions as far as size and speed were concerned.

At this point, we can now easily return to the general theory: Even in the most complex cases, it is only a subjective impression that a learning system can receive a really new piece of cognitive information from the environment. In reality, nothing is taken in, but only an interaction with the environment occurs in which the required amount of physical information is always interpreted by already present semantic instructions.

What must take place if learning is not as we have just described it? Let us assume the exact opposite as being true, since this is also the most widely spread view of the nature of learning, and see, what on closer observation, the logical conclusions are. We therefore accept, a least for the moment, the idea of an environment being able to instruct the organism capable of learning, consequently to convey to it in physical way the information about what it ultimately has to do in order to make the most for itself of any concrete situation. As soon as we accept this idea even provisionally, we lead ourselves into the greatest difficulties which could even not be overcome by a renewed acceptance of radical Lamarckism. If the environment has the power to inform a living organism about what would have to be done in order to behave correctly, it would mean nothing else than that there would be no compelling reason for the organism to limit the assimilation of this kind of vital information. In fact, just the opposite should happen: any living system that—owing to our fictitious interference— gets the unexpected opportunity to receive the relevant units of information from the environment should immediately begin to concentrate this vital knowledge on itself. In this way, it would not only drastically increase its chances of survival, but would also ultimately reach the fantastic situation of absolute immortality since every piece of information necessary to counter any threatening situation could be taken at any time from the environment. This conclusion may seem somewhat exaggerated at first sight but it is indeed the only logical consequence that can be drawn from the still widespread idea that living systems are capable of assimilating real cognitive, i.e., semantically significant information from their environment. The exploitation of such a possibility would have the consequence that a biological evolution, i.e., one that would depend upon random experimentation with genetic changes, may no longer play any essential role. That would mean that even certain protozoans or at least some very primitive metazoans for which learning has already been proved could evolve without resorting to Darwinian evolution—an absolutely ridiculous idea.

What can we conclude from all this? A more exact examination of learning shows us firstly that, for the present, a purely physical instruction from the environment must be excluded for reasons of probability and secondly, therefore each learning system must already, i.e., *a priori* as meant by Kant, know exactly in what way it has to react to very definite spatiotemporal relationships between what are often very complex stimuli. Learning is therefore in no way a random and, for this reason, a particularly creative and innovative behavior but rather nothing other than the expression of diverse complex strategies applied by living systems to solve vital problems in the most efficient way. It is already the term learning "mechanism" which indicates that we are not dealing with a chance result from an imaginary physiological lottery but rather just the opposite, namely with a highly specific and, above all, a biologically purposeful pattern of behavior. With this last insight, we have turned full circle back to the biological theory of evolution that requires that a real evolutionary and therefore cognitive step forward can only be achieved by an absolutely undirected change in the structure of the whole system. An evolutionist and simultaneously also an epistemological interpretation of learning must therefore, as a consequence, read completely differently to what it has done so until now. In this interpretation, learning is still an extraordinary complex kind of behavior but it now has nothing to do with acquiring something new. What, up to now, has been thought to be characteristic and unique of learning, i.e., the acquisition of new knowledge, of new cognitions, must now be excluded with certainty. Learning as a behavior without any real learning effect—can that be right?

Current biological research into the mechanisms of learning shows us exactly which direction the new investigations into learning are going (Shettleworth 1994). Much earlier, inspired by the great theorists (to name just the most famous: I. P. Pavlov, Clark L. Hull, J. B. Watson, Edward C. Tolman, B. F. Skinner, E. L. Thorndike) generally valid rules which should form the basis of all learning behavior were searched for. Nowadays, we are concerned with throwing these all too idealistic and reductionist notions overboard and replacing them with systematic research into the enormous diversity of things that up till now have been very vaguely lumped together as learning. This was initiated above all by those experimental psychologists who dared for the first time to take a serious interest in a biology of learning. Instead of hastily campaigning for an ideal taken from physics with which forcibly to construct the general, if not universal, laws of learning, they began to be much more interested in the detailed biological systematics of the corresponding kinds of behavior (Bitterman 1975, 1988; MacPhail 1982; Shettleworth 1993a, 1993b).

It should be stated here that physics was so strongly considered to be the model science because of its strictly quantitative nature that researchers in the most varied disciplines attempted to emulate physics in order to be labeled "scientific". As an example of an experimental psychologist involved in this movement, we can take the influential learning theorist Clark Hull, who made a what is nowadays an amusing attempt at devising, using Newton's famous *Principia Mathematica* as a model, a mathematical-deductive theory of learning. The title of his main work, published in 1940, was, *A Mathematico-Deductive Theory of Rote Learning*. Exactly as Newton interpreted heavenly bodies as point sources with purely quantitative dimensions, Hull ignored all the qualitatively different characteristics of his experimental animals in order to formulate his three basic laws of organismal motion, in almost perfect equivalence to Newton's famous Laws of Motion:

1. At the beginning the animal is at *rest* or is making *random movements* (cf. Newton's Law of Inertia).
2. New habitual actions are acquired by *reinforcement* (cf. Newton's Law of Dynamics).
3. Every action is automatically accompanied by an opposing reaction called *reactive inhibition* (cf. Newton's Law of Reaction).

Possible differences between species in their behavior were interpreted, again in accordance with Newtonian physics, as being simply quantitative changes in the constants of the universal equations of learning. The actual, mostly highly complex behavior of animals in their natural surroundings was irrelevant to the theory and the only important things were the countable movements of, in principle, interchangeable experimental objects. An interesting overview of the infancy of modern mechanistic behaviorism was given by Hilgard and others in *Theories of Learning and Instruction* (1973).

What initiated a fundamental change in direction of research into learning, however, was a piece of work that was carried out in that old style of behaviorism which operated exclusively in the laboratory. It was the experimental psychologist John Garcia, who, by a chance and therefore—as he nowadays quite readily admits—purely anecdotal observation and not an exactly scientific measurement of the behavior of his experimental animals, discovered that the theories of learning current at that time could not be the last resort. During tests which were quite commonplace at the time to find out the tolerance level of laboratory rats to radioactive radiation, he noticed that the animals developed without any obvious reason a marked aversion to the drinking water offered to them during the tests. Garcia's spontaneous conclusion, which we now know to have been exactly correct, was that the rats must have mistakenly associated the nausea caused by the radiation with the water which smelled of the plastic container from which they had to drink. When they encountered water from such a container they demonstrated what Garcia graphically described as "telltale signs of disgust" (Garcia 1991, p. 34). They jumped away from the drinking spout, pressed their heads against the ground, and retched as if trying to get rid of a terrible taste from their mouths. For the behaviorist theory of learning, such a finding could only be a random artifact without any further significance since the stimulus from taste of the water could not possibly be associated with the nausea caused by the radiation because there was absolutely no real connection between the two. Garcia and his coworkers were at first rewarded for their discovery not with praise but abuse for their unscientific methods, at least until the moment when this observation ridiculed by colleagues was quantitatively proved in a series of tests devised by him for this purpose. He used saccharine in water together with a low dose of radioactive radiation and it produced exactly the same effect, namely, the rats from then on hated this sweet-tasting water. Only after these data were published in this now-acceptably scientific form (Garcia, Kimeldorf, and Koelling 1955) did experts in the theory of learning begin to puzzle over the consequences that this absurd result brought with it.

What had gone wrong with the design of the experiment? Why had the animals behaved in such an illogical way? A quick look at the ecology of rats can immediately clear up this misunderstanding on the part of the experimental animals. The animals were completely correct in their decision to avoid this suspicious water since any other reaction, given their ecological history, would not have been sensible. Rats, during their phylogeny, probably have never had anything to do for a longer period with high doses of radioactivity that can produce nausea—atomic power stations have not existed long enough. However, they have always been confronted with dangerous substances which can cause internal injuries, e.g., plant toxins, products of putrefaction and fermentation, when it seemed wise that the irritation caused by this substance should be imprinted as reliably as possible in order to prevent the error of ingesting the substance being repeated. The resulting nausea must have conveyed to the animal that, in future, it should avoid the food it had eaten last. Garcia's rats consequently did nothing other than apply an proven behavioral strategy to a situation for which the animals could not have been prepared for by evolution. Because they felt nauseous and only had the plastic-tasting water available, since they could not perceive the radioactive radiation, they had to conclude with evolutionarily proven logic that the water was the real culprit. This is easy to understand since not a single species, even humans, has yet had to develop a sense for perceiving radioactive radiation. This also is possibly the explanation of the special cognitive difficulties we have in our dealings with atomic power. Perhaps, this will change in the near future for other reasons: recently, a small mouse was discovered for which the insanity of Chernobyl up till now has not brought about a single lethal mutation (*Science* 273: 313; in spite of the radioactive contamination the entire fauna around Chernobyl has profited from the absence of human habitation; Williams 1995). A similar situation to that of the rats falsely associating radioactive radiation with taste and smell is the case of human seasickness where, just as erroneously, our nausea brought on by our disturbed sense of balance in a heavy sea is associated with the delicious cuisine on modern ships. The result being that even the most extravagant captain's reception is not very well-attended although it is included in the price. That seems easy to understand but what bearing does this have on our conception of learning?

The significant point in Garcia's surprising discovery—and retrospectively, this is how we should see this observation—lies in the fact that the animals cannot react in any other way. Their biology usually helps them, in most cases, but damns them, in certain rare situations like the one previously described, to behave exactly in that manner and no other. When this is indeed the case and, as a consequence, this story about the rats remains not an individual case, then we automatically have a big problem with a physical theory of learning which is concerned only with abstract positive or negative stimuli and mechanically reacting test objects rather than real living organisms. Which means it can have no idea about the actual concrete problems that animals as well as humans have to overcome in their varyingly long and mostly complicated lives.

Encouraged by these first sensational observations, Garcia personally took control of the further investigation of this phenomenon. The results of his efforts unearthed far more interesting information than one could have expected. It soon became clear that the behavior of rats virtually never followed the general laws of learning à la Skinner or Tolman, but time and time again a very specific strategy of selective learning was utilized. For example, as already pointed out, nausea as a rule was strongly associated with the taste of food at the time it was offered or even hours after, whereas the smell only in a weaker form and visual and acoustic stimuli even less. A painful experience such as a short electric shock, in contrast, did not cause a taste fixation but turned out to very effective at imprinting visual and acoustic stimuli. The natural logic of this behavior and not just for this species is plain: only substances which have been ingested can normally produce nausea but, of necessity, not objects simply heard or seen at the same time. In contrast, only the latter (e.g., predators, biting and stinging insects) cause muscular sensations of pain whereas these cannot be caused by something which has just been eaten. At least something has now become abundantly clear, namely that experimental psychologists would be well-advised to seek tuition in biological behavioral research if they wish to continue to consider learning as their legitimate subject of study.

The results of such studies and numerous others with similar experimental approaches were extremely surprising and led to a completely new theory of learning being formulated as a working hypothesis for the future. In the Constraints Theory of Learning, we are no longer concerned with those attempts, preordained to come to nothing, to reduce problem-solving strategies to atomic units of behavioristic conditioning. Nowadays, we are concerned with just the opposite, namely with investigating the totality of these strategies which determines what intelligent behavior in animals and humans is (Shettleworth 1972). Such a biological approach to learning which takes into account the complex total structure of behavior is, on one hand, especially interested in the specific limits or constraints of learning and, on the other, in its diverse qualitative levels which, up till now, have been a taboo subject for behaviorism, but are inseparably linked with intelligence. One thinks here of the long-avoided problematic of levels of consciousness that has only recently been treated scientifically once more (Griffin 1991; Schleidt 1992; Heschl 1992b; for a new philosophical approach, see Dennett 1993). A biology of learning which has only just begun (Angermeier 1984), now emphasizes primarily the natural frame conditions of what seem to appear flexible and apparently boundlessly moldable patterns of behavior (see introduction to Marler and Terrace 1984: *Does a General Theory of Learning Have a Future?*). The starting point and simultaneously a catalyst for this change in direction of the development of this branch of research was the exponentially increasing inability of the traditional theories of learning to explain even approximately the multitude of new results concerning intelligent behavior of animals and humans. Above all, it was the numerous results from ethological research on the complex behavior of animals in their natural environment which gave rise to the rightful doubts about the general applicability and therefore validity of the behaviorist axioms.

However, even within their own camp, there was some dissatisfaction among psychologists involved in experimental learning research with their stereotyped routine laboratory studies, always with the same animals (mainly rats, pigeons, rhesus macaques) under perfectly unnatural conditions. This dissatisfaction had to do with the all-too simple and increasingly unbelievable interpretations within the framework of the conventional theory of learning. The revolutionary work by Garcia is only one example among many. It quickly became clear that those few very abstract principles of association based on the early investigations of Pavlov (1849—1936) which took place round about 1900 were insufficient to be able to set up a really comprehensive and general theory of learning. Increasingly numerous exceptions occurred which of course contravened the rules until it was no longer possible to say what was the norm and what was the exception because of the abundance of conflicting results. As a way out of this confusing situation, people now began, in compliance with the doctrine, to shove these unpopular exceptions to the classical principles of association into a drawer labeled "biological constraints of learning" and to concentrate on those cases of conditioning that seem to support the theory. Among these constraints, there were considered to be two essentially different categories of phenomena. On one hand, after the disclosures by Garcia, one knew that the association of stimuli is often much more specific and limited in form than the theory of learning forecast and, on the other hand, one discovered still further forms of learning that in no way fitted in with the behaviorist concept. Perfect examples of the latter are learning within the framework of imprinting and songbirds learning to sing—both forms of behavior can only be with difficulty or even cannot be reconciled at all with a universal theory of learning. The discovery of the unusual effect produced by only certain combinations of stimuli (*selective learning*) as well as solving the puzzle of formerly unknown learning mechanisms (*specialized learning*) must ultimately and inevitably lead to declaring the theory to be scientifically bankrupt no matter how beautifully it is formulated.

However, the title for the new theory, "constraints" theory of learning is a rather unfortunate choice and has a great disadvantage. It only emphasizes the negative aspect of the biological influence and therefore suggests that only the most general conditions may be genetically programmed in some way and hence let one erroneously think that there is still some creative freedom of learning that must exist, at least within this limiting framework. The latter view is an extraordinarily wide-spread one since it seems to be a compromise reasonably acceptable to both sides. Nevertheless, the long-overdue integration of biology and psychology of learning is diametrically contrary to such a compromise as already recognized at least partly by some psychologists. Sara Shettleworth, for instance, at the end of a comprehensive overview on "Biological approaches to learning", comes to the following conclusion about how "future directions" of psychology should look like:

Articles on learning pay more attention to what role learning might have in animals' lives, while biologists studying animal behavior have become more knowledgeable about principles of animal learning and cognition. Throughout this chapter have been mentioned numerous examples in which biological and psychological approaches have been integrated, and at least as many possibilities for future integrative research. Such integration will have been achieved when some future edition of this handbook does not need a separate chapter for "biological approaches" (Sara J. Shettleworth 1994, p. 212).

The fact that the underlying biological mechanisms can be viewed as constructive and hence do not only systematically limit the learning behavior of animals and humans but that they also structure and coordinate all the important details is still very hastily overlooked by most psychologists. The latter point, however, is the most significant one in our debate on the evolutionary status of learning behavior and it can now be decided in a clearly positive way. The main result of the long history of research into learning is not so much to be sought in the almost trivial knowledge that each kind of learning, including that of humans, always has to remain somehow limited, but much more in the fact that learning, in contrast to genetic mutations, which can bring about a real cognitive advance, is nothing more than a well-directed, built-in, problem-solving strategy. Learning is therefore only an expression of already preexisting biological, i.e., evolutionarily through natural selection tested information.

The latest results of research into learning have increasingly proved themselves to be a corroboration of a view of learning from a consistent evolutionary theory that admits that learning is very much concerned with complex behavior but has absolutely nothing to do with the possibility of individual cognitive gain. One of the first theoretical attempts to view learning behavior under such an altered, evolutionary perspective was the idea put forward by Harley (1981) to consider the existence of so-called evolutionarily stable (ES) learning rules. An interesting continuation of this new way of looking at learning was given in a recent article by Tracy and Seaman (1995) in the abstract of which one can find the revealing introductory words:

Suppose a strategy for learning the optimal behavior in repeatedly played games is genetically determined (Tracy and Seaman 1995, p. 193).

In order to make the paradoxical *Conclusio horribilis* of this chapter, that learning has nothing to do with real learning, that is with an increase in cognition, we will take an abstract example from the human field. Let us imagine, that we have been put in a situation not of our choosing where we are forced to learn the solution to what for us represents an absolutely new problem, e.g., how to operate a new machine, learning a new method of calculation, or a new game, etc. To understand how much with regard to cognitive information must already be presupposed, we only have to examine more closely the following simple questions and their answers:

How can one recognize the learning situation as such?
⇨ One must already know *in advance* how to recognize it.
How can one recognize the problem to be solved is one as such?
⇨ One must already know *in advance* how to recognize it.
How can one take the first step in the direction of solving the problem?
⇨ One must already know *in advance* which step to take in which situation to solve which problem.
How can one recognize whether this first step does indeed bring us any nearer to solving the problem?
⇨ One must already know *in advance* how to recognize the relative success of the first step.
How can one take the n-th step in the direction of a possible final solution to the problem?
⇨ One must already know *in advance* in each case which step to take at which point in time.
How can one recognize if the n-th step really does bring us any nearer to the solution of the problem?
⇨ One must already know *in advance* how to recognize the additional success of the n-th step.
How can one recognize the complete success of one's handling of the problem?
⇨ One must already know *in advance* how to recognize this ultimate success.

A similar procedural schema must be also valid for all the many of our so-called unique personal experiences. Subjectively, this is perhaps the most difficult aspect of a consistent evolutionary interpretation of learning, namely to grasp that although we all have the strong impression that our own personal lives are filled with a continuous series of unforeseeable personal experiences even the exact and most detailed knowledge about the respective particular circumstances must have already been present in the semantics of our genome. It is in fact as if everybody's concrete life would have already been written down and this in a paradoxically both at the same time abstract as well as concrete manner. To follow this unusual view it might be perhaps helpful to call into mind that any of the countless details of our experiences are nothing but completely autonomous attributions of our sense sense organs and the central nervous system to either the inner or the outer world. This is also the case for the strong feeling of whether we or others (= historicity in general) have something like a unique personal history, i.e., for all our conscious experiences and things we imagine since both memories of the past as well as plans for the future are always purely internally made constructions of our genetically instructed brain where, of course, a great deal is due to external environmental influences in the sense of purely physical things on offer or—much more frequently—elemental challenges, but have nothing at all to do with a cognitive instruction by those same "sense-less" influences (in the words of Woody Allen: "Tradition is the illusion of permanence"). Looked at epistemologically, this means that living organisms are a kind of cognitively closed solipsist Leibnizean monades since all the existing knowledge of the individual is internally constructed, and this not only in a merely philosophical sense as is already well understood by the rightly more and more successful trend of radical constructivism (theoretical founders: Maturana and Varela 1980; recent treatments in: Glasersfeld 1996; Müller, Müller and Stadler 2001; Foerster and Pörksen 2001; for impressive pathological cases of unusually constructed "realities", see Sacks 1985), but also—and this is hard for many people to accept—in a very concretely materialist, i.e., genetic sense. Hence, learning, as well as experience in general, means nothing more than already possessing in advance the complete knowledge of what is to be done to solve a certain problem of survival (and, ultimately, there exist only problems of survival). Immanuel Kant was therefore correct in a very biological sense in his clear-sighted criticism of the possibilities of experience since without the *a priori* preconditions of cognition which he postulated evidently nothing happens (Kant 1781/1999). Moreover, David Hume had already come very close to finding a solution to this question when he said near the end of his epistemologically most-stimulating investigations almost as an aside that our learned habits are not much more than a kind of instinct although possibly a bit more complicated than in our nearest relatives in the animal kingdom (Hume 1777/1975). In considering the status of learning behavior in relation to the unified theory of evolution we have tried to work out here, we can supplement the critical voices of these two skeptics with another very important element. Not only is a certain minimum number of special *a priori* precognitive pieces of information essential to be able to learn and experience something new and also a certain number of Hume's general Laws of Experience Activity that allow the learning organism to recognize things about the world instinctively and correctly, but far beyond that any differentiation between preexisting *a priori* knowledge and knowledge to be learned *a posteriori* has become invalid when viewed from evolutionary theory. What Kant and Hume were still convinced of, that human individuals in trying to understand the world, by learning are capable of making a cognitive advance can now be excluded with certainty. *A priori* and *a posteriori* can further be used as purely temporal expressions but their relevance as important epistemological criteria in relation to individual learning behavior has been reduced to zero. However, these classical terms possibly could gain new epistemological relevance in relation to the description of real evolutionary changes: Within such a new perspective, *a priori* would describe the situation before a random genetic change and *a posteriori* could be used to refer to the amount of cognitive gain that has possibly been achieved by the creature that may come into existence after this change. Interestingly, such an approach would come very close to the molecular definition of the knowledge K the biochemists Kuhn and Waser (1982) have given, even though in a symmetrical manner (see chapter 6).

Let us go back to learning theory: The simple grasping reflex in newborn babies remains temporally *a priori* with regard to the much more complicated first true eye-controlled grasping movement of a 5 to 6-month old child (Bower 1979, Piaget 1936, White 1970) but this no longer means that reaching later stages of behavior includes a real cognitive advance. The difference that was long considered significant, namely that between innate instinct and that behavior called "learning" has now disappeared completely. The popular conception that the term "innate" must logically mean "present at birth" and therefore erroneously equates a purely temporal statement (when did the behavior first appear?) with a definition from epistemology or genetics (where does the information for this corresponding behavior come from?) has caused great confusion in the many arguments about this topic. Above all, the rearing of animals for experiments with deprivation of experience which is hardly pleasant for the animals as well as being scientifically irrelevant was a direct consequence of this confusion. However, Lorenz himself could have put an end to this nonsense since his idea of an, as he first called it, "innate schoolmarm" (at the time, this was thought as an ironic attack on Lehrman, cf. Lorenz 1985, p. 280), and later, in a more technical manner "innate teaching mechanisms" (Lorenz 1977, p. 88) on which meanwhile a significant part of the new kind of learning research is based (cf. Gould and Marler 1987, Marler and Terrace 1984) had already pointed to the right direction. Unfortunately, he not only believed in an important categorical difference:

... here it was shown that we must distinguish what is instinctive and inherited and what is transmitted and acquired as two fundamentally different things. If we want to make it clear how different the roles are played in various kinds of animals, we have to observe the behavior of animals reared in isolation so that the influence of transmitted information can be excluded from the start (Konrad Lorenz 1935/73, p. 199).

but he also remained convinced of a basic epistemological difference between "innate" and "acquired": Innate and acquired are not defined by reciprocal exclusion but by the method of the *input from which the relevant information is taken* which represents a precondition for every adaptive change (Lorenz 1985, p. 280). So, ultimately, he kept on believing in a strict reciprocal exclusion of phylogeny and ontogeny, namely as strictly separate or, as he said, "fundamentally different" and independent sources of information. In contrast to Lorenz who never surmounted the old dichotomy, Peter Marler and James Gould made the deserving attempt to elaborate further a truly "new synthesis" (Gould and Marler 1987, p. 85) of biology and psychology by being among the very firsts of having grasped the profound and, as we still will see, absolute, i.e., 100% causal dependence of the latter from the first:

The emerging picture of learning in animals represents a fundamental shift from the early days of behaviorism. ... It is now understood that ... the animal is innately equipped to recognize when it should learn, what cues it should attend to, how to store the new information and how to refer to it in the future. Even the ability to categorize and perform cognitive trial and error, a process that may be available to the higher vertebrates, may depend on innate guidance and specialization. ... The idea that human learning evolved from a few processes, which are well illustrated in other animals, to fit species-specific needs helps to bring a new unity to the study of animal behavior and a new promise for understanding human origins (James Gould and Peter Marler 1987, p. 85).

So what has remained of those supposedly important differences between "innate" and "learned" if any at all? Not much, in principle, only a completely insignificant one. Various complex ways of behaving solve variously complex problems, there is not much more to it. The blinking reflex is mainly involved with the protection of the eye from external sources of danger, e.g., dust, water, rapidly approaching objects, whereas the solving of somewhat more difficult orientation problems such as, for instance, reaching for "unreachable" food (Heschl 1993b), the finding again of "invisible" objects (Etienne 1984), or even the construction of a purely imaginary "reality" (Piaget 1950/99), call for strategies a little bit more clever than a short muscular reflex action. Not to mention the solution of tricky mathematical problems. But what is so astonishing about this trivial difference?

9 Edelman's Errors

The intellect is the greatest unknown
in his head, ... If the idea, to get some
idea, was a good idea, I now have absolutely no idea of an idea.

Johann Nestroy

The true problem concerns the origin of new cognitive structures.

Jean Piaget

Before we concern ourselves with further questions about the problems of real cognitive progress, it is necessary to discuss a thinkable alternative to the evolutionary interpretation of learning presented here. This alternative states, at the level of the neurons, a continuation of the neo-Darwinian theory in relation to the explanation of the human intellect. The new theory has been named neuronal Darwinism by its founder Gerald Edelman a former immunologist. The idea concealed behind this catchphrase is as simple as it is attractive. Edelman claims nothing more than it is legitimate to compare the fate of single neurons within complicated neural networks with that of single individuals in the Darwinian *struggle for survival* and this not just formally but to consider them being identical (Edelman 1989).

His considerations on this subject begin with the supposed insight that Darwin's theory of evolution was unable to supply us with a well-founded explanation of the human capacity to think and to learn and it had therefore become necessary to complete Darwin's great work with this in mind. Let us assume that this view is indeed correct, we would then be presented with the rewarding task, we would hope, of producing a kind of selection theory for neurons. First, we would have to solve a not exactly easy problem since we must accept that the number of nerve cells in most vertebrates after a certain stage in development remains relatively constant and subsequently that there is a substantial drop in number due to aging and to wear and tear and therefore can have very little or nothing to do with the acquisition of cognitive abilities. In other words, the great host of nerve cells is a group of cells that are rather successful at surviving with their duration of life considerably exceeding that of all the other cells of metazoans (e.g., blood cells, cells of the immune system, germ cells, etc.). Even if we accept, for the sake of the theory, that neurons are really involved in a Darwinian struggle for survival, we would have the greatest problem with the death of individual cells which has to be understood as the result of a breakdown caused by their environment, since we know in the meantime from tissues with a much greater turnover in the number of cells that this dying, as a rule, is anything but passive and instead, seems to be a well-controlled molecular process of directed self-destruction. The latest investigations into the fascinating phenomenon called *apoptosis* (Wyllie 1980) or, if epithelial cells are concerned (Puthalakath et al. 2001), *anoikis*—genetically programmed cell death—present a quite different picture of extremely complex processes of well-controlled cell extinction and survival in metazoans:

The wonder of multicellular organisms is that each individual cell seems to know what to do, where to be, and how to behave. Such remarkable self-organization relies in part on the surprising alacrity of cells to commit suicide, a process called apoptosis, should they stray or be misplaced from their normal somatic compartment and so become deprived of the requisite social signals needed for their survival. Nowhere is this phenomenon more evident than in epithelial cells, which derive much of their positional information from their association with their neighbors and with the extracellular matrix (Abigail Hunt and Gerard Evan 2001, p. 1784).

In contrast to the normal necrotic cell death, a pathological form of cell death due to acute cellular damage caused by external influences such as mechanical or chemical injuries, apoptosis and possibly other forms of directed cell death (Joza et al. 2001) represent a kind of genetic self-destruction program where the presence of specific signals at the cell membrane (Nagata and Golstein 1995) induces a kind of purposeful suicide (more details of this process can be found in Steller, Nagata, Golstein and Thompson 1995; for the connection with the metazoan evolution, see Ameisen 1996; for the importance to a normal development, see Song, McCall and Steller). The decision to commit suicide follows within a few hours of the instruction and the remains of the dead cell are consumed by phagocytosis by specialized cells of the immune system which means that many of the still energy-rich biogenic substances of the destroyed cell are recycled to some extent. A very detailed analysis of this process in a leech with the flattering name *Caenorhabditis elegans* brought to light the participation of 14 different genes (Ellis, Yuan, and Horvitz 1991; Hengartner and Horvitz 1994).

In the meantime, what has been shown by this gene program is that it is something very old and wide-spread since, in very different taxonomic groups, such as worms, insects, and vertebrates, identical sections of which have been found in the DNA. The analogy between the extinction of cells within a metazoan and the extinction of a genetic line through Darwinian selection is therefore not just a difficult one but also a quite misleading one since genetic lines of descent are, as a rule, not there with purpose and preprogrammed intentions to maneuver themselves directly into evolutionary extinction. Yet let us return to Edelman's ideas in spite of all these basic contradictions and let us look at them more precisely, what then remains are changes to the brain which can perhaps somehow support these ideas.

What does indeed change in part dramatically, in contrast to the comparatively stable number of nerve cells in a newborn vertebrate, during ontogeny, is both the strength and number of connections between the neurons. This can, in humans for instance, mean an increase in the weight of the brain by a quarter in babyhood until ultimately four times its initial total weight in an adult (average: 1400g, i.e., 2% of body weight; Deacon 1992). The thesis of a real Darwinian selection in neurons must be qualified right at the start since obviously, such a selection, if occurring at all, is confined mainly to the formation of the synapses. Edelman naturally noted this and he thus restricted the application of his model of a neuronal Darwinism to the maintenance or selective reduction of nerve connections. Although the basic nervous circuitry of the most important connections in the central nervous system (CNS), consisting of brain, spinal chord, and sensory (coming from the sense organs) and motor (going to end organs e.g., innervation of the muscle fibers) nerve paths, in a more or less direct way derive their structure from the effect of genetic instructions (Tanabe and Jessel 1996), a great number of new connections are formed during the course of the embryonic development and to some extent right up to adulthood and these new connections can later undergo the most varied changes which can be demonstrated quantitatively (Kaas 1995; Levitt 1995; Neville 1995).

Here, at least, Edelman appears to be correct since the term "selection" is constantly used even in technical jargon to describe the development of functioning neural networks. Essentially, what is meant is changes in the number of branchings and concomitantly the potential strength of the neural connections (axons, dendrites) between the affected regions of the brain. Not a few neurophysiologists go as far as to say that these refinements and restructuring of the basic neural connections correspond to learning processes such as we know from the higher levels of behavior and can therefore can be viewed as more or less one and the same and thus investigated under identical perspectives. Consequently, Edelman's neuronal Darwinism can be constructed from existing opinions and therefore enjoys great popularity among neuroscientists. The reference to learning is especially important to our discussion and extremely informative, since it does indeed indicate a close connection that must exist between two completely different levels of physiology. Some theorists have claimed all along that learning is random behavior and from its environmentally determined results new information may be acquired and neuronal development can therefore, as recently surmised by some neurobiologists, be a kind of learning since the genetically given basic structure of the nerve system can obviously be refined by environmental influences. What exactly does happen during the neuronal development of a living organism?

Calibration or fine tuning, but not in the sense of Piaget's Lamarckian "accommodation" (Piaget 1971), is the best term for all these numerous processes that, from single neuron up to whole organ (for the functional necessity of the chick embryo to "train" its legs when still enclosed within the egg, see Wu et al. 2001) or even whole-body phenomena (e.g., long-term effects of weightlessness), take place during the development of the basic neural structures into a fully functioning nerve network. A concrete example may help us to visualize this better. In order to see in reality, that is to obtain fairly good internal pictures of the outside world („pictures" of course meant metaphorically since in our brains it is dark, unusually quiet, and rather scentless), we need not only two intact eyes with their light-sensitive receptors in the retina (fovea: approximately 160,000 cones/mm^2 for color vision) but also at least three further structures in the brain. Firstly, the activity of numerous neurons in the eye itself is needed to transmit the excitation from the sensory cells at the periphery to the inquisitive center. This task is carried out by the two optic nerves, which from their ontogeny are outgrowths from the brain and not part of the sense organs. The first station in the brain itself where the fibers of the optic nerve switch over to other neurons lies in the geniculate body of the diencephalon.

This goes on further until we reach that first central position in the nervous system of vertebrates where a first neuronal equivalent of the environment can be produced. In the primary visual cortex the route from the external stimulus to the central representation ends for the time being, which now, and that is what is so special about it, is theoretically available to the whole organism as up-to-date information. During embryonic development a basic circuitry of eye-geniculate body-visual cortex is laid down through the agency of a multitude of genetically coded messenger substances. This is insufficient, however, to produce a fully functional visual system at birth, since the wiring to the various destinations is still much too inaccurate to guarantee reliable functioning, especially in the geniculate bodies and the primary visual cortex where a topological representation is necessary. Consequently, a subtle fine tuning of the whole system is required. As a rule, this takes place in such a way that only correctly functioning connections are not just not reduced but considerably strengthened while incorrect connections are broken down or at least strongly reduced. Another clear proof of Edelman's selective neuroDarwinism? The details of the processes involved tell a different story.

In order to make it possible for the brain to analyze the right and left field of vision simultaneously to give us binocular vision with all the special advantages of high three-dimensional resolution, we must ask ourselves the question of how such an exact and permanently reliable topological, i.e., spatially perfect attribution of the two sensory fields can be achieved. Because of the rapid growth of the embryo in the early phases of development, there are continuous changes in almost all the spatial relationships so that a rigid and geometrically predefined structure would not be up to the task. Here, living systems must take a completely different route. They are architects and builders at the same time. They simply calibrate themselves by using certain internal and external parameters to make the corresponding corrections. In the case of our eyes, the process used goes as follows. Neighboring neurons in the retina fire salvos of impulses even before birth at definite intervals in the direction of the visual cortex that, however, can only be reached via the intermediate station of the geniculate bodies. These salvos of simultaneous neural excitations allow the geniculate bodies to define the respective regions for the left and for the right eye. The method is, when we assume a highly uniform speed of the impulses in the axons involved, relatively reliable since only the simultaneously excited neurons in the geniculate body can originate from one and the same eye.

The end-result of such a process of self-structuring is a clearly recognizable morphological subdivision of the geniculate bodies into parts for one eye or the other as well as for the narrow region that is innervated from both eyes. The subsequent, even more difficult step consists of transferring this first neuronal allocation structure as much as possible unchanged to the primary visual cortex, however, with the additional condition now that the relatively small regions for the left and right eye must be realized in adequately corresponding, i.e., the same relationships as from the field of vision, as columns side by side. What is meant here is that the same scene in the environment that is perceived by both eyes at the same time must indeed be represented neuronally at a *single* position in the visual cortex even though separated from each other. For this purpose it is now essential to call on the perception of real objects of the outside world for the fine adjustment of the visual cortex since the blind firing alone of the embryonic neurons cannot be an aid to perfect organization (for the recent discovery of both normal "forward" and "reverse" signalling and hence a potentially more precise formation of neuronal connections, see Cowan and Henkemeyer 2001). It is therefore no wonder that the acquisition of the ultimate columnar structure of the visual cortex (once left eye, once right eye, and so on) can only take place when the newborn baby experiences real seeing for the first time.

If this self-calibration through the possibility of seeing is hindered artificially by unforeseen disturbances or is only allowed in one eye, then corresponding deficiencies or one-sided abnormal development in the representation of the field of vision is the result. The structuring of the primary visual cortex as clearly differentiated columns does not take place at all or only one-sidedly to the benefit to the still intact eye. If such a drastic reduction of the visual faculty remains for a relatively short critical period after birth, the lack of this calibration shows itself to be partly irreversible and can result in considerable shortcomings in later visual processing (Crair, Gillespie, and Stryker 1998). In spite of this reduced flexibility in the interconnection of the basic structures of our brains, a permanent and continuous updating and fine adjustment of the complete wiring takes place. During this kind of on-line adjustment a certain regular change between self-induced spontaneous activity and externally induced excitation shows an obvious evolutionarily tested solution to this difficult morphometrical problem of self-calibration (see Katz and Shatz 1996). That such self-calibration processes do indeed take place throughout our lives in all our active sensory and motoric systems has, in the meantime, also been proved at the levels of both behavior and perception. So, for example, the influence of an additional *Coriolis force* during artificial rotation leads to corresponding compensatory reactions of the whole body that are able to reestablish the original precision of the most various types of movement (Lackner and DiZio 2000). However, internal influences can also be important, as for instance the amount of glia cells in different regions of the brain which seems to have a yet unknown impact on both the formation and plasticity of synapses (Ullian et al. 2001).

According to the extent of the displacements of activity brought about by a change in the environmental conditions, one can expect, within certain limits, corresponding, preprogrammed, morphological changes in the relevant neural structures. A good example of this are injuries or other disruptions of function caused by a poor supply of blood to nervous tissues such as by a stroke— in such cases, the organism attempts with the remaining intact tissue to produce a kind of substitute structure which to a certain degree can replace the original function (Nudo et al. 1996; Sober et al. 1997; Xerri et al. 1998). A somewhat artificial and strange experiment with an adult owl monkey (*Aotes trivirgatus*) has recently demonstrated this kind of flexibility for a comparatively normal change in activity. The poor monkey was "allowed"—in fact, he had to be well rewarded for doing so—to keep a disk moving for one hour a day for three months with the limitation that he could only use the second and third and occasionally the fourth finger to do so. After a successful scientific autopsy of the participant the results were more than clear. The regions of the somatosensory cortex corresponding to the preferred fingers had expanded at the cost of the others (Recanzone et al. 1992). Something similar has recently been discovered in violin players who of their own freewill undergo a comparable torture—the extensive training of certain fingers is meant. So the four fingers of the left hand, which have to do the main work in Rimsky Korsakov's "Flight of the Bumblebee", are clearly more strongly represented cortically than all the other fingers, whereby, the extent of this change correlates very obviously with the age that the violinist began to play (Elbert et al. 1995).

Let us now return to the theory. If the so-called neuronal Darwinism wants to be a serious alternative to a real Darwinian interpretation of learning which is propagated here, then this calibration process in the brain should proceed in a completely different way. First of all, every single connection between nerve cells then should originate in a fully random way. This would mean that in each individual organism a completely new spatiotemporal pattern of more or less chaotic proliferations of nerves should be expected whose equally random connections would thereafter undergo a real Darwinian selection process. Both views can hardly be correct since neither are the numerous new endings and connections formed in a truly random fashion in time and space, nor do poorly functioning synapses subsequently die owing to lack of adaptation to their neuronal environment. Instead they are reduced as a result of a proper genetic program. The neurobiologist Carla Shatz phrased this fact quite succinctly:

The precision of these connections are one of the most astonishing things about the brain. None of it seems left to *chance* (retranslated from C. Shatz 1992).

This is unaffected by the seemingly anatomical "jumble" of the cortical connections (Braitenberg and Schultz 1989). It is indeed the axons, extending purposefully in the correct direction of their well-defined target structures, be they sense organs or other brain structures, which develop only when they arrive at these structures and in addition only here do they multiply their branches. Neither do the axons perish from a simple lack of some essential resource but if it is necessary that they disappear, then they do it by themselves dissolving according to definite and relatively well-known molecular rules (Barinaga 1995, Marx 1995, Thoenen 1995; Acheson et al. 1995; Frank 1997; Colman, Nabekura and Lichtman 1997). This means that what is taking place here is nothing other than a well-controlled morphological rebuilding process in which the whole structure of the brain with its neurons survives and very definitely not a Darwinian struggle among some mutated neuronal monsters over the privilege of the human intellect.

Some of the most important mechanisms that form the basis of these apparently Darwinian processes have, in the meantime, been investigated down to the finest molecular details. The most well known are probably the processes at the so-called Hebb-synapse, where the strength of the connection is only permanently fixed if both neurons that are involved at the connection are simultaneously activated at regular intervals and these should be not too long. When a pre-synaptic cell stimulates a post-synaptic cell, the latter must be in an active state so that the connection between the two can experience a long-term potentiation (Nicoll, Kauer and Malenka 1988). In principle, what lies behind this is none other than the fascinating, effective, and purely neuronal realization of Hume's *principle of coincidence* with which, as already suggested, a not inconsiderable number of fine adjustments can be made. However, the dissolution of a synapse is more than just a passive not-being-strengthened because of a lack of neuronal synchronization. Synapses have long been found in various structures of the brain (e.g., hippocampus, cerebellum, visual cortex) and their strength of transmission was not just not increased by asynchronous activity of the neurons involved but drastically reduced. One assumes with great probability that, for this reason, a concrete molecular mechanism for such an activity-related weakening of synaptic connections must exist.

The Hebb-synapse is not the only possibility that exists for describing simple "learning" behavior at the neuronal level. Normally, the Hebb mechanism is operating in cases of so-called *implicit learning* whereby the associated stimulus is repeated many times and only slowly learned. Another possibility is one that comes more into play in cases of relatively complex and rapid learning. *Explicit learning* takes place. in contrast to the subconscious learning of things in implicit learning, with the participation of conscious processes of imagination so that qualitatively novel phenomena such as recall, memory and ultimately fantastic abilities such as logical and abstract thinking can appear on the evolutionary scene. Characteristic for this kind of learning is the noteworthy fact that, in contrast to the rather slow inactive familiarization, even the very first confrontation with a certain situation can already lead to a clear-cut result. This learning then comes very close to the purely researched phenomenon of the subjective aha-experience, known to us as the creative moment of a new observation or even discovery. Different to the Hebb-synapse, we do not find an automatic synaptic strengthening of two nerve cells simultaneously activated and excited one after the other but rather a third so-called modulatory neuron is brought into play to stabilize the learning process. In this case, the post-synaptic nerve does not need to be activated at the same time as the exciting neuron in order to induce a strengthening of the synaptic connection. This task is undertaken by the modulatory neuron which controls the learning process with its own directly attached synapse, the synapse of the main connection. When the pre-synaptic and the modulatory neurons fire at the same time, the connection is strengthened even if the post-synaptic neuron is wrapped in silence at that moment. This outward seemingly unimportant modification of the simple Hebb connection is of great consequence for the whole of the nervous system since now a high degree of parallel interaction between various and even widely separated regions of the brain has become possible. It appears that here in this conscious area an impulse-for-impulse-synchronization accurate to milliseconds plays a decisive role (Singer 1995) and not as Edelman recently speculated, alone the number—as large as possible (Edelman's somewhat cryptic estimation goes: "Millions of neuronal interactions", Ahuja 2000)—and "complexity" of neural interactions (Edelman and Tononi 2000). The first time such a modulatory variation of inter-connection was discovered by Kandel and Tauc in 1965 in a well-known sea-snail (*Aplysia*) where, however, it has still to do with purely implicit learning.

The two variations of neuronal strengthening mentioned so far present a highly interesting connection between the development and differentiation of the brain on one side and its function in cognitive processes themselves on the other side. We can even dare to set out the hypothesis that ultimately the mechanisms are one and the same, naturally with exceptions as far as the spatio-temporal parameters of the occurrence and the extent of the effective restructuring are concerned. This indissoluble connection between form and function becomes immediately clear if one observes more closely what has to happen for a short-term synaptic change to become permanent, i.e., from an initial tentative association to become a long-term memory. Experiments with mammals, but also with the brightly colored sea-snail *Aplysia*, have shown that the consolidation of learning takes place in well-separated phases. Firstly, a so-called short-term memory retains the learned item by means of a long-term reinforcement, by which is meant an increase in synaptic efficiency that lasts for minutes, hours, or even weeks and this can be adequately described with what was previously stated about the varying strength of synaptic connections. With the transition to a real long-term memory, other additional processes come into effect in which the continuous transition between morphological structure and physiological function cannot be clearer. New genes are activated for the expression of new proteins and finally there is a growth of new synaptic connections. Recently, a particular significance of the removal of permanently inhibiting influences on memory storage has also been recognized. The genes that here play a central role have now been termed *memory suppressor genes* (Abel, Martin, Bartsch and Kandel 1998) analogous to the already well-known *tumor suppressor genes* (Knudson 1971).

At this point, let us deal briefly with the general question of the how of a causal interrelationship between form and function since there is a misleading view that certain functions, in particular those which have to do with behavior, can exist to some degree independently of their morphological substrates and therewith achieve degrees of freedom that go beyond their mere biological-material limitations. Often in this connection, use is made of the very popular though completely wrong comparison of biological *hardware* which, as every computer freak can appreciate—and who is not one nowadays—can run very varied *software*. As applied to the human intellect, this would mean that our respective biology is a hardwired neural structure while the actual interesting thing about it is that using it in thinking can be viewed as being softwired and therefore practically infinitely variable. If one really wishes to use such technical, modern-sounding jargon, then a living organism can best be understood as hard- and software combined. To present this even more pictorially, the organism is a computer that operates itself, develops its own programs, manages all the necessities for existing as safely as possible for a long time, and finally—otherwise its evolutionary chances will not last for long—founds a family of its own by producing real "smallware" as offspring. However, it is such a non-trivial test of real material survival and not just a purely formal Turing test (Turing 1950) which has to be passed by all that which has taken place since the invention of the computer under the pretentious name artificial intelligence (AI) or even artificial life (AL) (Heschl and Peschl 1992).

For the moment, however, it is sufficient to point out that the historical dividing of the whole of biology into subjects some of which concentrate above all on the biological hardware (anatomy, morphology, histology, systematics) and some almost exclusively on the software (physiology, biochemistry, molecular biology) is merely the result of the historical fact that the beginnings of a first scientific approach to nature—for a long time the school subject biology was given the romantic name "natural history"—can be found in the dead collections in the museums of natural history consisting of innumerable native and above all exotic creatures. This led to the paradoxical situation where some morphologists because their valuable objects were dead wanted to understand their functioning, i.e., simply being alive, as being a difficult if not insoluble philosophical problem. The so-called body/soul problem in this connection represented the climax with respect to this natural philosophy since an attempt was made that right from the start was doomed to failure—an attempt to do away with a historically determined terminological separation which simultaneously may not be really abolished—because, as every gullible person knows, nothing can exist when it is forbidden. Several hundred years ago, the French thinker, René Descartes (1596—1650) gained a lot of credit in this field as he philosophized over the seat of the mind in the pineal body. Today, we should keep in mind that modern biology has not yet succeeded in bringing any comparable phenomenon to light that could give us the right to believe that the software of life could exist independently of its hardware. It would be a rather absurd finding since we would be confronted with the parody of viewing life as being independent of living systems. Or to express this in the words of biomechanics which is particularly experienced in dealing with living examples:

Structure without function is a corpse and *function without structure* is a specter (S. Vogel and Steve Wainwright 1969).

We now come to a preliminary conclusion to our short excursion into neurobiology. What is special about the developmental processes that occur in the brain? Very simply, it consists of the organism already knowing how it must proceed. It knows exactly where it must send which axon from which position to which target structure. It also recognizes exactly when its axons must have reached the well-defined area of these destinations and it knows that here it first involves at least some branching of the end of the nerve fibers. Then it must already know how it can differentiate between correct and incorrect connections and then decide to strengthen the former and to break down the latter. Not a mention of Darwinian selection in this area of well-coordinated ontogenetic programs. But in fact, this would make no sense. The nerve cells all have identical genetic material so therefore from the elimination of the "more poorly" functioning cells one cannot expect any advance or evolution. Even worse, by applying evolutionary theory, we can expect just the opposite to an Edelman competition, since the cells are genetically identical. Hence more than in identical twins (Segal 1988) which still can lead an independent life we must expect nothing less than a maximum of *cooperation* and that is what our brain, at least as long as it is functioning more or less properly, lets us see clearly every day in the truest meaning of the word.

As soon as we talk about physiological mechanisms, whose coordinated molecular processes can be investigated, we can no longer speak of randomness. Real physical chance can only be described from its general conditions, i.e., can only be described indirectly since, by definition, one cannot make any reliable prognosis about a random process. So, for example, the fall of dice, or the result as the number of spots can be described with expressions from the theory of probability without being able to forecast the next event reliably. In this way, the mathematical theory of probability is nothing more than an attempt to form a general theory of chance which, from itself, excludes any prognosis being made about the underlying processes. The neuro-physiological phenomena show us, however, that this kind of chance in research into cognitive performance is out of the question. Just the opposite, modern brain research is aimed in a completely different direction, namely at the increasingly better understanding of the basic molecular mechanisms and their integration into an evolutionary framework of an exponentially expanding discipline:

Should this be valid on the molecular level, should the same molecular processes in the development and differentiation of the brain proceed in part as in learning, then learning research would bring together cognitive psychology and molecular biology, as well as the human organism in a much more comprehensive manner than ever before. With such a uniform research initiative, mental processes would be demystified much faster. These investigations would then find their correct environment—in the evolutionary oriented edifice of biological thinking (retranslated from Eric Kandel and Robert Hawkins 1992).

The last point in this interesting scientific prognosis can, in principle, be decided today since evolutionary theory gives the only correct answer and it states that individual and therefore our own personal learning has nothing to do with a real progress but rather can only be the finished product of preceding phylogenetic and hence genetic evolution. For the immune system, which for Edelman was still the paradigmatic system for his later-developed neuronal Darwinism, this was very recently confirmed in an impressive way. Even in this field, it was established that what took place was just the opposite of Edelman's concept of a Darwinian selection of randomly produced antibodies which produce the relevant antigen through their adaptation to the enemy they must combat (be it chemicals, bacteria, or viruses). Organisms which possess a complicated chemical defense system, as do practically all *metazoans*, do not acquire their immunity selectively against a specific pathogen by a Darwinian adaptation process to it—this would indeed be a new acquisition in the course of ontogeny. Instead, they show that their immunity is only due to the necessary information for their defense against highly specific invaders being already present as "well-thought out" (or even better, well-proven), i.e., phylogenetically evolved instructions in a limited set of certain gene combinations. The end result of this finding says nothing more than that the supposedly *acquired immunity* is not, as generally believed previously—today, one speaks of "adaptive immunity" (cf. Encyclopedia 2000) —, acquired by an individual as a new characteristic during its lifetime, but rather that due to the long phylogenetic evolution of each individual metazoan has long been present in a finished form (Tough, Borrow and Sprent 1996; cf. Wedemayer et al. 1997). Therefore, nowadays, one differentiates, with good reason, between immune selection and true natural selection since although the initial conditions for variability are partly similar (somatic mutations, germ cell mutations), the selection processes involved are fundamentally different—namely internally prepared in one case through the controlling function of so-called innate immunity (Dempsey et al. 1996, Fearon and Locksley 1996, Fearon 1999, Brown 2001; for the difficult "Art of the Probable: System Control in the Adaptive Immune System", see Germain 2001; for its age of about 700 million years, see MHC Sequencing Consortium 1999) versus externally "unprepared", i.e., environmentally induced in the other case (Barbas III et al. 1997, Heschl 2001). In the meantime, it has already been proved that newborn mice possess a fully competent immune system even if it is still low dosed (Ridge, Fuchs and Matzinger 1996; Sarzotti, Robbins and Hoffman 1996; Forsthuber, Yip and Lehmann 1996). In the words of two researchers directly involved into the new discoveries about the true nature of the immune system:

The essence of innate immunity is the detection of molecules that are unique to infectious organisms. This capability allows the innate immune system to guide the selection of antigen by B and T lymphocytes, and the secretion by helper T lymphocytes of cytokines that promote an appropriate host response to the infection. Therefore, mammalian innate immunity is not merely a vestige of ancient antimicrobial systems that have been made redundant by the evolution of acquired immunity. Rather, it dictates the conduct of the acquired immune response. ... Here we show that cellular and soluble components of innate immunity provide instruction that enables the acquired immune response to select appropriate antigens and the strategies for their elimination (D. T. Fearon and R. M. Locksley 1996).

However, the most direct proof that the adaptive system is closely dependent on the instructions of the innate systems comes from the existence of two mutant strains of mice, first, the RAG knockout mice which are unable to rearrange the genes that are central to adaptive immunity and, second, the STAT-1 knockout mice which lack those y-interferon mediated pathways that are required for the interplay between the adaptive *and* the innate response (Pardoll 2001; see also Du Pasquier and Litman 2000). Finally, to close the circle by coming back to neurobiology: concerning the seemingly unresolvable problem of a both determinist and nevertheless flexible neuronal specification, it was Jacob Monod who some years ago first foresaw that even the complicated dynamics of human brain development can be explained in principle by recurring to preexisting genetic information:

We know that there are about 10^6 genes (today: 10^5) which one could theoretically use if one considers the quantity of DNA contained in the cells. In fact, only a fraction of the DNA is reserved for structural purposes, in such a way that there are no more than 10^4 structural genes available. In every cell, you have to recruit a certain proportion of these genes to assume the normal functions that every cell has to assume. 10^3 genes, that is about what is needed to guarantee the biochemical functioning of an elementary cell such as a bacterium. Now, we need 10 structural proteins to label a particular cell. We can choose these 10 structural proteins from among those 10^4 nonutilized. Consequently, we have the choice between something like 10^{40} labels, which is more than is necessary to construct the CNS (central nervous system). How do we then make the choice? 10^{40} is only slightly different from 2^{130}: hence we need 130 bits or binary decisions to arrive at a label, by utilizing only a small fraction of the genome's information to make the correct choices, even for the specification of an extremely complex organ (translated from Jacob Monod 1979, p. 291).

10   On the Heritability of Jazzophilia

The old problem of the relation between heredity and environment in
evolution ... has, I think, largely been
cleared up by the recognition that the
capacity of an organism to respond
to environmental stresses during development is itself a hereditary quality.

C. H. Waddington

We will find the individual genes,
even those for complex character
traits which are affected by several
genes.

Robert Plomin

If learning has a lot to do with complex behavior but nothing to do with real cognitive growth, we then must ask ourselves what is the purpose of our attempts, often with a great expenditure of time and money, to determine the quantitative extent of so-called heritability of various kinds of behavior. Heritable is that part of behavior that is supposedly genetically fixed and therefore transmitted to the offspring. What remains of the very same behavior is then consequently ascribed to the determining influence of the environment so that, in the most simple cases purely in an additive manner, the sum of the two factors decides the actual expression of a type of behavior. In many cases, it has been shown that the kind of interaction between the genome itself and the environment can influence the form of a type of behavior. Nevertheless, in all these cases, we start out from the concept that in principle it is possible to differentiate between these two forms of an instruction to the organism. The intrinsic scientific purpose of this division is that a biological theory of behavior must be particularly interested in the genetically mediated part since this alone is of significance for evolution. Everything else caused by various environmental influences and partly by purely coincidental factors is devoid of any such special significance since the empirical refutation of Lamarck's ideas is an established fact and that there can be no inheritance of acquired characters. The efforts of the biologists to peel away the, for them, all-important evolutionary part from all the observable characters is therefore easily understood and should, therefore, be viewed without any of the many supposed implications for an eventual moral judgment of human behavior. The strong need to pass the latter has produced some odd quirks in the field of experimental psychology and this is no coincidence. The perennial debate over nature versus nurture is so overburdened with ideology that no attempt should be made here to describe the all too well-known details (cf. the "sociobiology-battle" described very vividly in Segerstrale 2000). What is of much greater interest within the framework of an evolutionary theory of cognition must be related to the basic methodological justification of such a determination of heritability. Should this turn out to be criticizable or even refutable regarding its axioms, it would have inevitable consequences for what has been the customary splitting up into nature and nurture and the polemic that has been associated with it in the open debate.

Let us begin with a very simple example from biology to unveil the essentials of the method of heritability determination. In a standard work on evolutionary genetics (Maynard Smith 1989) the following data, obtained from fruit flies of various origins, are given as typical source material for further determinations:

Wing length of *Drosophila subobscura* (standard units):

Population	Temperature during development	
	15 °C	25 °C
Scotland	130,2	109,8
Israel	20,3	99,7

On closer examination, it becomes apparent that the wing length of this tiny little fly is dependent in a simply additive way on the genetic constitution (Scottish or Israeli flies) as well as on the thermal environmental conditions during the breeding of the animals at different temperatures (15 or 25 °C). On one hand, a temperature difference of 10 °C causes an almost identical change in the length of the wings (20.4 and 20.6 units). On the other hand, the difference between both populations independent of the breeding temperature is 9.9 and 10.1 units. The situation seems in fact to be extraordinarily clear and allows only one interpretation to be made. The total variation V_{Phen} with regard to the wing length (phenotype) of *Drosophila subobscura* is composed of two independent components, a hereditarily(H) determined variation V_H and an environmentally(E) determined variation V_E. Consequently the following equation must be valid: $V_{Phen} = V_H + V_E$, whereby this formula can be related to any phenotype or biological character. And indeed, a better result could not have been envisaged even by the most optimistic experimentalist: 30.5 = 20.5

+ 10.0, what more can one say. The principle behind this division into two parts says nothing other than that all measurable differences whether found in behavior, physiology, or in "rigid" morphology (see wing length) must be genetically or environmentally determined. Only the former are regarded as worthy of consideration in the context of evolutionary theory, whereas, the environmental influences are judged as being troublesome. They will therefore be eliminated immediately for the further discussion for clarity. In the model, this means even more consideration is given to the remaining purely genetic influences as they are regarded as decisive, whereby, in such simple cases such as the wing length in *Drosophila*, the effects of genes on certain external characters are viewed as being additive. For an efficient determination of the total variance, which can even be split further into various sub-variances (real additive variance: see the example; *dominant* variance: an allele A dominates the expression of a character over allele a; *epistatic* variance: a nonallelic gene suppresses the effect of A or a) a great number of data are necessary on large populations or, at least parts of them, since otherwise a statistical treatment of the topic makes no sense. Therefore it is unnecessary to emphasize that the single individual in such population-genetic analyses cannot play a decisive role or that no concrete statement can be derived from them. The fact that this is so often what is attempted during the various controversies will be dealt with in greater detail somewhat later.

For the moment, it is interesting to see that the purely genetic model as used for evolution-theoretical determinations is quite inadequate for the prognosis of single empirical events. This fact is not just a little astonishing since originally we proceeded from the premise that only genetically determined influences can play a role in biological evolution. For example, there is the remarkable observation that even isogenous, i.e., through inbreeding largely genetically identical ancestral lines can throw up clear variance in their characters. Since V_H tends towards zero in these cases, this variance must be a consequence of the environment alone, i.e., V_E. Agreed, a minimum of variance remains even with identical genotypes *and* identical environmental conditions. When we accept that slight and uncontrollable environmental differences no longer play a role, all that remains is the possibility of special system-internal cycles having an effect during the course of ontological development. Since, in the meantime, the appearance of states of deterministic chaos have been demonstrated in living systems (Markus et al. 1984; Hess and Markus 1985; Schiff 1994), this kind of variance can still be explained. Independent of this, the nature of living systems lies in the taming of entropy and therefore reducing the danger of disorder-producing physical chance (Hess 1997). However, let us return to genetics: in addition to all this variability, there is an even more astonishing result, namely that inbred lineages can display greater variance in their characters than the original mixed population from which they are derived by complicated breeding processes. This demonstrates the necessity of also including a certain degree of influence by non-additive effects of special gene-environment interactions since, interestingly, the environmental influence on the phenotype seems to be greater on inbred populations than on heterogeneous ones.

Without wanting in any way cast doubts on the internal logic of heritability determinations or to attempt to refute the applied calculations on the basis of their underlying algorithms,—it would be a fruitless undertaking since the purely formal correctness is undisputed—we want, however, to make an attempt at investigating the significance of such models more closely with respect to the status of learning behavior as far as evolution theory is concerned. Let us go directly to the crux of the matter and try to integrate learning itself into population-genetic considerations which are based on the concept of the inheritance of characters. Learning behavior is, how could it be any different, conventionally understood as being a character whose existence is due to external factors which are to be found in the environment. In other words, the effect of genetic influences can be virtually set, axiomatically and universally valid, at zero ($V_H = 0\%$) which means that the environmental influences must be at a maximum in our formula of heritability ($V_E = 100\%$). Consequently, the empirically determinable variance of one kind of learning (e.g., maze learning) would be caused alone by the environment and this completely independent of any genetic factors. Therefore, every learning behavior should be immediately removed from all evolution-theoretical considerations. But then, what good is a population-genetic investigation into the learning ability of a certain species if this kind of behavior, as a character now already in many respects taken to be non-biological, is completely determined by the environment? One can recognize at once that such an approach must be doomed to failure since any close reference to biology is consciously excluded. Only radical environmentalists would agree with such a view of learning and continue to claim that learned behavior—and ultimately even independent of the species being observed—represents exclusively the result of an instruction to the living organism by the environment. People such as Hull, Watson, and Skinner and other dogmatic learning theorists were indeed not very far away from this viewpoint. In the meantime, a kind of tacit compromise has been established where it is assumed that learning is indeed somehow influenced by genetic factors at least peripherally but the relationship between environment to influence of heredity was a specially variable one. To some degree, this new development can be explained by experimental investigations such as those by Garcia into species-specific limits of certain kinds of conditioning processes but also by the rapidly increasing influence in this field of the behavioral biologists who, from the beginning, were much more oriented towards evolution theory—a field which, up till recently, was the sole domain of behavioristically oriented human psychologists. Above all, it was the evolutionary theorists themselves who had to be interested in the integration of learning behavior into the theory. Therefore it was not astonishing that, with time, the once strongly resisted concept of a biologically influenced limitation or even detailed pre-structuring of learning has become accepted at least in its initial stages. There was no alternative since all the many new results pointed in one and the same direction. Learning could then be understood for the first time as a special biological character with distinguishing features, as a result of instructive environmental influences but—and this is the interesting point—that could not be entirely independent of genetic factors.

With the knowledge that learning is not quite as simple as the early environmentalists and learning theorists would have liked it to be, namely completely independent of any genetic influence, an extraordinarily keen interest began to manifest itself in what is the actual proportion of the inherited versus the acquired in learning by animals and humans. In humans, however, these investigations were always enveloped in controversial ideological arguments so much so that it was often difficult, if not impossible, to sort the chaff, i.e., what was ideologically read into or added, from the wheat, i.e., the real empirical results. Environmentalists were obliged by their scientific confession to take the side of the importance and the indispensability of environmental influences in the expression of any human behavioral character, whereas the evolutionists dedicated themselves to their creed which primarily consisted of reducing the environmental factor to a negligible quantity. The disputes were, as a consequence, rather vehement and have remained so to some extent right up to the present day (Segerstrale 2000). Behind all this, one found and still finds the political sympathies and opinions which, when labeled "conservative" swear on the inborn capabilities and talents of their electors and when labeled "progressive" put all their hopes into a hopefully increasing number of ones who can both learn as well as be taught.

Scientific investigations into the heritability of behavior did not begin only after the downfall of that extreme theory of learning from the early twentieth century which was categorically based on learning representing a knowledge-increasing instruction from the environment. One such classic example study took place as long ago as 1934, when the American Paul Wilson introduced the first really scientific and systematic method, that of using twins. The observation of twins and in particular the comparison of twins developing from one egg cell (monozygotic) and twins from two egg cells (heterozygotic) promised to deliver more exact information on the actual influence of what was inherited from the parents on the behavior of the offspring. The complicated double helix structure of DNA was at that time an unknown and was only decoded in 1953 by Watson and Crick. The discovery that genetic material was made of a nucleic acid had, however, been made 25 years previously in 1928 by Griffith who was working on a bacterium (*Pneumococcus*). Twins from one egg cell and therefore genetically identical were expected to demonstrate behavior more similar to each other than twins who were not genetically identical. This assumption had been more or less confirmed some time before by researchers such as Dahlberg (1926), Holzinger (1929), and Siemans (1924). However, their work proceeded from the assumption that it was valid to consider the environment of the twins investigated as being perfectly identical, above all when they had been brought up within the same family. Wilson was the first to examine this assumption more closely and ascertained something astonishing for that time. First, he noticed that, even between separately born siblings and siblings born at the same time, large differences exist in their environmental conditions. We should not be surprised that twins because of their identical ages experience very similar environmental conditions. Then it was established, just the opposite to the earlier assumptions, that the environments for non-identical twins were more varied than for monozygotic ones. Wilson explained this situation by saying that the different genetic makeup had led to a corresponding different active selection with regard to the environment by the non-identical twins and concluded that the actual influence of the two elementary factors can only be estimated with very great difficulty.

In the meantime, the research into the heritability of intelligence has degenerated into a regular boom and almost every year, sensational new results on the quantitative relationship between nature and nurture in humans are published. They all show more or less again and again the same important results: Intelligence or general mental ability g, operationally defined as intelligence quotient IQ, (1) in the first place has to do with the individually different and genetically determined speed of information processing (Fink and Neubauer 2001), remains incredibly stable across life span (McClearn et al. 1997, Deary et al., in press), and last but not least influences a great variety of various life-style and life-outcome variables (Brand 1996). A comprehensive overview of modern research into individual differences with regard to psychological variables is given in Lubinski, D. (2000). In 1990, an extensive American study (Bouchard 1990) appeared which is still today treated as the standard empirical work on the topic. In contrast to many other previous investigations, this was dedicated to the study of the early childhood of twins living separated from each other, regardless of whether they were monozygotic or were from two separate eggs. In a ten-year investigation, more than a hundred such pairs were subjected to a great number of intensive psychological as well as physiological tests and the results were analyzed statistically. The result in brief was that approximately 70% of the variability of the intelligence quotient was demonstrably linked to genetic factors. However, there was worse to come; the most diverse data on personality, temperament, career and leisure-time interests, as well as social attitudes of monozygotic twins brought up separately were approximately the same as those raised together. As an explanation, the authors put forward the hypothesis that genetic differences cause psychological differences primarily indirectly in that they affect the selection of the effective surroundings of the developing child (see Wilson's assumption above). Effective surroundings can be understood as everything that within the framework of the interaction with the environment the child purposely chooses or perceives as such. In other words, the growing child does indeed learn entirely from experience as the radical environmentalists have always believed and may well determine the greatest part of the observed psychological variability, only the effective, i.e., the real experiences that the child makes are to a great extent chosen by him. The special way that children select their own experiences from the huge number theoretically possible can, in the opinion of the authors, only be explained by the constant influence of the genome. With this in mind, we can no longer speak of a purely additive effect on the expression of human intelligence, we must now emphasize the constant interaction between the genes and the environment. What clearly emerges here from this and similar investigations is how the classic gene/environment conflict has been severely modified if not eradicated completely. In conclusion, the authors speak of a new formula to elucidate their unequivocal results: nature *via* nurture and no longer nature *versus* nurture. This would mean that just the most heritable characters, intelligence in this case, must be accessible to an intervention from outside since the heritability in these complex areas of behavior is related to the kind of interaction with the environment and not, as believed previously, to the predetermined realization of an unchangeable end product. In a paper by Martin et al. (1986), this was succinctly expressed:

Humans are explorative organisms whose inborn capabilities and tendencies help them to select from the profusion of possibilities and stimuli what for them is relevant and adaptive. The results from versatility and learning therefore do not hinder, but rather reinforce the effect of the genotype on behavior (N. G. Martin et al. 1986).

It therefore only remains to explain briefly what the other 30 or at least 20 per cent are about—between 5 and 10% of the variance in heritability investigations is blamed on some unknown interference, a kind of experimental background noise—since they are needed to bring us up to the 100% needed for our full intelligence. If 70% of our most noteworthy species character, namely that of being excessively intelligent, is influenced genetically, then the rest must at least be able to impart that longed-for degree of independence, known as freedom of will or freedom of spirit. Even here, there are rather a lot of most varied problems. In principle, it is at exactly at this point where everything begins all over again with the concomitant criticism of the previously mentioned basic concept of addition. To be as brief as possible: there is a lot to be said for the present concept of a quantitatively variable heritability of ways of behavior being quite simply wrong in spite of the undoubted occupational therapeutic effect of the associated calculations. Without claiming completeness, the most important points of criticism will be dealt with more closely in the following.

The sharpest, though purely technical, criticism of this concept comes from statistics itself and concerns the most commonly used method of calculation, the classic two-way analysis of variance (ANOVA), the starting point for *multiple regression calculations* which are applied on a large scale. As Douglas Wahlsten was able to show recently from just a few informative examples, this kind of statistical treatment is simply not suitable for excluding those gene-environment interactions which could throw doubt on the validity of the traditional additive model (Wahlsten 1990). That means that, in many cases, one can expect to obtain absolutely fictitious numbers on the additive effects of genome and environment, which, on the other hand, gives good reason to severely criticize the concept of heritability itself (for a devastating criticism of the "the obsession with heritability", see Sarkar 1998). The previously described study of the behavior of twins brought up separately is a beautiful example. Although the authors have obtained, to all appearances, a secure result of 70% heritability for intelligence, they do not go so far as the obvious claim that this would mean this 70% would be present in a static preformed way and then be augmented somehow by environmental influences in an additive manner for the remaining percent. Just the opposite, the study concludes strictly speaking with an alternative hypothesis, namely, that only the specifically individual kind of interaction with the environment can be genetically programmed. Obviously, it is well understood here that one cannot possibly speak about a character such as intelligent behavior that to a large extent should originate without any interaction with the environment.

Another severe shortcoming of most investigations carried out by psychologists into heritability , different from those by behavioral researchers in biology, consists of specific behavior patterns not being recorded in their structural entirety with regard to the population but mostly only the hypothetical end results of some complex behavior to get easily manageable measurements for their calculations. Here, one can ask oneself whether, in actual fact, the one and the same behavior is being investigated. It is exactly here, in the area of cognition, that it is difficult to answer affirmatively. Let us take an example from the field of memory research whose testing by simple questioning about items to be memorized is indeed a quantitative matter, easily made use of, but what in fact lies behind a correct, or an incorrect answer, or even no answer at all disappears without a murmur in the physical ideal of being statistically calculable. Let us descend one or more steps in organic complexity and remind ourselves of the simple example at the beginning of the chapter in which Drosophila from different countries and perhaps with different mentalities develop wings of different lengths according to the temperature prevailing during ontogeny. We can have serious doubts that, e.g., the temperature-determined length x = 115.0 of any of these flies is really exactly the same as the population-determined length x = 115.0 of any other flies. Of course, viewed quantitatively there is no question that they are identical but is an insect wing with the exact length of x = 115.0 automatically the same as another wing also with the length x? Examined in physiological detail this is not very likely, even if seen from external appearances probably very little difference is recognizable. To go one step further, one can understand the variation of wing length as being a genetically laid down character which in various species exists in completely different forms. This is so for other groups of insects (e.g., butterflies, beetles) which can show different variation patterns of one and the same character.

This example may sound rather artificial but the problems which lie behind it are not . Let us search for another example from the field of behavior. As an ethologist, one thinks immediately of the very adaptive abilities of some land-living vertebrate quadrupeds under certain circumstances to solve the problem of the surprising plunge into an aquatic environment by the skillful application of the behavioral pattern of swimming. How does this look as far as humans are concerned? It is well known that newborn babies are able to swim well or at least lie face down on the water and splash around without any great problems with breathing. However, as soon as our swimmer is turned on his back the situation very soon becomes dangerous. This shows that this form of swimming is probably a kind of rudimentary four-legged walking in water. However, this fantastic archaic ability is lost very quickly and is only regained very much later as proper swimming with considerable problems in learning the technique. What interesting insights can a heritability calculation give us? Quite simply nothing apart from abstruse and contradictory results. The inborn ability to swim and then its disappearance as well as its later reemergence would be itself a problem. In addition, if we simply draw a line through the reckoning and—what is ethologically justified—regard the swimming of newborn babies as not being really swimming, we would still have problems with this category of behavior, namely consciously acquired swimming. Each person has his own personal style of swimming and this does not facilitate its being easily plotted as a point on a continuous scale. Simultaneously it is then questionable whether it is meaningful to compare the swimming of one person with the swimming, or not swimming of another and if, for example, it is justified to calculate a mean value. We are naturally exaggerating somewhat in order to express the problem that lies behind it more clearly. It is obvious that the concept of an additive connection between the genome and the environment can only be related to quantitative characters and only within this context can it be applied meaningfully. I would argue strongly that such isolated, purely quantifiable characters do not exist at all in nature. It appears more plausible that it is man's large capacity for invention that cuts out so-called quantitative characters artificially from complex biological structures so that they can be used in routine calculations without the disturbing accompaniments. These methods can be very successful in certain other areas of application such as physics and modern technology but their success within the field we are discussing here is modest.

The following point lead on without a break from the preceding one. All heritability calculations are based on purely intraspecific comparisons which means that one confines oneself for methodological reasons to within a single biological species. As soon as one leaves the area of a certain species, the concept becomes really contradictory and, within a very short time, no longer applicable although, in principle, there should be nothing against a comparison being made between species. Since members of each species are related to members of *every* other species and, in addition, many species have only minor genetic differences to each other (e.g., Felidae, the cats), it is whimsical, to say the least, to restrict the application of heritability calculations solely to within a species. At least no convincing argument has been put forward that would forbid it. Such a comparison of species must inevitably lead to partly abstruse results since purely quantitative data on characters, which are often not homologous to each other, i.e., are not due to genetic relationships, are meaningless. A vivid example from the field of morphology would be, like the one from *Drosophila*, the comparison of wing length in birds, insects, and bats. From the field of behavior, one could, for example, take the ability to swim and make a comparison between carnivores, hoofed animals, and primates. The result would naturally be complete nonsense because for a supposedly identical form of behavior completely different heritabilities would be calculated. Swimming within these three groups of animals certainly does not represent homologous behavior, i.e., ancestrally similar behavior (see Green 1994) so that a paradoxical result would not be a surprise to a zoologist.

In general and in conclusion, one can state that, from all the results from research into heritability, their diversity must remain inexplicable because a real causal reason for it can never be given, that is for why this particular percentage of heritability or why that percentage environmental influence. So the finding that the preference for or rejection of jazz music appears to be 45% heritable is very amusing but it still cannot be explained. This is not much different for the heritability of 51% for attitudes to capital punishment (pro or contra), a figure of 36% heritability for or against mothers going out to work, 7% heritability for being for or against coeducation, 44% with reference to the British royal family, with 25% the truth of the Bible, and with 8% pajama parties (all values from Martin et al. 1986). It could not be demonstrated any better how absurd such physically "exact" calculations are. On the contrary, as soon as one even begins to think more closely about what they could mean, the more one meets contradictions that turn out to be insoluble. Let us take the example of jazz: one would assume that the contact with and therefore the appraisal of this kind of music, whether positive or negative, can only be purely environmentally determined since there are quite a few countries where jazz is simply not played. One should try to determine the heritability of the attitude to jazz in these countries, it would, without doubt, be a frustrating undertaking. In any case, when a relatively high percentage of heritability can be established, it is often an easy task to imagine a situation in which just the opposite is valid. So, for example, let's assume that the body weight of a person is, under normal conditions, approximately 75% genetically determined (a value between 70 and 80% is always more or less acceptable, that means politically correctly chosen). One can easily picture extreme circumstances (e.g., famines) when the weight reached by an adult is totally under the control of the environmental conditions and the genetic component shrinks to a very small percentage. Astonishingly, this fluctuation in the index of heritability for one and the same character is not understood as being a challenge to, let alone a refutation of the concept of the additive effect of inheritance and environment. Just the opposite is the case. For instance, Hans Eysenck (1916—1997), who was one of the most famous in the field, stated in a book with the provocative title "The Inequality of Man" the following on the factor heritability (h^2):

Characteristic for a population is the improbability that h^2 is constant, it can vary from time to time and place to place. If, for example, one made education, in the sense of greater equality, available to everybody, the effect of heritability on g (= "general mental ability") would increase in the same proportion as the environmental influence is reduced; if one were to limit education to fewer beneficiaries, the opposite effect would be achieved (retranslated from Hans J. Eysenck 1975. p. 103).

Eysenck took on the wearying task, with a veritable flood of figures, of making the dominant influence of genetic factors over environmental ones plausible but at the same time not being able to offer a credible causal theory to explain it. The citation shows what experimental psychologists actually mean when they speak of heritability. The factor h^2 means here nothing more than the statistical variability in the probabilities of a certain population encountering certain environments and says as such absolutely nothing about from which source—environment or genome—the information for the development of a certain character is thought to come. The example with education imposes itself and demonstrates the actual weaknesses of this concept. As Eysenck correctly stated, the influence of heredity h^2 on the manifestation of *general intelligence g*, whose possible approximate location in the brain (Cortex latero-frontalis) was only determined recently (Duncan et al. 2000), must measurably increase if an obvious limiting environmental factor, as for instance the accessibility to education, is made easier. The connection is trivial: If from 100 potentially intelligent pupils, we assume all 100 were able to go to identically good schools, then the remaining differences in their intelligence would push the calculated heritability to great heights because the environment as the cause of these stubborn differences must be ruled out. The factor h^2 for g increases, the heritability of intelligence apparently becomes greater. When, on the other hand, from 100 potentially intelligent pupils, 80 of them are, for some (flimsy) reason or other, prevented from receiving a good education, it would mean that the variability of intelligence in the grownups naturally was determined by the environment since the limiting environmental conditions are the main cause of the measured differences. Consequently, the influence of h^2 on g decreases, i.e., the heritability of intelligence becomes less. Now comes the crucial question: what exactly is h^2? It can be many things but one thing it is certainly not and that is a measure of the heritability of biological characters. What h^2 means, in fact, is nothing more than the *probability of concurrence* of an already existing inherited character with an environment that is either beneficial or detrimental.. If the environment is detrimental, or does not foster the development of the character at all or only insufficiently so, then the apparent heritability h^2 must decrease and this even in the case when a perfect genetic heritability for the corresponding character would have been proved. It is therefore rather trivial to recognize that the absence of bodies of water can hardly bring about such a recognizable swimming behavior (the only "exception": sparrows taking a dust bath). It would also not be very surprising if the renunciation of an organized school system leads to only a few first-rate mathematicians being produced. It is therefore logical to think that the previously mentioned 100 individuals all possessed the genetic makeup for high intelligence and pass it on to their children without it being simultaneously essential that all these 100 people show a high intelligence actively in their own lives since especially adverse conditions in a negative environment for intelligence can prevent the latter taking place. As Eysenck (1996, p. 12) demonstrated by means of a special ranking list showing the relationship between occupations and intelligent quotients, there seems to exist a row of optimum social niches for people having different levels of intelligence. The average IQs were as follows: patients in mental homes = 57, casual laborers = 82, untrained workers = 87, trained workers = 98, craftsmen = 109, specialists or technicians = 117, managers = 132, academics = 153. Naturally this must be an oversimplification of the real relationships. For people forced to grow up in an environment which is wrong, so to speak, one can generally predict that they will not find the advantageous conditions needed for their intellectual development. However, the reciprocal situation can also happen: the person who would like things simple suffers just as much in a highly intellectual environment as someone who prefers the complicated and lives in a world of ignorance.

Hereditary characters must, however, as they ought to be understood in a strictly scientific sense, be real biological constants and therefore, in one and the same individual, can in no way vary in dependence on location or time and, consequently, the associated environmental conditions. It stands to reason, if one imagines that heritable must ultimately mean that certain genes or whole groups of genes must exist, e.g., those responsible for the development of general intelligence g, which, of course independently of any special environmentally determined impetus, are able to pass on the realization of g to the next generation. It is therefore biologically not possible to imagine that a real character is 30, 40 or even any other number of percent genetically determined since we only have to know whether a certain gene is present in the actual genome or not. But again this is a triviality: A 50%-existing sequence of nucleotides can hardly be thought of as an empirical concept to be taken seriously. Either a piece of genetic information is 100% available and can therefore be transmitted in the course of reproduction or such information simply does not exist. Genetic "heritability" that varies according to the environmental conditions is nothing other than a direct contradiction to the basic theses of modern evolutionary theory.

The popular game with the heritability of different characters is an uninteresting and ultimately confusing approach (cf. Devlin, Daniels and Roeder 1997) since the more or less random coming together of certain gene sequences with certain environmental circumstances cannot, per se, be of consequence but rather the reproductive success alone of the individual who carries this gene sequence is what is relevant. A pseudo-heritability of this sort, as it is commonly calculated again and again for the most impossible human characters (e.g., 45% for jazzophilia or jazzophobia; see Rushton 1989, p. 511), says nothing about one of the fundamental questions in biology, namely where does the information come from and does that form the basis of a biological phenomenon? A comparison of various species would be more likely to supply us with an answer to this question if we are still set on using the heritability method. A comparison of various primates, for example, would soon bring to light the fact that the ability to grasp higher mathematics cannot be taught, even under the most advantageous conditions, to chimpanzees, to Barbary apes, and categorically not to marmosets. This would also be a decisive empirical test for the apparently obvious statement that the environment itself can give rise to new characters. That most professional heritability researchers consider it illegitimate to cast even a descriptive glance over the boundaries between species will continue to make it difficult to reach a consensus. A final important point of criticism for the additive heritability concept concerns the question of the underlying mechanism, the explanation of which would ultimately decide whether the concept is valid or not. Arthur Jensen in defense of his very controversial treatise entitled *Environment, heredity and intelligence* (1969) was

surely not completely in the wrong with his opinion that the quantitative results of heritability research cannot be applied directly to the individual since
we are talking about a purely statistical parameter for whole populations and
not a description of single cases. The controversy became very heated over
Jensen's statement that the poor performance in intelligence tests by the black
population of the USA could not be explained by environmental factors alone.
Jensen stated:

Heritability is a population statistic which describes the relative size of the genetic component (or the assembly of genetic components) in the variance of the character in question
within the population. In relation to a measurement or a character of an individual, it has
no sensible meaning. An individual measurement has no variance by definition. There is no
possibility of separating the IQ of a specific individual into its heritable and environmental
components. If we were to say this individual has inherited 80% of his IQ and has obtained
20% from the environment, this would naturally be nonsense (retranslated from Arthur
Jensen 1973, p. 80).

However, that cannot be the final answer since if it is illogical to say that the intelligent behavior from single individuals is 70% inherited and is determined
to 30% or even less by the environment, then one may certainly not ask what
the relevance is of all these kinds of calculations when the relationship to empirical individual cases is such an intricate one. It appears to me that if someone insists from the determination of, as an example, the average length of
the human nose, no conclusions can be reached about the actual length of the
nose of an actual person that otherwise would be described by the usual species-specific characters. Naturally we do not expect our nearest neighbor to
have a nose which is exactly the average length which anthropologists have determined by great efforts from data collected from all over the world. We
would, however, in all probability be talking about a nose length within this
scope and not in dimensions of a completely different order. In addition this
value would continue to be a declaration of length and not, for some unfathomable reason, suddenly mutate in the direction of sizes of other modalities.
This means that the average nose length of *Homo sapiens* is not normally concluded as being a mysterious measure for any other hidden character, just as
we cannot conclude from the hereditarily determined variance of a population, a suddenly inconceivable size for the single individual. Unless one would
like to deliberately prevent this consequence being drawn. Thus it seems as if
many heritability researchers have a kind schizophrenia in that, on one hand,
they are keen to prove finally the genetically determined heritability of human
intelligence in physically exact quantitative percentages, and on the other, they
take great pains to emphasize that in no way does it conclusively mean that this
same intelligence, in fact, has to be really genetically determined. What is meant by this case is the fictitious reader or the other person taking part in the

conversation whose highly intelligent performance may not be decreased whatsoever by the hint of an overstrong genetic component. The previous citation from Jensen is symptomatic for the ambiguity in this approach.

Fortunately, there are at least two cases of behavior that have been so well analyzed that they allow us to judge to what degree the heritability concept is entitled to make statements about the underlying mechanisms. The species in question are both marine snails—the already mentioned sea hare *Aplysia* and the species *Hermissenda*—and here, examples of so-called "innate" and so-called "learned" behavior were investigated right down to the finest molecular-biological details. Individuals of *Aplysia* look a little bit like real hares with the long ear-like processes on their heads, as well as with their considerable weight of up to 5 kg, although they have never been seen to move like them by hopping. For *Aplysia*, it was demonstrated for the first time how a whole family of genes originating from a supposed prototype, control the complex innate behavioral pattern of egg laying (Scheller and Axel 1984). For *Hermissenda*, in contrast, it was proven that, by the learned association of light and turbulence stimuli in the surrounding water, firstly on a prespecified wiring system of nerve cells and secondly, founded on a laid-down sequence and the neuroanatomical and ultimately biochemical changes based on it (Alkon 1983). In other words, the actual difference between both kinds of behavior is most easily described as a difference in the existing structural complexity but is in no way just confined to having something or other to do with the difference between genetically and environmentally determined information. In fact, the underlying mechanisms show very well that the corresponding biological information must have already been present in both cases. Otherwise, neither could *Aplysia* have been able to carry out its egg laying nor could *Hermissenda* have made the association between certain light stimuli with certain turbulence stimuli. *Hermissenda*, under favorable conditions, moves towards the light, i.e., the surface of the water in order to graze on its usual food source, hydrozoan polyps. However, it reduces this positively phototactic tendency as a meaningful adaptation to turbulent conditions. When the sea remains rough for a long time, it can withdraw to deep regions for weeks on end. This proves at the experimental level what we had already surmised about learning behavior somewhat earlier (chapter 8). Precisely defined rules of animal behavior are always already in existence. Up till now, the finest empirical proof that learning has nothing to do with the new acquisition of information comes from a completely different kind of animal. For the very first time, a gene for certain learning and memory abilities was identified in *Drosophila*, the favorite fly among all modern behavioral geneticists. Because of this discovery, the fundamental and irresolvable conflict of gene versus environment has to be thrown on the rubbish heap of outworn concepts (Levin 1992). At this point, we can already dare to make a prognosis about the future of heritability calculations. With the increase in similar investigations such as the one into the *Drosophila* learning and memory gene, this increasingly theoretical branch of research consisting of nothing more than psychometrics will rapidly sink into oblivion, where it has belonged for some time, since the more information we have on the exact causal connection between genome and complex behavioral patterns, i.e., the ones that have been learned , the more uninteresting and confusing such calculations will appear. The partially continuing existence of the heritability concept expresses nothing more than our present relatively large-scale ignorance in this area.

In conclusion to this chapter, let us turn to the main statement of a cognitively interpreted theory of evolution and attempt to integrate its most important statement, i.e., learning has nothing to do with cognitive gain, into population-genetic thought. As we have seen, this statement directly contradicts the meaning of all psychometric heritability calculations insofar as it shows that no such calculation can determine quantitatively the real source of the semantic information. What is repeatedly calculated with great enthusiasm is simply the probability of a material realization of already previously existing genetic information under given environmental conditions. Only thus is it at all understandable and ultimately trivial to expect to obtain completely different results according to the environment and species being investigated. Our genetically determined ability to learn to swim can only be demonstrated with great difficulty without the presence of water. That these calculations have absolutely nothing to do with the actual mechanisms which underlie complex behavior, shows moreover the impossibility, for example, of actually imagining the interaction of 70% genetic and 30% environmental influences. In which moment in an instance of behavior would the genetic influence cease to exist and how would this behavior be able to continue on the basis of a merely environmental influence. This kind of foolish question can be posed again and again but one would never be able to produce an answer. An evolutionary interpretation of the heritability concept which views the variation and increase of biological information as an important component of evolution must, however, postulate the following consistent result of all its calculations: 100% of the information in living systems must be autonomous, i.e., of genetic origin (Heschl 1992a). This must be so since even a fraction of Lamarckian information, which could be assimilated directly from the environment would be a direct and insoluble contradiction to the basic axioms of modern evolutionary theory which still are (and will remain to be) mutation and selection. As far as the role of environment is concerned, it consists exclusively of making a selective evaluation of the organisms as they change with time. Darwin himself had a very suitable term for this influence: "natural selection" (Darwin 1859).

To make the difference between the mere probability of realization of a phenotype already laid down in the genes and the acquisition of really new information even clearer, we would like to emphasize that without exception each and every biological character is permanently influenced by the genes as well as by the environment in its development and its ultimate manifestation. That does not mean, however, as is so often erroneously interpreted, that information novel to the living system is purposely adopted. The environment continually influences every single organism in an extremely selective way but that in no way means it has the power to instruct living creatures in a purposeful manner on how to master their problems of survival. The complex composition of the food with which we stuff ourselves in great amounts everyday tells us little about how we should digest it biochemically and a problem in mathematics does not tell us how we can successfully solve it. If that were so every person, indeed every animal could solve it every time. The still wide-spread concept that the environment can teach us as the only chosen species is, in this respect, much more misleading than Darwin's well-meant comparison of natural selection with the known process of artificial selection in the breeding of domestic animals. If the principle of environmental instruction was to function in a single case, we would have no need to concern ourselves with evolution in the future but rather with a supernatural theology of immortal creatures (for an entertaining treatment of the absurd consequences of immortality, see Danny Rubin's fantastic story "Groundhog Day" from 1994). Modern behavioral genetics has, for a long time, been talking about completely different but fundamentally more interesting things:

Behavior genetics is no longer concerned with simply testing for a genetic component to variation. Genetic studies of discrete behavioral variants within populations examine processes that maintain different strategies within a population. Genetic analysis of quantitative traits can lead to speculation about the history of selection on behavioral traits. The study of genetic interactions between traits indicates possible constraints on evolutionary change and provides information on the physiological and developmental basis of behavioral variation (A. A. Hoffmann 1994, p. 42).

The advanced behavioral geneticist, who is involved with this material on a daily basis, is already talking, as a matter of course, about something apparently paradoxical such as the "nature of nurture" (Plomin and Bergeman 1991), with a, in the meantime already independent "genetics of experience" (Plomin 1994) as the first very promising results. Empirical research has known for quite some time an apparently completely paradoxical "learning through instinct" (Gould and Marler 1987). Here, two at first sight completely different independent research areas ending up with a very strikingly similar statement. And one day, one can already take bets on it, this kind of genetics, at present considered unconventional, will be able to confirm in all its complicated details what we here have merely predicted from evolutionary theory. In actual fact, a few scientific papers have already indicated the basic direction in which future research will develop. The following, among others, have already been substantiated empirically:

1. The influence of genes on the intellectual development does not, as would be expected because of the constant increase of the cultural influence, diminish with age but, just the opposite it can be proved to be the determining factor in behavior right up to the end of life (McClearn et al. 1997).
2. Social intelligence is also determined by genetic influences. A first example are that certain kinds of disturbances of these influences (Turner's syndrome: all or part of one X chromosome is missing) by an identifiable gene locus on the human X chromosome (Skuse et al. 1997).
3. Finally, it is already possible, for a certain part of variability of the IQ within a certain range (2% = 4 IQ points) to determine a specific gene locus called *IGF-2R* (Plomin et al. 1998).

It appears as if Plomin's announcement at the beginning of this chapter will not remain an empty promise. However, true behavioral genetics in the future will mean the analysis of extremely complex molecular cascades interwoven in a multiple manner which can potentially lead from every point of the genome to the most sophisticated patterns of human behavior. Thus we should not go too far at this point, this is still the very beginning, no more but also no less.

11 A Superfluous Law of Evolution

The Law of the Inheritence of acquired characters:
Everything that the individuals gain
or lose through the influence of the
circumstances which its race has
been exposed to for a long time, and
consequently through the influence
of the predominant use or continuous nonuse of an organ, is passed
on to the offspring by reproduction
provided that the acquired changes
are common to both sexes or the pro-
ducers of these individuals.

Jean-Baptiste de Lamarck

In this book we will and should not just ridicule Lamarck, after all, he was the first person to grasp the true meaning of the concept of the evolution of living things a long time before Darwin did. We can make the statement here that with reference to the crucial question of biological inheritance he was completely right even if in a very paradoxical way and even if it was somewhat against his real intentions. His so-called (individually) acquired characters can indeed be transmitted to the next generation and will do so permanently since these characteristics are not truly acquired, but rather the necessary information for them has always been present in the genes. That means nothing other than that what Lamarck postulated, that which has been argued about and has been steeped in legends right up to the present day—the inheritance of acquired characters—is not just confusing and troublesome but superfluous as well since the supposedly novel information concerned with this process already exists in the genes and therefore does not logically need to be stored there once more. For this reason, there is no more need, as erroneously assumed for *Homo sapiens* again and again, of the supposition that a very special kind of cultural or historical evolution could take place which is independent of and, in some cases, against the principles of biological evolution. The exact opposite is true because Lamarck postulated a mechanism which, strictly speaking, at any rate is possible within the framework of the Darwinian theory of mutation and selection and therefore has no more need of a separate explanation. What we can do with Lamarck's concept of acquired characters with a good conscience is to delete the term "acquired" since learning and other apparently new acquisitions have nothing to do with really new acquisitions in an evolutionary sense.

We should now formulate this connection more clearly in order to avoid misunderstandings as much as possible. Lamarck's famous-notorious inheritance of acquired characters is based on metazoans, in the course of their more or less long lives, being able to gather experiences which afterwards as new information can be somehow incorporated into the genome to be preserved for evolution. Such a direct and, above all seemingly directed passing on of new genetic information would naturally mean a great acceleration of the evolutionary process since the complicated route of the randomly controlled trial and error method due to mutation could be circumvented. The original reason which led to this simple and intuitively convincing concept, however, lies in the fact that with the first germination of ideas about evolution, the question of how the great number of apparently perfectly adapted life forms could have come about. For the pioneers of evolutionary concepts such as Buffon (1707—1788), Lamarck (1744—1829), Geoffroy Sainte-Hilaire (1772—1844), Goethe (1749—1832), and Erasmus Darwin (1731—1802) the grandfather of Charles Darwin the concern was to primarily explain this evidently strong purposefulness in the natural system of living organisms and in their evolution.

The best known story of evolution in this connection is the one about the ever-increasing length of the giraffe's neck. Lamarck assumed that it had been brought about by the perpetual striving of the animals to reach their food, the green leaves in the crowns of trees. Because the construction of the animals' bodies was so well adapted to their behavior, it seemed apparent that the purposefulness of the behavior of individuals was adequate to explain this trend towards having a longer neck. An alternative possibility was not put up for discussion at the time and unfortunately, this fact is always forgotten. Even Charles Darwin himself found nothing extraordinary or wrong about Lamarck's interpretation and so the latter did not, for a long time, stand directly in the way of Darwin's actual discovery of the effect of selection. In fact, just the opposite, Lamarck's mechanism of inheritance and Darwin's principle of selection rounded off the picture of evolution at that time to some degree in an optimum way. Lamarck's animals struggled heroically to adapt to their surroundings, somehow handed down genetically the adaptation to their offspring and were therefore superbly equipped for facing the merciless selection of the Darwinian kind. That such a combination of purposeful variation and purposeful selection must produce an evolutionary short circuit through which, in shorter and shorter time periods, increasingly better adapted and ultimately immortal organisms would arise escaped even the most perceptive critics of evolutionary ideas. We are not saying here that Darwin's concept of natural variation did not indicate a completely different direction since it clearly placed more emphasis on a more random natural basis. The following citation even proves that Darwin himself was not far away from the discovery of the mutative random character of biological variability, even though he still hesitated to draw the necessary consequences for his theory:

The case, also, is very interesting, as it proves that with animals, as with plants, any amount of modification in structure can be effected by the accumulation of numerous, slight, and as we must call them accidental, variations, which are in any manner profitable, without exercise or habit having come into play. For no amount of exercise, or habit, or volition, in the utterly sterile members of a community could possibly have affected the structure or instincts of the fertile members, which alone leave descendants. I am surprised that no one has advanced this demonstrative case of neuter insects, against the well-known doctrine of Lamarck (Charles Darwin 1859).

It was therefore more that the highly critical Darwin did not or perhaps even did not want to recognize the well-hidden Munchhausen components in Lamarck's attractive ideas where, ultimately, living organisms could pull themselves from the evolutionary swamp by a lock of their own hair when it was demanded by the environment. Even Ernst Haeckel (1834—1919), the German champion of Darwin's theory of selection, fell into the same trap although he explained to us the historical and scientific connections with regard to the most important representatives of evolutionary thought better than any academic philosopher:

Just as I did, right from the beginning, Spencer categorically opposed the germ plasm theory from Weismann which denies this important factor of phylogeny and wants to explain the latter exclusively by the "omnipotence of selection". ...; since Charles Darwin was just as firmly convinced of the fundamental significance of progressive heredity (note: = inheritance of acquired characters) as his great predecessor Jean Lamarck and as Herbert Spencer. I had the good fortune to be allowed to visit Darwin three times in Down and each time we exchanged our unanimous views on this main issue (Ernst Haeckel 26. 3. 1898, translated from Heberer 1968, p. 420).

Times have changed and, in the meantime, it has been demonstrated that a directed evolution such as Lamarck imagined is simply not possible since only mutations, i.e., undirected random genetic changes, can provide the starting material for natural selection to become effective. At that time, August Weismann was therefore *completely alone*, what is readily forgotten, in being correct with his germ plasm theory. What, however, has become of Lamarck's individual acquisitions, have they already disappeared from our modern concept of evolution? Interestingly, no and this, as we will see in a moment, still causes the same severe theoretical problems even today. While the interpretation of the evolutionary mechanisms has been changed decidedly by the discovery and investigation of the molecular components of inheritance, the interpretation of the activities of the individual remained at the historical level of Lamarck since, still today, learning is considered to be an individual advance though now absurdly no longer useful for biological evolution. It is exactly on this point that it has been simply forgotten to criticize both Lamarck and Darwin since our previous discussions have clearly shown one thing, namely that learning has nothing to do with cognitive progress, and also its result, i.e., that what seems to have been learned, represents neither an adaptive improvement nor a new acquisition of the metazoan organism. As soon as one introduces such a consistent evolutionary interpretation into the still ongoing discussions about Lamarck's version of evolutionary theory, for once and for all, the devilish Gordian knot of the apparent Lamarckian errors is unraveled.

In fact, many examples of this Lamarckian error still exist and closer examination shows us what among the fundamentals is confused. Let us take probably the most spectacular experiments which were often carried out to substantiate Lamarck's controversial ideas and are attempted even up to the present day. It does not concern the complex affairs of passing on learned abilities from one generation to the next but rather some basic physiological regeneration processes which, in a Lamarckian perspective, represent the simplest step of an adaptive modification, or in short, an apparently new acquisition by the individual. Mostly, these experiments concern the intentional removal of part of an extremity or some other important organ of an animal whose condition in the next generation is examined with great attention and hope to see if there is any noticeable change. The newly acquired character should, according to the considerations of convinced Lamarckists, somehow be able to influence the genetic equipment of future generations. However, even the creator of the germ plasm theory, let himself be seduced into carrying out some quite bizarre procedures to fend off this erroneous belief:

At the same time, experiments with mice proved that cutting off their tails, even from both parents, does not result in even the slightest shortening of the tails of the offspring. I have carried out such experiments myself and that for 22 consecutive generations without any positive success. Among the 1592 young, produced by tailless parents, there was not a single one with a tail defective in any way (August Weismann 1904, vol. 2, p. 56)

The many dissected and in this way mutilated experimental animals, whether they were primitive invertebrates (e.g., insects, snails) or the far more attractive vertebrates (e.g., toads, salamanders, mice) showed and still show only one thing in great clarity. Exactly that what was erroneously considered to be a biological character, i.e., the amputated tail, the cut-off extremity, the finger that had been removed, etc., is of course not transmitted from one generation to the next, but in its place, everything that this special kind of experimenter could not recognize as an inherited character, namely the already existing capacity for the regeneration process itself. Interestingly, this was not recognized before more than 30 years ago by the influential embryologist and developmental biologist C. H. "Wad" Waddington:

> The old problem of the relation between heredity and environment in evolution (what is often, though somewhat incorrectly, spoken of as the problem of Lamarckism) has, I think, largely been cleared up by the recognition that the capacity of an organism to respond to environmental stresses during development is itself a hereditary quality (C. H. Waddington 1968b, p. 20).

However, Weismann had already expressed a similar view some decades before:

> ... is the capacity for regeneration ... an occurrence of adaptation, even though one with its beginnings in antiquity, which is based on a special mechanism, and does not appear everywhere to the same extent and at the same strength? We became acquainted with facts previously which should have pointed us towards this opinion. ... New experiments ... put these facts beyond doubt (August Weismann 1904, vol. 2, p. 3).

In direct contrast to Weismann, Waddington took this completely correct insight and drew an absolutely absurd conclusion from it, namely, the conclusion that what he considers inherited anyway, as the consequence of the mechanism of a so-called genetic assimilation, could become an innate character. In plain language, this means he claimed that what is already something anyway can still become it:

> Further, the demonstration that a combination of this fact with that of developmental canalization leads to the occurrence of a process of genetic assimilation, by which the effect of an "inheritance of acquired characteristics" can be exactly mimicked, has removed the whole heat out of this ancient discussion. These developments of biological theory have, however, I believe, considerable philosophical implications, since they show how it is possible for characteristics such as mental or perceptive abilities, which originally arose in interaction with the environment, to become in later generations "inborn" to the extent of being independent of any particular environmental stimuli (C. H. Waddington 1968b, p. 20).

What more can one say? First, all inborn characters are in constant interaction with the environment, so that there is no difference to supposedly "acquired" ones, and second, this would still be pure Lamarck—although with an impressive sounding technical term, namely "genetic assimilation" to make it appear modern. Nowadays, we know, however, that information that was already in existence—from the perspective of the organism in question this means: how do I heal my ruined foot, how do I regenerate my injured tissues, etc.—is passed on without any problems, whereas that strange "information" which was artificially thought up by an experimenter, such as the act of mutilating a part of the body, of course cannot be transmitted to the next generation. What good would it do? Going further than that would truly be a miracle since, in such cases, even the conscious intention of the experimenter would have been passed on! In principle, those people were expecting that the experimental animal should reach an age comparable to his mutilated ancestors and then—with wise foresight of things to come—do to itself what the evil experimenter had in mind. One can see even at first sight what here—to the misfortune of many experimental animals—tragically was not correctly recognized. One believed certain unfavorable environmental conditions (e.g., mutilation of the tail) to be the acquisition of a new character and did not see the genuine biological character (i.e., healing of the stump after injury) which ultimately is the only important thing. Again, we can put this fact succinctly with the words of Weismann: "There is no inheritance of mutilations!" (Weismann 1904, vol. 2, p. 56).

A very similar situation exists in all those cases in which from making use or not making use of an organ—for Lamarck the most important driving force of morphological evolution—the structure of the organ is altered whether by a simple reduction or an increase, but perhaps also by a complicated restructuring of the relevant tissues. In this instance, the best-known example is the case of animal muscle tissue, where the end-result in humans can be either, by excessive exercise, an attractive (or less attractive) muscleman or, during enforced inactivity (e.g., when a limb is set in a plaster cast) a complete atrophy of the musculature. Many such changes are reversible to a large extent and no doubt represent an adaptation to the prevailing conditions. For Lamarck, it simultaneously represented a purely individually acquired character which, to the advantage of the descendants, was to be passed on. Again it is easy to show what should happen here: naturally the well-known muscular arms of the village blacksmith are not passed on so that his baby son comes into the world with arms like a miniature Popeye the sailor and instead of grabbing for his pacifier he grabs for a hammer. This would not necessarily be an advantageous thing to pass on, in fact it could be a danger to the person looking after him. The "real" Popeye, of course, did not acquire his huge muscles through body building but by consuming large quantities of spinach and they shrank back to a normal size very quickly. In this case also, we must not exclude that scoffing spinach to excess with all its welcome consequences can be transmitted directly to the offspring. The genetic information involved here and which, without any Lamarckian problems, can passed on immediately to next generation and obviously already exists in the blacksmith father and similarly in other fathers, but not necessarily identical, consists of quite simply being able to increase the size of the muscles by the corresponding demands to a certain size, or to maintain them at a constant level, or to make them smaller, as the case may be. The relevant heredity program can be changed in many ways but only by random mutation processes. One can, for example, imagine that the dimensions of the, let us call it the standard muscle heredity factor, can take on various values which, even in fully untrained children, can bring about a clear difference in build. The specific kind, the dimensions, and the course of the muscle building (or degeneration) can take on many forms and lead to very different results. Now it is not difficult to see that everything down to the last detail can be inherited, the building up of body musculature (cf., "Schwarzenegger gene": Dickman 1997) as well as the physical condition that one needs for it (*ACE* alleles: Montgomery et al. 1998). Had Lamarck thought out his concept to its logical conclusion, he would have—as far as this would still somehow concern him—saved himself a great number of posthumous, and largely unjustified insults. How far the argument over supremacy in the world of biology in the 19th century between France and England led to an exaggeratedly negative portrayal of Lamarck as a more or less an anti-person to Darwin who was strictly speaking a Lamarckist, ought not be judged here. However, it is rather astonishing that in our modern more liberal era the classification "Lamarckian" is used as a term of abuse in biological circles, just as the word "communist" has become nowadays.

Taken all together, regeneration and healing processes as well as organ changes through use or nonuse during ontogeny makes a lot of sense, it fulfills a biological function and is therefore evolutionarily adaptive, but it has nothing to do with any kind of new acquisition by the individual (for the present state of knowledge, see special issue of *Science* 276 [1997], 60—87: Frontiers in Medicine: Regeneration). The individual, so highly regarded by us, is, as we have already come to understand, generally excluded from evolutionary progress, which by itself is enough to show that, in a purposeful and lawful way, strategies are brought into play which had to be struggled for previously at the actual evolutionary level of the phylogenetic germ line by random mutation against the rigors of natural selection. Here, we can successfully apply our principle of random cognitive acquisition and consequently establish theoretically why Lamarck was so mistaken and simultaneously was unintentionally right. Individuals always use purposeful, i.e., not random, methods to solve specific problems and this is exactly the fact which tells us that, in all these cases, they have nothing to do with real evolutionary progress. At the same time, nothing stands in the way of the real genetic transmittance of these individual, often very varied behavioral strategies, to the respective offspring.

In a way, Lamarck was badly done to because acquired characters are indeed inherited even though for the simple reason they quite do not represent anything newly acquired.

Essentially, the so-called Baldwin effect is based on the same logical error: the genes only partly determine a given behavior, what is still lacking is then learned by generation "zero" and hence the new environment or niche, which as a result can be occupied by the species, selectively favors in the following generations the emergence of that *genetic* information still lacking which—and here comes the inconsistency—has already been acquired in the learning process of the first generation. In this regard, statements such like "behavioral flexibility enables organisms to move into niches that differ from those their ancestors occupied" (Deacon 1997, p. 322) are simply wrong and do not alter the situation very much. Instead, for the modern evolutionary treatment of phenotypic plasticity its genetic determination has already become a matter of course (Agrawal 2001).

Lamarck's last excusable mistake was not so much that he postulated a novel evolutionary mechanism opposing that of Darwin but rather that he, and Darwin too, could not recognize that, in reality, the ontogenetically developing individual metazoan must remain excluded from evolutionary new acquisitions. The normal picture of Lamarck as the direct opponent of Darwin's concepts must be ascribed to a kind dramatic masterpiece by the following generations of evolutionary theorists since both of these researchers were in absolute agreement in one point, which only later became the most disputed, namely the inheritance of acquired characters. One of the best and simultaneously most amusing proofs of this can even be found in Darwin's famous "Origin of Species":

Not a single domestic animal can be named which has not in some country drooping ears; and the view suggested by some authors, that the drooping is due to the disuse of the muscles of the ear, from the animals not being much alarmed by danger, seems probable (Charles Darwin 1859).

If we were able to bring both personalities face to face in a scientific confrontation today, they would be truly amazed at how different the two concepts of the mechanisms of evolution have become. Essentially, the only point where, to all appearances, Darwin actually contradicted Lamarck concerned the acceptance of spontaneous generation for the origin of the simplest living organisms from nonliving material (for a detailed history of the Victorian debates over spontaneous generation, see Strick 2000). However, let us now return to the actual problem. That special information of the individual which, according to the erroneous concept of the Lamarckists, including Darwin, was somehow to be stored in the genes does not need to be transferred there in a complicated way for the very good reason that it already exists in the developmental instructions of the genome. Most of the experiments to support Lamarck's concept were based on the well-meant belief that something what the free individual had developed in his life could somehow react on the physiology and especially on the genetics. From this vague idea, only later did the apparently more earnest problems develop with regard to the reality of

Lamarckian evolutionary mechanisms and their confrontation with the Darwinian principles of genetic chance and natural selection. In Darwin's time, this was not yet possible because no concrete knowledge of the genetic basis of evolution existed except for that of the Austrian recluse, scientist, and abbot, Gregor Mendel (1822—1884), who, when his works on his experiments with peas were discovered, became the posthumous founder of genetics. It is astonishing that even today some aspects of Lamarck's theory of the acquisition of acquired characters as a conceivable mechanism of evolution, though often hidden and not immediately recognizable as such, still haunt professional circles. Again, Weismann is still the best source of knowledge about this fact:

It is always the old stories, some expounded unchanged, others in a new version (August Weismann 1892b, p. 520).

Even in recent times some professed Neo-Darwinists, obviously without noticing it, continue to pass on certain hidden elements of Lamarckism. Currently, the most widely spread interpretation of the problems of Lamarckism, or as a solution of the understood version says a Lamarckian inheritance of acquired characters has not asserted itself evolutionarily because most of these characters in no way represent an improvement for the organism and therefore such an inheritance more probably brings about a disadvantageous effect and would be selected out. Maynard Smith, for instance, an influential evolutionary theorist unsuspected of any kind of Lamarckism states this view exactly with the following words:

We would like also to know why the genetic mechanism is like that. If a tape recorder can be designed to transmit information in both directions (note: play back and record), surely a genetic mechanism with a two-way flow of information—from phenotype to genotype as well as from genotype to phenotype—could have evolved. The answer is that most phenotypic changes (except learnt ones) are not adaptive: they are the result of injury, disease, and old age. A hereditary mechanism that enabled a parent to transmit such changes to its offspring would not be favoured by natural selection (John Maynard Smith 1989, p. 12).

Only from this rather purely technical reason, is it further argued, is the so-called central dogma of molecular biology universally valid, in which an information transfer from the genotype, i.e., from the nucleic acid DNA (double helix) and RNA (simple strand), to the phenotype, i.e., the proteins (complex 3-dimensionally folded amino acid chains), is allowed but the realization of a reversal of this coding direction is strictly forbidden. Whereby, the central dogma here is no other than an attempt at a molecular biological foundation of Weismann's non-inheritance of acquired characters. In other words, it is the physicochemical conditions at the molecular level alone that forbid any instructive reaction from the program executing protein back to the information-storing nucleic acid. This argumentation seems obvious at first glance in all those cases where, by special regeneration processes, negative influences which arise from injuries, illnesses, and aging are counteracted to some extent although, as indicated in the previous examples, it is definitely not—as Maynard Smith erroneously assumes—the injury itself which represents the character but solely its successful healing. Consequently, we should not interpret a clinical picture of an illness as a negative and therefore not to be heritable new phenotype. Here again, the functionally correct reaction of the organism is confused with the problem posed by the environment since the cause of the illness itself (e.g., injury, virus infection, poisoning by various organic or inorganic substances, harmful oxidation in the cells, etc.) ought of course not be inherited.

However, the appropriate reaction of the body that helps it survive, very much so. Only the latter results in the corresponding clinical picture and the only relevant information behind it is that which, in a multicellular animal, in every single one of the many million copies of its basic genetic program is held ready at all times for such eventualities. A separate discipline, called "evolutionary medicine" is in the meantime become involved with this topic whereby, for the first time, the healing process itself as well as the no less important attempt at healing by another person—from former medicine man to the modern doctor of medicine—are coming under scrutiny in an evolutionary perspective (Nesse and Williams 1994; Fábrega 1997). To understand the full significance of such an enterprise, we should not forget that our bodies are full to the brim with individual genetic information. If we forget for a moment the wateriness of our earthly bodies—we are approximately 65% water—we should remember that of the remaining solid substances up to 15% consists of nucleic acids. This means that if we assume an average dry weight of about 24.5 kg (from a total of 70 kg), there is about 3.7 kg of highly concentrated information (Lehninger 2000).

With aging, one can very easily get the idea that there is no good reason for it since getting old—as everybody involved in the process will readily confirm—is not such a pleasant experience that one values it as a specially wise feature of Mother Nature. In particular, the natural ending of life, the unavoidable death of the single individual is not celebrated as wonderful invention of the evolution of multicellular animals. Ludwig von Bertalanffy (1901—1972), referred to as the "holistischer Organizist" from Vienna, was the first to point out the evolutionary price of the origin of metazoans being the obligatory death of the individual. In fact, it is not difficult to show that the whole of mythology and religious thought is, in principle, either directly or indirectly, derived from the very understandable desire to achieve personal immortality (Topitsch 1988). However, even if it is indeed difficult to see this as a consolation, the death of the individual does have an evolutionary purpose:

Only from the standpoint of usefulness, can we understand the attribute of natural death, we agree that the germ cells must be potentially immortal like unicellular organisms but that the cells that make up the bodily tissues can be transitory and how, in the interests of their performance and one-sided differentiation this somehow has to be. … There can be no doubt that death as an attribute is virtually contained in the organism and it is intentionally so, the ultimate goal of the development which begins with the egg cell, with the release of the gametes, i.e., with reproduction reaches its high point and then is followed by a rapid or slow decline until the natural demise of the individual (August Weismann 1904, vol. 1, p. 274 and vol. 2, p. 298).

As it is with dying, so is it with growing old. Far from being the purely physical result of degeneration processes which ultimately—comparable to the situation of an overworked machine—leads to the technical collapse of the living robot called the body, death and aging are definitely functional aspects of often very different stages of life cycles where more or less everything has its own place. For example, for *Homo sapiens* it is known that his life is split up into a large number of various stages, where a stage—similar to the succession of those in insects—is followed by the next often linked to a change in the purely physical area (physiognomy, body shape, sexual characters, etc.) as well as in the appropriate behavioral strategies practiced (Science Special Issue [1996] Patterns of Aging. *Science* 273: 42—74, 80). If that were not enough, a gene was discovered recently which—very similar to that found in mice (Kuro-o et al. 1997)—is held responsible for the manifestation of the Werner-Syndrome, an extremely accelerated aging process in humans (Yu et al. 1996). Fortunately, however, the opposite also seems possible since there exist genes on chromosome 4 (Puca et al. 2001) which are able to produce "Sardinia's Mysterious Male Methuselahs" (*Science* 291, 2074—2076; for the potential absolute limits of the human lifespan, see *Science* 291, 1491—1492). From the fetus via the baby right up to granddad or grandma, a person goes through several stages with often very different aims and functions whereby the exact succession of the individual phases is naturally nowhere near as rigid as the dramatic structural changes during the ontogeny of holometabolous insects (cf. the paradigmatic case *Lepidoptera*: egg => caterpillar => pupa => butterfly). Nevertheless, it appears that even in the case of the vertebrates, where some groups (newts > salamanders > frogs) are admired for their developmental flexibility (a famous example is the neotenic axolotl, *Ambystoma mexicanum*, which, under normal conditions, retains its larval salamander morphology, but under artificially changed conditions as for instance removal of water is still able to develop the complete adult body), that there are definite genetically programmed intervals of approximately 90 minutes of growth or reconstruction to maintain the rhythmic momentum of ontogenetic development (for a review, see Pennisi 1997). At the same time, it is therefore clear that these changes in our appearance as well as our behavior follow a set pattern that cannot be simply dismissed as signs of old age if one wants to make an evolutionary adequate and impartial judgment.

The personal death itself, however regretful we may find this subjectively, represents a very interesting aspect of an evolutionary interpretation of the human individual genesis (Rose et al. 1991). As we have seen, the ontogeny of metazoans must remain excluded from real evolutionary progress—which means the same as cognitive progress—for selective-functional reasons since random somatic mutations almost always represent a danger to the organism (a somewhat more complicated situation exists in the immune system where "directed", i.e., genetically controlled somatic mutations play an important role in antigen defense; cf. Heschl 2001). The information-theoretical connection to this subject was first defined in developmental biology by Lewis Wolpert (1978) and M. Apter in cooperation in a theoretical article in 1965 under the title "Cybernetics and Development". This also contained a first profound criticism of the overzealous application of Shannon's information theory, since "it is more correct to think of instructions instead of information, ... (since) instructions do not exist at a special location but rather the system behaves as a dynamic whole." Wolpert summarized this view three years later at a in a sense historical IUBS Symposium under the aegis of C. H. Waddington (title: Towards a Theoretical Biology):

We suggested that the development of this complexity was more easily understood if the egg was considered to contain the programme for making the organism rather than the complete specification of the organism to which it would give rise (Lewis Wolpert 1968, p. 125).

Resulting from all this is the first logically compelling argument for the necessity of the death of the individual from the perspective of the theory of evolution. To make possible a real evolution of a multicellular species, there must come a more or less predestined time to the individual when as it is so aptly put "your time has come" to make way for possibly more advanced types, in the final analysis one's own offspring changed by genetic mutations and new combinations. In the terse words of Stephen Stearns : "The price of the specialization in the germ line and the soma is death" (Stearns 1992, p. 183). Our personal feelings may tell us that this sounds rather unpleasant, however, the biological connections speak a very clear language. During the last few years it has become increasingly clear that the natural death of the individual—we are obviously not talking about being eaten by an escaped lion or run over by a car—is not a simple biochemical capitulation of our own physiology to the stress of the daily struggle for survival but rather a well-defined kind of genetically controlled expiry date. Naturally that does not mean that our genes have nothing else to do but be concerned about a chronologically exact limit of our individual life spans. This explains why the unconditional genetic determination of the potential length of the life cycle is not excessively high and lies, in all the species investigated so far, below a value of "heritability" of 35% (Finch and Tanzi 1997). It is sufficient for this purpose when we learn what the direct (Eizinger and Sommer 1997; Hodgkin, Plasterk and Waterson 1995; Jazwinski 1996; Yu et al. 1996) and indirect (antagonistic pleiotropic effects, increasingly frequent mutations: Williams 1957; general decline in hormone production: Lamberts, van den Beld and van der Lely 1997) effects on the respective genetic constitutions ultimately mean as to what our personal life expectation really is. The facts are still, e.g., thread worms live on average for 15 days, fruit flies for 40 days, mice 27 months, and humans 72 years (Finch and Tanzi 1997). Only one thing is decisive here and that is, with very few exceptions—some grass rhizomes are estimated to be 15,000 years old (Nooden 1988)—in practically all other metazoans, no effective selection towards an infinite individual life span has taken place:

Thus the evolutionary answer to the question "Why do we age?" has three parts. First, we have a clear separation of germ line and soma. Second, the force of selection declines with age until past some point, determined by factors discussed under the evolution of reproductive lifespan, older organisms are irrelevant to evolution. Third, under these conditions two sorts of genetic effects become possible, (a) the accumulation of more mutations with (note: negative) effects on older age classes than on younger ones and (b) the accumulation of antagonistically pleiotropic genes that benefit younger age classes at the expense of older ones. Maturation is the point in the life history past which these effects should start to occur and before which they should never be seen (Williams 1957) (Stephen Stearns 1992, p. 200).

In agreement with all this is that, on the level of cell reproduction, there is evidence that the somatic lines can only divide a limited number of times and have to manage with this. In the germ line, on the other hand, special genes are activated which have to bring about the potential "immortality" of their line (Ehrenstein 1998). So, for example , nerve cells essentially reproduce themselves very rapidly only in the early stages of embryonic development in order to maintain a more or less neural status quo. This, on closer observation, is no bad thing since—in contrast to the sluggish plants—an animal's power of movement should be fully operative at birth since in most species this is a very dangerous time for them. Therefore the number of nerve cells remains relatively constant for the whole lifetime (Morrison and Hof 1997) or at least does not increase in total through regeneration, which in the long term would be a precondition for living forever.

Let us return to Lamarck once more. It is completely inaccurate to believe that most phenotypical changes are not adaptive and therefore their transmission to the next generation would not be favored by natural selection because they are the result of injuries, illnesses, and aging. An injury, the pathogen, and the environmental factors which accelerate aging are obviously anything but adaptive and should not, or, more logically, cannot be passed on to the offspring, whereas, the ability to repair an injury, overcoming a serious illness, and even the purposeful aging process can be passed on genetically as a useful attribute and will become permanent. Even these few examples show us how careful we must be when we want to speak of nonadaptive characters of biological systems. Mistakes are made in this respect even by convinced Neo-Darwinists.

Even more paradoxical and therefore more amusing for the present topic is the situation in relation to the evolutionary interpretation of learning which we dealt with in chapter 8. In this case, it has not escaped even the most hard-boiled reductionists, that it is anything but easy to treat them without further ado as really disadvantageous, i.e., nonadaptive characters of biological systems. One agrees consequently what appears to be a reasonable compromise, which consists of accepting as valid the results of learning processes as the only really significant Lamarckian exception with regard to normal metazoan evolution. Interestingly, even influential theorists such as Maynard Smith, Gould, Dawkins, and many others have agreed with such a clearly non-Darwinian interpretation. Thus it seemed obvious that both learning and the transmission of its results over the generations could no longer be described by applying the usual principles of Neo-Darwinism. Hence, a very special kind of evolution, namely cultural evolution had to take place to remedy this severe problem and simultaneously to at least stay in any kind of conceptual contact with evolutionary theory by using biological nomenclature. However, that in fact an actual denial of evolutionary theory had been born was only realized by a few biologists. If evolutionary theory is in fact to realize Darwin's challenging claim for a really comprehensive explanation of living systems, then it cannot capitulate so quickly and easily to a phenomenon that no doubt represents a topic for research on a specific form of biological behavior, but nevertheless remains an intrinsic part of evolutionary biology. Richard Dawkins who, like many others before him, also applied himself to the evolutionary explanation of complex cognitive phenomena recognized this difficult situation and attempted to find a way out of the dilemma. In contrast to the majority of evolutionary epistemologists associated with the school of Lorenz and Campbell who continued to cling to either an insoluble duality or a chronological parallelism of biological and cultural evolution, Dawkins at least made an effort to find a solution within the Neo-Darwinian paradigm. The meme theory he conceived of the cultural development was thought to be a kind of logical progression of the already well-tried theory of biological heredity. For Dawkins, memes are something like the basic intellectual units of what is normally viewed as human culture (*mimem* [Gr.] = imitation; *memoria* [Lat.] = memory; *même* [Fr.] = same; Dawkins 1976). The extent of these, i.e., of everything which can be understood as a meme, however, is very encompassing and simultaneously very difficult to discriminate from other phenomena. Strictly speaking, very widely varying thoughts and ideas are involved here which are ultimately capable of being expressed in words and therefore are mental concepts that can be passed on, i.e., from simple descriptions of classes of objects and events such as "house", "mouse", "hole", "live", "eat", and "hunt" up to far more abstract concepts not easily accessible directly with our senses, e.g., "natural selection", "inheritance of acquired characters", "electrical induction", "logical deduction", etc. The meme therefore includes more or less everything from any people of a society with a cultural tradition that has been thought and passed on.

Whereas during genetic evolution, tangible genes struggle for survival in multicellular bodies, in the cultural history of modern man, ideas and thoughts are increasingly taking over the role of the once dominating evolutionary replicators and therefore weakening the influence of the genes more and more. In the last instance, Dawkins has even in mind a kind of historical takeover by which the heavenly hosts of the ultimately most successful meme („God", "life after death", "theory of evolution"?) because of their incomparably high speed of propagation and replication can become the dominating units of information of the whole of human development. In the logical consequence of this concept, he then comes to the conclusion that one day in the future, computers will be designed as carriers and increasingly, as then autonomous developers of the newest memes, could take over their own evolution, a Hollywood-like idea which accordingly was taken up by a broad spectrum of the public with comprehensibly great enthusiasm for the currently most popular toys if not soon "contechnics" of *Homo sapiens*.

The idea of memes may sound very convincing and fantastic but does any heuristically valuable concept lie behind it which could help us to understand better the biological and the, if not "evolution", then at least cultural development of the human race, above all more satisfactorily? Dawkins himself announced, however, very honest skepticism in this regard after he received strong criticism from colleagues in the same field over his meme theory, some of whom accused him of betraying a good thing, namely genetic evolution. Still in 1976, with great enthusiasm and conviction with a whole chapter to itself with the title "Memes, the new replicators" was situated at the conclusion of his selfish gene bestseller. In 1987, just a decade later, he wrote in "The Blind Watchmaker" significantly only on one short page described as *science fiction* about his original meme concept. Although the whole invented scenario was not devoid of a certain originality and in the meantime had gained not a few adherents and supporters, it was bound to come to nothing because of a basic deficit. What we are dealing with here, as Dawkins himself finally admitted for the first time in "The Extended Phenotype" from 1982, is at most simply a superficial analogy which was conceived rather to illustrate the standard evolutionary concept of gene selection pictorially. In the meantime, it looks as if the meme theory is more suitable to hindering than to promoting a serious evolutionary view of cultural development in humans. Obviously, the main problem lies in that in the final reckoning we have yet again an explicit Lamarckian effect of the phenotype upon the so-called replicator. Dawkins speaks of a "Lamarckian chain of causes" in this connection, which in the meantime has been accepted with contagiously memetic enthusiasm by a—to self-fulfil the prophecy—constantly increasing number of diverse social scientists, orientationless philosophers, and, of course, ubiquitous AI-people (for a highly infectious possibility of infection, see the *Journal of Memetics— Evolutionary Models of Information Transmission*: www.cpm.mmu.ac.uk/jomemit/). Here, yet again, the uniform nature of the most important mechanism of evolution, i.e., the basically random character of the processes of change in the evolving systems, is called into question. This again would mean that evolution and culture would have to be strictly separated from each other—similar to the fruitless debate over the "true" connection between mental and material phenomena during the last few hundred years. If we accept this view, we would land in exactly the same place where most philosophers, scholars from the arts, and from the humanities have already set up their permanent camp. Why have we then spent all this time on this excursion into the world of evolutionary theory when at the end of it we discover that we have turned full circle? The philosopher Daniel Dennett put forward a somewhat personally biased criticism of Dawkins' story of memes, but for all that, it is nonetheless quite to the point:

I don't know about you, but I'm not initially attracted by the idea of my brain as a sort of dung heap in which the larvae of other's people ideas renew themselves, before sending out copies of themselves in an informational Diaspora. It does seem to rob my mind of its importance as both author and critic. Who's in charge, according to this vision—we or our memes? (Daniel Dennett 1993, p. 202)

However, instead of simply accepting that the human brain is in fact a real dung heap as renowned star philosopher Dennett despite his initial criticism seems to do („once our brains have built the entrance and exit pathways for the vehicles of language, they swiftly become parasitzed and I mean that literally...", p. 200), we can assume that Dawkins is simply wrong: if one takes his idea of memes really seriously, then the whole thing would have to function exactly as Dennet described it above and in no other way. Hence ultimately, Dawkins, as a confirmed Neo-Darwinist, must fail with his surprisingly Lamarckian theory of culture and memes since, like many others before him, he based his ideas solely on the purely physical information concept which, by its mere reference to the number of possible states of a system, allows all kinds of information carriers to be generated or projected into the world wherever and whenever they might be useful. Using such a perspective, it would be again possible to separate the carriers of the information, the so-called replicators, from their simply phenotypical effect on the body. Dawkins even went so far as to degrade the whole organism to being a phenotypical vehicle segregating it from its selfish super-replicators, the genes without being aware that, at the same time, what determines the semantics of the genetic information, i.e., its inseparably functional significance to the living process, would have to be lost. This is exactly the weak point of this and all similar concepts which simply take the fashionable information term to indiscriminately and often falsely apply it to everything possible. So suddenly, at one time, any structured arrangements of the inanimate world become important static pieces of information which are held to describe or explain it and it is often not noticed that this fascinating information is only attributed to the things themselves through our involvement with them. Thus, in principle, Dawkins meme theory, viewed heuristically, is not much more valuable than any astrological or otherwise esoteric interpretation of the meaning of the universe and its inhabitants. Astrology naturally has a meaning but only in relation to the thinking of the people using it, in whose heads all this mysterious information can be found.

Interestingly, the whole group of Popper/Lorenz-influenced evolutionary epistemologists—and not a few prominent people on both sides of the scientific-philosophical divide belong here—have worked out a theory of cultural development that is practically identical with Dawkins meme theory and therefore just as enticing but at the same time also just as wrong. What is meant here is Popper's famous sub-worlds of our world which we can be sure is already more than complicated enough without these. But not so for the great Karl Raimund Popper. World No. 3, in Popper's choice of words, doubtless the most valuable of all the worlds, includes all that which by Dawkins comes under the catchword of "meme pool", a term coined in analogy to the corresponding biological term "gene pool". It concerns the so-called knowledge in an objective view, under which Popper understood the totality of cultural inheritance coded as philosophical, theological, scientific, historical, literary, artistic but also technological knowledge laid down in specific material substrates of World No. 1 (e.g., tools, machines, books, works of art, music). Characteristically, Popper, exactly like Dawkins, did not realize that in all these substrates no real semantic information exists, let alone anything so un- and supernatural as "objective" knowledge. On its own, this was not sufficient to demolish the persistent popularity of his beautifully simple and impressive model of different "ontological" layers of the world (Kant on ontology: "the science of the properties of all things in general"). In fact just the opposite, as can easily verified, most scientists who have been involved with epistemology at some time in their careers return to Popper's Cartesian concept for the reason, among others, that it so convincingly conforms to what so many scientists see as their Holy Grail—namely objective knowledge. Associated with the somewhat more elevated, if not already transcendental World 3 are Popper's Worlds 1 and 2 which should set the things a little bit in motion so that thinking and active creatures can come into play—the latter being Popper's compared with his beloved "objective" world not so popular "subjects" World 2 consists of individual and hence always subjective states of consciousness, whereas World 1 of purely material objects and states which ultimately adds a suitable brain so that everything can function properly. There is no way in which such a scheme can ever function in a practical manner since its application to the one and only world we in fact are confronted with requires exactly that what was previously excluded with such painstaking care. A single small example should be sufficient to illustrate the dizzying spasmodic thinking of what results. If we allow ourselves to become captivated and attempt to consider things scientifically, in which one of these three worlds do we actually exist? Are we in World 2 of conscious states or are we rather in World 3 of objective knowledge? But then, what has happened to World 1, to our most highly prized brain whose lowly material origins we can hardly deny? It is to be feared that we must exist simultaneously in all three Popperian worlds or, even worse, we have to accept that, without our material brains, we cannot think out our actual concrete thoughts at all. The supposed objectivity of our thoughts, however, bring us up against an even greater problem. If they can be situated in our brains but also in books, how can they then belong exclusively to the noble association of World 3? They may have some interaction with other worlds but how can they become part of World 2 in another person when that must be strictly forbidden for them? Or may they do this on occasions perhaps with special permission? What happens to them with their highly valued objectivity, have they perhaps been exchanged for a shabby subjective fantasy? On the other hand, they have to become even more objective if they appear in another brain in World 1. Or is our understanding at that point completely wrong? At least, we may not expect that any convinced devotee of Popper would agree with such a blasphemy.

We have, of course, allowed ourselves to dramatize the situation somewhat and we could easily extend this ad libitum but I think that it is already more than clearly sufficient to demonstrate the uselessness of such categorical splitting or rather: hairsplitting of the world. On closer examination, the philosophical nonsense behind becomes immediately evident. If we separate the world into only two (e.g., body and soul) or more, as the philosopher, the lover of wisdom, so elegantly says, ontologically separated worlds (e.g., Plato's ideas, Kant's categories and the world itself, and today Popper's worlds), in order to scrutinize their precise interaction afterwards, we have given ourselves a task that will occupy our valuable thinking force for at least a lifetime but with no prospect of ever finding a solution to the insoluble problem. However, perhaps that was exactly the intention! This becomes evident if one looks for instance at Nicolai Hartmann's (1882—1950) famous "Aufbau der realen Welt" (The Structures of the Real World; cf. Werkmeister 1990) appearing in 1964 where this German philosopher conceived a truly fundamentalist "doctrine of categorical layers of existence" (subtitle). Looked at epistemologically, there is a kind of schizophrenia at work here where one and the same phenomenon observed under varying conditions are often subject to completely different interpretations. The situation is even reminiscent to a certain degree of the egocentric stage of early intellectual development in children when an object or subject observed under different circumstances (e.g., after a change of perspective) are understood as being separately existing things (cf. 3-mountains-task: Piaget and Inhelder 1948) or changing numbers of them (cf. conservation of number: Piaget and Inhelder 1969). Amongs others, this is also illustrated by that amusing phase of language development where the 2—3 years old child who wants something special to eat does not say "I" but rather "Kevin" or "Anna" and then expects the chocolate to be handed over immediately. At this age, the small child still sees himself only as it is designated by the family and in no way already as a unity of his own self with his own designation. This changes slowly with time and the experience of the child (Loveland 1984; for a similar developmental pattern in great apes, see Miles 1994, Patterson and Cohn 1994). It appears that the unification of previously more or less strictly separate areas of experience is one of the specific characteristics of human development to more and more efficient, that is generally applicable cognitive functions.

Finally, before finishing the chapter, let us return once again to our preferred one world theory of life, i.e., that explicitly materialist theory of evolution as it was devised by the late atheist pastor Darwin. One thing remains clear: the physicochemical conditions described by the central dogma speak unequivocally against any Lamarckian feedback from the organism as a phenotype to its genotype, the genes, but why exactly is this so? The central dogma does not speak against Lamarck just for this reason, we can now be precise since with the latter, a purposeful, i.e., a random mutation-independent evolution would be accepted which attempts to solve something that does not need solving any longer, namely the source of the knowledge of the individual. Since from a simple physiological regeneration or healing process up to the learning and application of mathematical algorithms (Heschl 1994b, 1996) in humans, it is always exclusively purposeful and therefore, of necessity, previously (but not necessarily consciously) known strategies which are used. And this explains why the ontogenetical development of the individual, no matter how impressively it may present itself to the observer, has nothing to do with a real acquisition of information.

That the nucleic acids and not the proteins represent the relevant molecules of heredity should be considered a more or less an evolutionary coincidence or perhaps has a lot to do with their special molecular peculiarities and interaction with each other and with other types of molecules. In principle, as already discussed earlier (chapter 5) one could even imagine just the opposite, namely a situation where the proteins or any other suitable class of molecules would be the carriers of heritable information and the nucleic acids the conscientious executors of the evolutionary tested instructions of the proteins' programs. This would not, in any way, change the basic mechanism of any kind of organic evolution, namely the unavoidability of changing their underlying structures by some kind of undirected and unpredetermined genuine random processes. To use the somewhat well-worn comparison of the tape recorder once more: it is not an embarrassing slip-up of evolution that the magnetic tape of information can only be played in one direction or, better expressed, only in a circle (cf. "circular causality"; Haken 1989, p. 25). It is just the opposite, even the smallest possibility of a Lamarckian directed recording of supposedly pre-existing semantic information from outside would immediately reduce any kind of evolution ad absurdum or to its logical antithesis which is the inanimate world. Within the latter, every change is both "adaptive" and "useless" at the same time since there are no true dynamic systems to be conserved.

Evolution must therefore, no matter how annoying this may be to us as rational beings, love nothing more than to work on the sensibly organized with really chaotic methods to be able somehow to achieve success. The really convinced Lamarckists, however, have believed again and again that something like inheritance of acquired characters was not only possible but also *necessary* to explain evolution completely while the Neo-Darwinists, who rightly dominate the field of evolution, have unfortunately failed so far to notice that the mysteriously acquired Lamarckian information in learning behavior is in no way an insuperable problem that as a special case of a so-called cultural evolution which works by slowly evolving cognitive mechanisms of imitation and instruction (for a phylogenetic treatment, see Tomasello 1999) has to be rigorously separated from biological evolution. The fear, or rather the social taboo that lies behind this need for separation, which simply forbids the biologist from viewing the evolution of organisms including humans from a uniform evolutionary perspective was formulated by Dawkins all too clearly in his otherwise very refreshingly irreverent book *The Selfish Gene*:

If a species is to be expected, it must be for good particular reasons. Are there any good reasons for supposing our own species to be unique? I believe the answer is yes. Most of what is unusual about man can be summed up in one word: "culture" (Richard Dawkins 1989, p. 189).

Our own answer to this apparently trivial and yet most important question of the modern theory of evolution must read quite differently. It is a clear "no" and yet again "no", because today, in contrast to earlier, we can substantiate such a claim by adequate theoretical arguments. It looks as if we would be able to propagate our memetic thoughts and ideas in exactly the same way as our genetic heritage and this with a much higher speed and efficiency than real genes are in a position to. In reality, however, our own thoughts have kept a surprising fact secret from us, and this perhaps for a good reason: their knowledge was always there inside us and did not ever come from outside somewhere nor from somebody above who does not have a body.

12 The "Wonder" of Language

It was the spoken word that made us
free. ... Until the Great Leap Forward,
human culture had developed at a
snail's pace for millions of years. That
pace was dictated by the slow pace of
genetic change. After the leap, cultural development no longer depended
on genetic change. Despite negligible
changes in our anatomy, there has
been far more cultural evolution in
the past forty thousand years than in
the millions of years before. ... With
language, we can store precise representations of the world within our
minds, and encode and process information far more efficiently than
can any animals.

Jared Diamond

With this citation, we have reached the position where we should subject the apparently most important reason for accepting the status of man being something special to closer scrutiny. It should be emphasized once more that by special position we do not mean a unique character like our upright gait or our mole-like (Fam. *Talpidae*) nakedness since, as already mentioned, every species has its own special position. If this were not so, there would be no sense in talking about species. What is meant here is the much more basic question of whether we as humans have the right to boast of being a genuine exception to the known mechanisms of biological evolution. Even the trained physiologist Jared Diamond, from whose book *The Third Chimpanzee* (1992)—meant is our own species because of its close genetic relationship with the chimpanzee and the bonobo (Diamond suggests to include the last two species into the genus *Homo*, resulting in the new species *Homo troglodytes* and *Homo paniscus*; for the detailed comparative genetics of the group, see Kaessmann, Wiebe, and Pääbo 1999)—the citation above was taken, finally capitulated to the mysterious wonder of human language. Exactly "like some other scientists who have speculated about this question" (p. 54), the all-too plausible answer to the controversial condition of *Homo sapiens* occurs to him. It has to be "the anatomical basis for spoken complex language" (p. 54), what else, that has made us into the unique creatures which we undoubtedly are. If this were not so, it would not be understandable how we could differentiate ourselves as sufficiently as we would wish from our nearest relatives who, when viewed genetically, are so close to us. Only a 1.6% difference in the gene sequences between us and the two species of chimpanzee (King and Wilson 1975; for a review, see Gibbons 1998) cannot possibly be the essence of the matter, there must be something more significant that not just separates such a unique species as *Homo sapiens* from the rest of the large family of primates but also from all the other living species. Diamond himself seemed not so fond of this all-too popular interpretation because a few lines later he had doubts about whether the aptitude for real language made such a very great difference between man and his nearest relatives as was generally believed:

However, virtually every claim of any animal behavior suggestive of elements of human language is greeted with skepticism by many scientists, convinced of the linguistic gulf separating us from animals. Such skeptics consider it simpler to assume that humans are unique, and that the burden of proof should be borne by anyone who thinks otherwise. Any claim of languagelike elements for animals is considered a more complicated hypothesis, to be dismissed as unnecessary in the absence of positive proof. Yet the alternative hypotheses by which the skeptics instead attempt to explain animal behaviors sometimes strike me as more complicated that the simple and often plausible explanation that humans are not unique (Jared Diamond 1992, p. 146/147).

In the following, we will try to be more consistent than Diamond. If we include all that has been spoken so far and make the attempt to assess language using evolutionary theory, there will be not much left of the wonder of language. In particular, the point concerning cognitive gain, and therefore the status of linguistic statements and interactions as evolutionary progress, will show us that the portrayal of symbolic communication as the unique character of the human species which allows it alone to circumvent the mechanisms of biological evolution is simply mistaken. It is better to say we are dealing with an illusion since we will see that maintaining this belief is merely stubbornly clinging to a cherished faith. Perhaps the special fascination of verbal communication is that subjectively we have a sublime feeling of real participation in the personal thoughts of our fellow human beings and therefore this at least helps us forget the less comforting and objective fact of our separateness. Closely associated with this is the idea, or even as far as our feelings are concerned the subjective certainty, that we humans can exchange real knowledge and above all substantially new knowledge among ourselves.

The difficult thing about language is that it is so difficult to talk about because this talk is itself dependent on the reflective possibilities of language.

That may well be, but the situation changes quickly when one treats it as modern researchers of language usually do in the meantime, namely to consider language to be just one of many other forms of behavior that has to fulfill an adaptive function within the framework of a comprehensive evolutionary theory of human behavior. The great man who, after the first comparable attempts of the famous baron Wilhelm von Humboldt (1767—1835) "Über die Verschiedenheit des menschlichen Sprachbaues und ihren Einfluss auf die geistige Entwicklung des Menschengeschlechts" [*On the Variability of the Structure of Human Language and Its Influence on the Intellectual Development of Mankind*], succeeded in bringing the investigation of language down from the ivory tower of philosophy of language to the lowly levels of scientific research was Noam Chomsky. What he brought to light with regard to the linguistic behavior of humans was immediately comparable to the contributions of Charles Darwin to the formulation of modern evolutionary theory. This parallel to Darwin is far more surprising than one would ever think if one is only acquainted with the career of Chomsky while he was young.

He became not only interested in politics at an early age but was also actively involved—"I grew up in the radical Jewish society of New York. That was in the time of the Great Depression." (Chomsky 1973, p. 165)—so that it was a roundabout route he followed to reach the special field in which he later became famous. Stimulated by their common political interests, he began to study linguistics under Zellig Harris, the head of the Linguistic Department at the University of Pennsylvania at that time. During the first few years, he tried to apply behaviorist methods to investigating the acquisition of speech. Like any other good behaviorist, he assumed that by applying a few inductive principles, it should be possible to simplify this unbelievably complicated phenomenon. However, in contrast to the still great host of learning theorists, he very quickly recognized that with these all-too limited reductionist methods—no matter what behavior, of mice or humans—that nothing could be achieved with simple associative principles of learning. He only succeeded much later in convincing other people of the value of his newly formulated theory of linguistics (Chomsky: "Even after I became convinced myself, I could not get my work published. Few people conceded any value to this work." 1973, p. 187).

Chomsky's inspiration or, as he himself described it, an idea that was born through a hobby,"I assumed that these generative grammars were only fun and my private hobby" (Chomsky 1973, p. 186), was the assumption of a generative grammar as a kind of deep-rooted structure of the human intellect that contains all the necessary instructions for the acquisition of any concrete language, be it English or Japanese. According to Chomsky the existence of such a universal grammar (UG), which is assumed to be common to all people, is necessary because children up to the age of four under normal conditions achieve a mastery of language whose extent could not be possible by simple associative learning processes taking place. Fully independent of purely selective learning processes in which only individual words and their meanings or various special rules of sentence structure are acquired, and within a relatively short period of time, extraordinarily stable structures are built up which go much further than what the experiences through our senses ever have to offer. One of Chomsky's favorite games, used by him to substantiate the correctness of his argumentation, is related to the principle of structural dependence of grammatical processes. This principle states nothing other, for example, than that children never form new sentence constructions such as questions, negations, changes of tense, relative clauses, etc. by merely changing the word order in a sentence, in order to create a new statement independent of the initial structure of the sentence. This would be the case if only simple rules of learning theory applied. A concrete example would be that those 4 year olds would never change the rather precocious statement "Chomsky made the first attempt at scientific linguistics" to form the question "Linguistics scientific attempt at, made the first Chomsky?" which is theoretically plausible and, together with the appropriate tradition, could be maintained for generations.

Even more variations would be theoretically possible but do not exist in any language. If, as Chomsky proved, learning Japanese is only slightly different from learning English, we must ask ourselves how this can be possible without some form of preexisting structure playing a role. If learning a language actually did follow the idealized and universally held principles of behaviorist learning theory, then first, children would have to be capable of learning everything theoretically possible and second, languages would then, with the complete loss of being translatable, very quickly drift apart structurally. A good example for demonstrating that this is not the case is furnished by the reduction in the diversity of languages unfortunately taking place in New Guinea—nevertheless still about 1,000 languages spread over a little more than 300,000 square miles. The speakers of these languages are capable of learning, exactly like every other human, a new, apparently completely different language and translating it into their own. Recent estimates state that during the next hundred years of rapidly advancing globalization, almost half of the 6,500 existing languages will disappear or, expressed more correctly, assimilated into each other (from *Atlas of the World's Languages*. Mosely and Asher 1994). However, this is most clearly demonstrated by small children, no matter where they come from, they possess the necessary pre-knowledge to learn to speak any one of the multitude of languages of our world and do it perfectly. It is fascinating for parents to observe how easily children between 3 and 5 years old learn another language in addition to their mother tongue without any of the laborious lessons in grammar. usually necessary for adults who even then only manage it with a great effort and seldom perfectly. What does remain untouched, however, is all those individual characteristics which everyone keeps independent of the respective language learned, whether it is to articulate certain kinds of sounds in a specific way or express a personal style of syntax or lexical meaning.

Chomsky's parametric theory of language states that every person is endowed with an innate set of universal principles (i.e., "universally" valid for all members of our species) with which he can assimilate any language. It is, however, completely clear to Chomsky, that the assumption of such a universal grammar represents a rough simplification since—as a child's learning of a language nicely shows—individual differences from the outset cannot be completely excluded, in fact they are the rule. Nevertheless, it is absolutely legitimate, as practiced day in day out in biology, to select the type of behavior common to all people from the abundance of individual ways of reacting when determining the typical characters of a species. Nowadays, despite a continuing theoretical debate about the nature of the so-called "typus" (cf. Riedl 1987a), it bothers no one that coughing, sneezing, or laughing are specific kinds of behavior of *Homo sapiens* although we, when we are city dwellers (which will soon be 80% of us), are permanently confronted with conspicuous new variations of certain behavioral patterns. *The* universal grammar will probably not be realized by any single human being of this world, like the fact that *the* pear or *the* mountain will never really exist, but this prevents nobody from understanding what Chomsky has in mind when he says:

Let us define "universal grammar" (UG) as the system of principles, conditions, and rules that are elements or properties of all human languages not merely by accident but by necessity—of course, I mean biological, not logical, necessity. Thus UG can be taken as expressing "the essence of human language". UG will be invariant among humans. UG will specify what language learning must achieve, if it takes place successfully (Noam Chomsky 1975, p. 29).

The universal grammar, sometimes also called "generative" grammar, since it can generate language is nothing more than a highly specific system of rules with which, during the unfolding of ontogeny, a theoretically unlimited though not endless number of sentences can be produced. What has to be learned by every child it seems is definitely not the language itself but at most the variable regional and historic peculiarities in pronunciation and meaning of a certain language idiom whereby apparently very different languages become nothing more than trivial variations of the same basic pattern. This explains, among other things, why the direct translation from one language to another is at all possible since this would be ruled out completely if the acquisition of a language occurred according to purely associative rules or took place randomly under the strongly varying environmental conditions of different cultures. What consequences can be drawn from Chomsky's influential theory for our topic? The central question to the possibility of cognitive gain can also be answered very clearly in the case of human language. Language and every other form of symbolic communication based on it has a lot to do with rules, regularities, and well-defined mechanisms but at the same time, and for the same reason, not the slightest with any kind of cognitive progress and passing on information. Nobody apart from Chomsky has managed to demonstrate so beautifully that language can only be understood by its adherence to the rules and hence completely systematic character. In other words, the subject learning the language, usually the growing child, always knows down to the last detail exactly how it has to proceed since in the whole of language development, chance plays no constitutional role—it would be disruptive. In fact, any random changes in this area would unavoidably endanger the coherence of the whole system of speech. As the theory of evolution has taught us, such a change would only have a chance of improving the whole system at all, if it occurred in the germ line, the only place where it is possible without endangering the coherence of the physiology and the behavior of the developing metazoan right from the start. Consequently it is this extraordinary adherence to the rules which eliminates what is often erroneously ascribed to language, namely the possibility of achieving cognitive gain independently of purely biological evolution.

One could object here that such an interpretation was overburdening and misusing Chomsky's theory for one's own ends since today, nearly everybody admits that a certain structural frame for speech acquisition, namely the famous universal grammar, could be inborn but that an abundance of special details must first be acquired as qualitatively new information by the person learning. That one differentiates between a fundamental deep-rooted structure and a superficial structure laid down at the periphery speaks in favor of this objection. While the former acts like a biological substrate on which language acquisition is built, at least the latter should include with it, as the remainder of the freedom of the individual, the possibility of real new acquisitions. In principle, this would bring us back to the old division of inborn versus acquired and we would yet again be embroiled in the tiresome argument about the inheritance of the various components involved in the functioning of language. Fortunately, however, resorting to such an outdated dichotomy is no longer justifiable and Chomsky himself clearly defined what many of his interpreters have not yet understood or perhaps have not wanted to understand:

The term "deep structure" has, unfortunately, proved to be very misleading. It has led a number of people to suppose that it is the deep structures and their properties that are truly "deep", in the nontechnical sense of the word, while the rest is superficial, unimportant, variable across languages, and so on. This was never intended. Phonological theory includes principles of language that are deep, universal, unexpected, revealing, and so on; and the same, I believe, is true of the theory of surface structures and other elements of grammar (Noam Chomsky 1975, p. 82).

What Chomsky here makes clear indirectly is nothing other than the empirical fact that the whole field of verbal communication represents a category of human behavior that is distinguished down to the minutest detail by well-defined and determined regularities. In contrast to many biologists who are frightened of the natural wonder of man, he did not shy away making use of the same assumption for other seemingly uniquely human areas:

Alongside of the language faculty and interacting with it in the most intimate way is the faculty of mind that constructs what we might call "common-sense understanding", a system of beliefs, expectations, and knowledge concerning the nature and behavior of objects, their place in a system of "natural kinds", the organization of these categories, and the properties that determine the categorization of objects and the analysis of events. A general "innateness hypothesis" will also include principles that bear on the place and role of people in a social world, the nature and conditions of work, the structure of human action, will and choice, and so on. These systems may be unconscious for the most part and even beyond the reach of conscious introspection. One might also want to isolate for special study the faculties involved in problem solving, construction of scientific knowledge, artistic creation and expression, play, or whatever prove to be the appropriate categories for the study of cognitive capacity, and derivatively, human action (Noam Chomsky 1975, p. 35).

On the basis of this statement, one would have thought that with Chomsky we had encountered the most unexpectedly consistent epistemologist if only he had, when confronted with the awesome, all-or-nothing decision for an evolutionary interpretation, at the very last minute, taken the bull by the horns and not run away from it. This was the Noam Chomsky who, taking up the position as a rationalist linguist and not, as one would believe from his nativistic theory of language, as a biologist fully convinced of Darwinism, considered the linguistic competence of the species *Homo sapiens* as a preexisting character. This was the man who saw not much more in the contribution of language to survival than he saw in the contribution of arms and legs to the special feature of human locomotion since our first *erectus* ancestors, this great thinker ultimately rejected all the consequences concerning the crucial point of the whole business, namely, the epistemological status of verbal communication. The whole thing begins with a more than a skeptical consideration of the influence of natural selection on the origins of intelligent ways of behaving in humans:

Like physical structures, cognitive systems have undoubtedly evolved in certain ways, though in neither case can we seriously claim to understand the factors that entered into a particular course of evolution and determined or even significantly influenced its outcome. ... Among the systems that humans have developed in the course of evolution are the science-forming capacity and the capacity to deal intuitively with rather deep properties of the number system. As far as we know, these capacities have no selectional value, though it is quite possible that they developed as part of other systems that did have such value. ... It would be a serious error to suppose that all properties, or the interesting properties of the structures that have evolved, can be "explained" in terms of natural selection. Surely there is no warrant for such an assumption in the case of physical structures (Noam Chomsky 1975, p. 58/59).

All that finally ends again, in relation to the essential question about the possible limits of human intelligence, with the ancient myth of the special position of the human mind:

Notice that these quite natural views on the scope and limits of knowledge set no finite limits on human progress. The integers form an infinite set, but they do not exhaust the real numbers. Similarly, humans may develop their capacities without limit, but never escaping certain objective bounds set by their biological nature. I suspect that there is no cognitive domain to which such observations are not appropriate (Noam Chomsky 1975, p. 124).

This obviously subjective and firmly held conviction that, in principle, the possibilities of the human intellect are infinite, held not only by Chomsky, but also by Piaget, Popper, Lorenz, Campbell, and many others, is repeated again and again and needs closer examination. At least theoretically it cannot be excluded that, as a result of the repeated application of specific rules of grammar on language, an enormous diversity of different sentences can be generated. In Chomsky's words—"to say it once more, a grammar is a system of rules and principles which generates an infinite class of sentences" (Chomsky 1977 p. 55). However, this does not necessarily mean that the associated cognitive faculty of humans can be assumed to be limitless. Independent of all the many different variables which, through the fluctuations in the environment and the social milieu, have an influence on language, variants of the universal basic pattern of the UG are predictable or can be generated when needed. For just this reason, the apparently infinite number of possible variants has nothing to do with cognitive progress since the recognition system, i.e., the living organism, always knows exactly how it is to proceed. Just the latter has to be ruled out if we wish to have the right to talk about real cognitive progress. In principle, the apparent infinity of language systems is directly comparable with the admittedly less well-founded infinity of the realization of instinctive behavior since animals do not hop around, or roar, or eat, or mate, etc. like preprogrammed robots in perfectly stereotyped ways. They produce well-determined patterns of behavior based on flexible, regulatory systems. Unfortunately, however, flexibility in behavior systems is often erroneously equated with a degree of indetermination in the underlying genetic program, a conclusion that is devoid of any empirical basis.

Let us return to Chomsky's surprising climb down over the integration of his theory of language that is oriented very strongly on scientific results into one of the most important basic theories in biology, the theory of evolution. His last minute turnaround, more anti-evolutionary than anti-biological, back to the philosophical myth of man as the wonder of nature is especially regrettable because he was so near to the correct view of things as far as the connection between evolution and cognitive progress is concerned. Right from the start, his well-argued "hypothesis of innateness" goes much further in all areas of human thought and intercourse than the later so-called evolutionary psychologists and epistemologists, whether active empirical scientists or simply interested philosophers, had trusted themselves to say about the claims of human intelligence being dependent on evolution. If that were not enough, Chomsky also provided us with a concise and, above all, absolutely correct description of the main issue of our investigation:

Returning to the main theme, suppose that we now select a problem in a domain D that falls outside of O's cognitive capacity. O will be then at a loss as to how to proceed. O will have no cognitive structure available for dealing with this problem and no LT (O, D) (= learning theory for O in the domain D) available to enable it to develop such a structure (Noam Chomsky 1975, p. 24).

Only somewhat later did he say:

Outside the bounds of cognitive capacity, an empiricist theory of learning (note: = learning through trial and error) applies, by unfortunate necessity. Hence little learning is possible, the scope of discovery is minimal, and uniformities will be found across domains and across species (Noam Chomsky 1975, p. 39/40).

The first of these two citations in relation to the analysis of the theoretical point of departure gets our full approval. A similar perfect description of the fundamental problems of overcoming cognitive boundaries was given by the first great language logician of modern times, Ludwig Wittgenstein, who, however, did not go much further than the starting point:

The book is aimed at drawing a boundary to thought, or rather—not thought, but the expression of thought: since in order to draw a boundary to thought , we must be able to see both sides of the picture (we must be able to think of what it is not allowed to think about) (Ludwig Wittgenstein 1921/1963, p. 7).

The consequences we can deduce from this today are of a very different nature to those of Chomsky, who in the final analysis introduced a generally vague theory of learning or of Wittgenstein, who with the well-known saying, "What one should not talk about one should keep silent", (Wittgenstein 1921/1963, p. 7) some time before prescribed a paradoxical kind of knowledgeable silence. Thereby, a conclusion consistent with evolutionary theory could not be easier to deduce. In the field in which one has no competence, one must remain cognitively blind since the possibility of forcing one's way into an absolutely forbidden area with just any kind of competence cannot exist. Beyond the boundaries of the cognitive competence of a creature, not as an unfortunate result, as still thought by Chomsky, but an epistemological necessity, we have to give the evolutionary anti-methods of random variation and natural selection the only possible chance of success. Contrary to Chomsky's assumption, the evolutionary mechanism has nothing to do with learning by the trial and error method since this also represents a directed cognitive competence just like any other imaginable. For this very reason, for every area that has been inaccessible up till now, only an undirected progress by genetic mutation is possible since—to formulate this very simply if not idiotically simply—naturally nothing can ever be known that *by definition* cannot be known.

In addition to this specifically evolutionary consideration, we can also formulate a system-theoretical treatment of the human power of speech which ultimately comes to the same conclusion. The connection between the power of speech and postulated cognitive gain through verbal communication can generally be viewed as the question about the possibility of a transfer of information from one biological system to another. Here, something like a break in cognitive continuity between physically separate systems comes into effect. This means that—in contrast to the quasi-perfect coherence within a living system which exists as long as the system survives—the external relationships between such autonomous systems, because they are under unpredictable influences, are subject to strong physical fluctuations. It may sound trivial but this fact allows us to speak at all of concrete systems that can be considered separate. These very special physical units also called "creatures" since they are alive (although nowadays it is better to say organisms) are simultaneously the finally decisive units or entities of selection in a systems theory of evolution. Already very simple considerations demonstrate the importance of individual organisms as the ultimately significant units on which natural selection can act. This is the reason why no single organ of a metazoan (heart, lungs, nervous system, etc.) can be selected in preference to the others since, in such a case, there would be an inevitable shipwreck—metaphors like "all in the same boat" or "share the same fate" indicate this connection even if this is occasionally overused in a confusing way by advocates of group selection (Wilson and Sober 1994). Coherent organization would be a somewhat more technical term for these autonomous systems, called living organisms, which are separated by physical discontinuities and are therefore automatically in competition with each other. That, in spite of this, fundamental system limits could be repeatedly pushed back during evolution, even if seldom, is shown beautifully by the origin of metazoans themselves. From previously independent individual systems, unicellular ancestors, in the course of time, completely new mega-individuals were formed in which the former gaps in the causal connection were bridged by newly created strong spatiotemporal coherences. The prerequisite for this, however, was the maintenance of genetical identity and thereby the cooperativeness of all the involved cells, now called somatic cells.

As far as the status of interaction between separate individuals systems is concerned, two things can be said. First, the interaction is permanent, i.e., no system can choose if it wants to interact with its environment or not—even a perfect noninteraction, if physically possible, would still be a special form of missing interaction—and secondly, it is cognitively neutral. The latter means nothing other than that the knowledge of each living system ends at its material boundaries and therefore everything that comes from outside demands a special interpretation. This special kind of interpretation must first be in a position to generate real semantic information recognizable as stable process structures from the purely physical information realized by the most varied material and energetic states. Formulated another way, this means that knowledge can only originate within the dynamically stable organization of a living system and what we like to consider an exchange of information between such systems represents, in reality, a purely physical interaction whose cognitive interpretation is dependent on the systems in question themselves. Living systems, in contrast to their accessibility to material and energy, are distinguished by their cognitive isolation. Nothing of this can be changed by the occasional wishful thinking that we humans are intellectually and inextricably joined to all other (or at least almost all) individuals of a social group and therefore the thinking of a single person cannot be a purely individual thing. Even when everyone in the world was of the same edifying opinion, all of them would be, i.e., each one of them, alone with it.

Back to communication between living systems. The neglect of systems-theoretical and epistemological facts of isolation in all those cases is not so serious where very similar or perhaps even almost identical systems interact with each other. Only in these special cases, can we actually assume that the semantic information obtained from the same physical information consisting of optical, acoustic, chemical, or other signals is more or less the same. In this way, one can, like Chomsky did for the reasons of a simpler manner of representation, neglecting all the existing individual differences, proceed from the existence of a universal linguistic grammar which allows us within the framework of verbal communication to exchange, thoughts, ideas, even whole theories with each other. However, this portrayal of the situation has the great disadvantage that one is very easily lured into the false assumption that real thoughts, ideas, etc. are being exchanged. Such notions of telepathic communication must be confronted with the fact that the purely acoustic realization of the spoken word "thought" can hardly contain the thought "thought". Just such a fallacy is committed in many popular philosophical articles on the nature of human communication where, e.g., the origin and the consequences of a Popperian world number 3, the world of objective knowledge in our intelligent books, films, and other "objective" media is discussed in all their details. That Chomsky himself did not fall into the trap set by this unconsidered and, viewed epistemologically, utterly incorrect idea and right from the beginning had a system-related concept in his head, is shown by the following citation:

As for the fact that the rules of language are "public rules", this is, indeed, a contingent fact. It is a fact of nature that the cognitive structures developed by people in similar circumstances, within cognitive capacity, are similar, by virtue of their similar innate constitution. Thus we share rules of language with others as we share an organization of visual space with them. Both shared systems play a role in successful communication (Noam Chomsky 1975, p. 71).

What ultimately remains from the great wonder of speech is that with this most dearly loved occupation of our species we are only dealing with a simple interaction like any other between living systems where the cognitive interpretation is determined fully independently of each other by the respective partners in the interaction. In many cases, these interpretations will, excepting those of monozygotic twins (see McClearn et al. 1997; Gottesman 1997) and other clones of nature, be very different from each other. So, for example, the communication between prey and predator with great probability does not lead to the same semantic messages of the interaction ("Look here, I am edible but please do not eat me" versus "Look here, I am a predator, I am sorry but I would like to eat you very much") just as it would not do so in the interaction between parasite and host (try to imagine a "dialogue"). As soon as there are even the smallest differences in the cognitive constitution, it becomes immediately obvious that equating purely physical with real semantic information leads to glaring mistakes in interpretation. In the case of perfect cooperation between identical systems, they are cleverly hidden but simultaneously let us misunderstand the precise nature of the interaction between separate living systems.

In the meantime, modern linguistics, which has, to a great extent with the help of Chomsky, left the old philosophy of language miles behind, has already found the final connection to the empirical methods and theories of a natural science (Wind et al. 1992). This is providing us with, when one considers the state of research only a few decades ago, a much more exact picture of the relationships in the case of the development and function of human speech. The more recent works of the linguistic researcher and, to a certain degree, Chomsky's heir, Steven Pinker are a vivid example of this. In his book *The Language Instinct. How the Mind Creates Language* (1995) Pinker summarizes very graphically the state of affairs in the case of language when he introduces that element which with Chomsky was unfortunately totally missing: the biological evolution of the ability to speak by the interplay of mutation and selection. Since, in his opinion the acquisition of speech had from only very little or nothing at all to do with a Skinner-type general learning process, the purpose of which has not been answerable for a long time, but to do with instincts which can clearly be compared with others which are largely innate, it is obvious to compare the evolution of the ability to speak with the evolution of all the other biological structures, and be it the evolution of the speech-trumpeting elephant's trunk. In some additional points, Pinker goes appreciably further than Chomsky in that he postulates the existence of an inner cognitive language called *Mentalese* that in every case precedes a grammatically determined verbal statement (compare the absurd campaign of the early "linguist" Wittgenstein [1960] against the acceptance of such an inner language). Such an opinion of a close relationship between thinking and speaking is new—still Chomsky sees both these brain functions as acting essentially independently of or, at most, parallel to each other. In contrast, Pinker decided for the first time to have a clear hierarchical order in which the "inner language" played a much greater role than previously ascribed to it. This is in excellent agreement with the early work of the great interpreter of Piaget, Hans G. Furth, who proved the existence of completely normally developed "Thinking Without Language" even in congenitally deaf children (1964, 1971, 1972). We know today that, among themselves, these children—incredible as it may seem—develop a language completely of their own which shows better than any linguistic jabber that language does not come from the outside (Goldin-Meadow 1998). As far as the crucial point of the question about the epistemological status of language is concerned, Pinker, in a discussion of an empirical work about the way that the correct conjugation of regular and irregular verbs functions comes to the following conclusion:

Although there is evidence that the memory system used in language acquisition and processing has some of the properties of an associative network, these properties do not exhaust the computational abilities of the brain. Focusing on a single rule of grammar, we find evidence for a system that is modular, independent of real-world meaning, nonassociative (unaffected by frequency and similarity), sensitive to abstract formal distinctions (for example, root versus derived, noun versus verb), more sophisticated than the kinds of "rules" that are explicitly taught, developing on a schedule not timed by environmental input, organized by principles that could not have been learned, possibly with a distinct neural substrate and genetic basis (Steven Pinker 1991, p. 534).

There is not much more to add except that—as already discussed when we examined Chomsky's ideas—all the possible purely associative mechanisms participating in the use of speech or acquisition of the power of speech are no exception to the fact that speech has nothing to do with any growth of knowledge or cognition. Independent of the specific kinds of learning mechanisms involved which actually are only competing for the best scientific description of the phenomenon speech, whether now a more universal grammatical structural development, or more behavioristic principles of learning, or a modern Pinker-type combination of both („rule-associative-memory hybrid model";

Pinker 1991) it is clear that speech represents nothing more than a repertoire of behavior developed completely in accordance with certain pregiven deterministic principles by variously linguistically talented individuals of the species *Homo sapiens* and therefore, from its very nature excludes any realization of cognitive gain. Nothing of this is changed by the astonishing observation that languages appear to develop so fantastically independently of their biological carriers, the human speakers, and this at such a speed that seems to leave the boringly slow biological evolution a long way behind. It is naturally very impressive to learn that two sister languages can develop to apparently mutual incomprehension within only 500 years or, to cite a somewhat different case, the complete replacement of an existing language by another. Appearances are deceptive, in the first case, there always remains a great deal of translatability and in the second, the genetic constitution of the acquirer of the language cannot be ignored. So, much to our astonishment and in spite of numerous personal contacts, it has not proved possible to arouse the enthusiasm of any of the still remaining intact groups of chimpanzees, gorillas, or orangutans for even one of our many marvelous human languages. This can, however, have much deeper-rooted reasons. Antoine Le Grand, a proponent of Cartesian ideas in England in the 17th century, reported the existence of "a certain tribe in East India which believes that, in their neighborhood, there were often apes who could in fact understand and speak a language but made no use of this ability because they were afraid that if they did they would be set to work" (from Antoine Le Grand: *An Entire Body of Philosophy, According to The Principles of the Famous René Descartes*). At the same time, however, this does not mean that our ape-like relatives cannot master some degrees of difficulty in *mentalese* such as shown by the recently studied more advanced cognitive abilities of bonobos (Savage-Rumbaugh and Levin 1995).

In how much detail our genetic constitution not only instructs the acquisition of linguistic abilities but also determines its particular details has been shown by recent investigations of the inheritance of certain forms of highly *specific language impairment* (SLI; Bishop, North and Donlan 1995). Various grammatical rules such as the formation of tenses, number, gender, cases, or person are from the sufferers systematically used incorrectly or not used at all. Comparable grammatical weaknesses showed 80% concordance in monozygotic twins, and even in twins from different zygotes 35% of the cases had the same speech impediment (Tomblin and Buckwalter 1998; cf. Dale et al. 1998). In an extended case study on the distribution of SLI in a family with 30 members, spread over 3 generations, it was proved that this deficit, with great probability was due to a fully effective dominant gene on a somatic chromosome (Gopnik 1990). However, whereas all of these investigations, as is typical for traditional behavioral genetics, were still indirect, i.e., purely phenotypic ones, a brand-new study has shown that there exists in fact something like a

Chomskyan master gene called *FOXP2* which seems to play a crucial role in normal speech development (Lai et al. 2001). Now it will be interesting to see if other primates, and in particular the great apes possess a comparable variant of that same gene.

Another topic which should be briefly handled is the very large area of research of the neural or neuronal basis of the power of speech. In this area, such a lot has occurred recently that it is almost impossible to give a suitable overview in just a few lines. A rather general neurobiological finding which in the meantime has been relatively well substantiated says that there must be at least three different systems involved in speech behavior. The neurolinguists Hanna and Antonio Damasio describe these systems as follows:

> We are convinced that the brain processes language by means of three interacting groups of structures. The first, a whole battery of neural systems in the right as well as the left hemisphere is present for the nonverbal exchange between the organism and its environment mediated by various sensory and motor systems—that means everything a person does, perceives, thinks, or feels.
> The second group—a smaller number of neuronal systems that are mostly situated in the left half of the brain—represents phonemes, phoneme combinations, and syntactic rules for the combination of words. When these systems are activated by the brain itself, they make the word forms available and generate spoken or written sentences: when they are, in contrast, stimulated by the spoken or written word, they conduct the first processing steps of these aural or visual language signals.
> A third group of structures, which are also mostly situated in the left hemisphere of the brain, mediate between the first two groups. These instances can call up a mental representation and stimulate the production of the word form, or they can receive words and cause the other parts of the brain to call up the appropriate representation (retranslated from Antonio and Hanna Damasio 1992, p. 80/81).

In contrast to earlier concepts, where one was more or less of the opinion that there must be a narrowly bound and isolated speech center, today it has been shown that there is a close intermeshing of the speech system with practically all the other cortical structures. Perhaps the most interesting thing here is the discovery of the third group of neuronal structures since it goes to the core of what is of great significance in verbal communication: the connection between purely linguistic representations (noun, verb, etc. as complex sensory or motoric events) and non-linguistic representations as the actual concept of our thinking (object and process representation, categorization, formation of analogies and metaphors, abstraction, formalization). That these mediation structures must exist in some form or other can be demonstrated very clearly with what we can describe as the "it's on the tip of my tongue"-syndrome which everyone experiences from time to time and which increases with age (James and Burke 2000). Here is meant that often embarrassing situation when, in the middle of a conversation, one cannot remember a name or a special word and, no matter how hard one tries, one is unable to recall the correct term from the internal dictionary, i.e., Damasios' speech system No. 2. One thinks of those situations such as oral examinations or interviews when all the listeners are waiting expectantly for the correct word to come out—it is obvious that our specialist knowledge is not blocked out, since we know exactly what we are talking about, neither is it our index of suitable expressions (the word seems to be waiting for its cue to come on stage), it is alone those instances of mediation so well-defined by the two Damasios. It is interesting that the only solution is to think of something else in such situations in which some circumstance (e.g., a disagreeable or a too much friendly conversational partner) has confused the mediation system—it may be looking in the wrong part of the dictionary. Usually, this allows a new attempt to be made. In any case, the cited saying is common in most languages (German: "mir liegt das Wort auf der Zunge", French: "j'ai le mot sur le bout de la langue") so that it is highly improbable that this phenomenon is abnormal or pathological.

Even more conclusive in this connection is the observations with people who have suffered irreversible damage to the brain area responsible, mainly the left temporal lobe, where more concrete expressions are mediated on the anterior inner pole and more general expressions on the posterior pole. In some cases not only the access to various areas of the dictionary units is made more difficult but also the grammatical structure of the spoken sentences is severely impaired. Characteristically, in these cases, the access to verbs and function words is simultaneously linked with such a grammatical dysfunction which indicates that grammar understood as a syntactic linking of words is not much more than symbolic operation in a purely verbal space. It may be here where Chomsky's universal grammar is located together with all its hypothetical deep- and no less deep surface stuctures of the human power of speech, which represents not so much an abstract speaking ability by itself but instead can be viewed as being perfectly identical with Piaget's universal action logic (see also Greenfield 1991, Deacon 1992, Weber 1998, and, more recently, Grodzinsky's BBS target article "language use without Broca's area" [2000]). In such a case, the known specific speech centers in the human brain should be considered less as centers of a pure speech awareness but rather as some kinds of swollen acoustic areas with appropriately (over)dimensional sensory and motor parts for handling the abundance of sounds to be heard and to be spoken.

The special location, or neighborhood, of these centers does indeed speak for such an interpretation. So for example, the Wernicke speech region, which is largely responsible for the sensoric storage of heard as well as spoken sounds, is immediately next to the primary auditory cortex, the already mentioned *Heschl gyrus*. Injuries in this region make themselves noticeable as impairment of the comprehension of speech (sensory aphasia). In contrast,

Broca's area, responsible for the syntactic realization of what is spoken, i.e., for the grammatically correct formation of words and sentences, can be found directly next to the primary motoric center for the region lips, jaw, tongue, and throat. If there is a breakdown in this region, as one would expect, there are symptoms of speech impairment (motoric aphasia) whereby it is typical that the normal, i.e., nonsymbolic functions of the muscles used in speaking (eating, drinking, and swallowing) remain completely intact.

There only remains an outstanding causal explanation of a long-known phenomenon, namely the so-called lateralization of the centers for the spoken as well as for sign language (Hickok, Bellugi and Klima 1996; for the probability that sign language preceded spoken language, see the impressive and since then unequalled results of Gardner and Gardner 1969 with chimpanzees), which, in nearly all cases, are found in only one of the two hemispheres of the brain, namely the left: Broca's motoric speech center in the 3rd left frontal convolution of the cerebrum and Wernicke's sensory speech center in the left parietal and temporal lobes. This is quite astonishing since, in no other vertebrate, has a comparably marked asymmetry of the two halves of the brain been found. Here we have, at last, a tangible morphological character in front of us which appears to differentiate *Homo sapiens* clearly from all other animals (for similar asymmetries in animals, see Güntürkün, Emmerton and Delius 1989; Bradshaw and Rogers 1993; Güntürkün et al. 2000; Güntürkün 2001). Unfortunately, this is not completely true: our nearest-related brother apes (and sister apes) must be included in this view because recently a proof of a clearly distinguishable Wernicke center has been produced in the common chimpanzee (Gannon, Holloway, Broadfield and Braun 1998). This explains on one hand how the cleverest apes in history (Washoe, Kanzi, Bakala, Goma, Congo and, last but not least, Judy) can understand so much, simultaneously without—if we disregard the astonishing cases of spontaneous two-and-three-word sentences in the bonobo Kanzi (Savage-Rumbaugh et al. 1986)—giving any more "sophisticated" commentary (Robert Seyfarth very aptly speaks of the existence of "half a language" in monkeys and apes, see Weber 1997) and, on the other hand, this makes it understandable how we humans in our development as children (and later learning a foreign language) begin with first learning to understand words and true sentences before we go over to actively speaking ourselves (Savage-Rumbaugh et al. 1993; probably, this is valid not only for primates, i.e. some kind of limited thinking without [much] talking or signing, which is nicely demonstrated by Alex, the speech-trained Kanzi of the grey parrots, cf. Pepperberg 1990, method from Todt 1975; note also the conspicuous difference in humans between hearing a song and singing it, between hearing music and playing it on an instrument, between contemplating a landscape and painting it and so on—all these asymmetries could possibly be caused by the difference between, in terms of proximate mechanisms, rather "selfish" perception and rather cooperative activities, the latter perhaps the true origin of spoken language; in favor of such an interpretation is numerical asymmetry, still existing in modern humans, between few highly gifted speakers and many listeners, few highly gifted performers and many spectators, few highly gifted writers and many readers). Yet again, only a very, very relative difference between speaking-thinking humans and apparently dumb apes (one may note the two different, but at the same time closely associated meanings of the word "dumb"; cf. McComb and Semple 1998). Be that as it may, we must proceed from the fact that each neuronal function should be primarily symmetrically organized since from the origin of the vertebrates both sides of the brain have agreed to separate processing of the sensorimotoric incoming and outgoing signals of the two symmetrical halves of the body. Then why this disturbing esthetic disfiguration of an otherwise harmoniously symmetrical central control organ? What can be the underlying reason for this unusual morphological peculiarity? At this point, we should take the liberty of proposing a somewhat daring but, as we shall soon see, a not too implausible evolutionary speculation. The starting point of my speculation is an observation that I was able to make about my two sons. I have long been interested in the ontogenetic development of pointing with the usually responsible forefinger at objects, a very characteristic way humans behave which in this special form cannot be seen in any of our nearest relatives. My thesis consists of this action which is universally present in humans and therefore unremarkable being perhaps more important for the intellectual development of small children than we have thought up till now. The actual initiator for a closer investigation of this reaction was a proper kind of pointing game that one of my nephews when 15 months old used to like playing with his parents and other relatives. He pointed at certain objects in various situations and silently demanded that the appropriate word be said to him. This apparently trivial way of spending the time obviously gave him a lot of satisfaction for not only the parents but all the other adults present reacted as if to a secret internal command and named the object in question (e.g., light) at which the little boy repeated the action again and again and he himself also repeated the word. I also observed the same game with other children so I took it on myself to look at things more exactly.

As both my sons (Oliver, Benjamin) reached the appropriate age—just to be sure, a few weeks before—I began to make regular video recordings in a half-experimental situation. The main filming location was at the daily breakfast table where for a few weeks the following typical scene was played out. Children just like adults are often very hungry first thing in the morning, and the object of childish desire, the drinking bottle with its delicious substitute mother's milk had been placed in a clearly visible position in the center of the table but somewhat out of reach. The distance between the little hand and the bottle was about half a meter. As one can imagine, the reaction of the unwilling star of this production was suitably violent but as the rest of the members of the family were able to (or were forced to) experience at close quarters also very informative. What began as angry shouting and groping in the direction of the milk bottle developed stage by stage from an initially decisive stretching and reaching for the object of desire, then more and more to a purely demonstrative grasping for it and finally to a real pointing at the solitary place for the fulfillment of the momentary desires. From what began with grasping with the whole hand changed with time to pointing at first with several fingers but finally more and more often to pointing with one finger alone at the unreachable but still very visible object.

In addition, the accompanying sounds vented by our highly motivated test subject also improved in parallel to the stage by stage improvement in the act of pointing. From the shouting and bawling, gradually more and more often came babbling and talking nonsense and finally something similar to word was produced. My older son Oliver even managed alone to discover anew the word "hunger" which no one had ever said to him. From a simple clucking, which in principle was the expression in anticipation of the long-overdue act of eating, developed a babbling noise which sounded like "nga, nga" and then finally to a very clear "nunga, nunga". In order to end the torment and to prevent my son from being irreversibly fixated on a strange and later possibly confusing version of the word, my son was allowed to learn the correct pronunciation of "hunger".

How can all this fit together? An evolutionary working hypothesis would be that the actual starting signal for the acquisition of human speech is an appropriate ritualization of the pointing behavior developed from the simple grasping behavior for objects and linked to a change in the accompanying sounds produced—namely away from the emotional arousal (mainly atonal noiselike sounds) to a symbolic, relevant statement (more tonal sounds, syllables, words) (even Thomas Kuhn commented on the cognitive significance of pointing: "Any one who has taught a child under such circumstances knows what a primary teaching aid pointing is", Kuhn 1974/92, p. 404). From simple grasping and manual handling of objects, the first "grasping" and finally naming of objects could have originated, whereby individual and social behavior receive their first important synchronization. This matches exactly with the fact that from the side of the considerate parents there is a program for this which promptly in such situations causes them to react with the correct answers (i.e., by naming the object, feature or process observed) and repeat this a thousand times if necessary. In addition, the pressure on a parent exerted by a child, perhaps one's own, who points with its finger at an object and simultaneously babbles something causes the adult to say the word in an exaggeratedly clear way and in a noticeably louder voice with a suitable change in frequency. It is known that adults sensibly speak in a much higher voice as soon as they begin to speak to small children, in fact more than a whole octave higher (*baby talk*: Ferguson 1964, Szagun 1980, Fernald 1985; *Motherese*: Oller 2000) and intone every single syllable very slowly and clearly in the learning to talk phase so that they offer their offspring the most optimal pattern to imitate (Kuhl et al. 1997). Children usually respond very positively to this kind of baby talk because, on one side, they can distinguish the transposed word much better and, on the other, the fear of the giant size of the adult is reduced by this playing down to them. As a rule, it is an easy for the growing child to harness grown-ups for their own purpose according to their level of verbal competence even when it is simply being an appreciative audience, even if only temporarily, for an exciting story for the beginning of long-term memorizing of the first words at the age of about eight months (Juscsyk and Hohne 1997). If the young gentleman or lady does not want to learn or can still not learn something, then he or she will not do it no matter how much trouble even the most loving parents take.

What is still missing from this picture is the current status of knowledge on the specialization of the cerebral hemispheres. Since most of us humans are inherently right-handed (9 from 10; but see Rosenbaum 2000) and use the forefinger of this hand for pointing more often than we do with our weaker left hand, it seems obvious that here an initially trifling effect can bring about that the left side of the brain, (which as morphologists have known for a long time, is responsible for the visual working of the right half of the field of vision due to the special, i.e., crossed wiring of the nervous connections) has to assign the right hand, which is largely seen to the right of the axis of symmetry, the first symbolic sounds and finally words. This admittedly rather unimportant fact could have furthermore caused a slight evolutionary tendency for the left half to bring into the role of the later speech hemisphere. In this connection, there is the highly interesting observation that only in right-handers (to 99%) do the speech centers lie in the left hemisphere while in at least a third of all left-handers, these same centers can be found laterally inverted in the right hemisphere. Further to this, a tertiary speech region, discovered relatively lately by Penfield and his coworker, lies in direct vicinity of the primary motor centers for the right arm and hand (Penfield and Roberts 1959). An additional support for the thesis comes from the investigations into human origins. The first enlarged frontal lobe was discovered in a 2 million year old skull of *Homo habilis* (nowadays *rudolfensis*) (KNM-ER 1479; Falk 1983), and it was on the left-hand side similar to Broca's area. From exactly the same epoch come the first archeological indications of the beginnings of handedness in the *Homo* line (Toth 1985). Given the enormous significance of object manipulation for the cognitive development of humans, this is a strong indication of a close evolutionary relationship between handedness of pointing and speech lateralization

(Corballis 1997) whereby a faulty development of this asymmetry has been associated with the origin of schizophrenia (cf. Maddox 1997). As a further proof for the close evolutionary relationship between handedness and language there is the fact that both linguistic abilities and skilled manual behavior patterns are specialized to the left side of the brain (Deacon 1992). And, finally, we should not forget that each of us was not able to start learn reading without using the right or left forefinger for fixing all those numerous senseless black letters on a white sheet of paper (for a social history of pointing and its pedagogic significance, see Treml 1996). Not to mention that we still use our forefinger, though a virtual one called a "cursor" to "communicate" with our by far most cherished new toy, i.e., the computer.

All this corresponds excellently with the fact that our nearest living relatives, the unusually intelligent bonobos, nevertheless symbolically reach for things and in specific directions with the whole hand and therefore at least show the first signs of being capable of real pointing. Although, for some evolutionary reason or other, it has never occurred to them to utter a sensible noise let alone a word, with this ability they are standing at the threshold to purposeful communication as shown by the next citation. For this purpose, one should hold a picture of a small, inquisitive pygmy chimpanzee called Kanzi before one's eyes who with his right hand is pointing in a certain direction and with his gaze is turned to the person looking after him, the mouth almost imperceptibly open as if he wanted finally to say something important. The legend for this nice picture goes such as follows:

Kanzi shows me where he wishes to go by gesturing. He began to gesture at about twelve months of age and continued to use gestures intermingled with lexigrams after he began mastering the formal systems (Sue Savage-Rumbaugh and R. Lewin 1994, legend to a figure on p. 142/143).

At the end of our excursion into the world of the child and his well-loved pointing game for controlling adults, I would like to mention a finding that even from the purely linguistic perspective presents us with a serious piece of evidence. It is known that there are only a few sounds that are actually found in all the world's languages. The most interesting of all these potentially ancient noises of our hairy ancestors is called "tik" which in most languages in some way has something to do with the meaning of "finger" („digit", "doigt", "dito", "dedo") or "thing" („thing", "ding") or—pointing out—"this" („dies", "tit [for tat]", "iets", "bit") (*Atlas of the World's Languages*. C. Moseley and R. E. Asher (eds) 1994). In this connection, the noticeable etymological association is significant. What does it say in the Bible? In the beginning, was the word. In Goethe's "Faust"? In the beginning was the deed. Both together could be correct, since at the beginning were fingers and things and soon after the first possibly awkward, perhaps wrong-handed pointing to the word (for the further development of the interaction between object perception and object naming in growing infants as compared with adults, see Leslie et al. 1998; Landau, Smith and Jones 1998).

Independent of the present approach, Ullin T. Place formulated a very similar thesis in "The Role of the Hand in Evolution" (2000) and developed it further to an impressive evolutionary scenario of the origin of speech through mutation and natural selection. This goes a great deal further than Chomsky and Pinker's substantially static speech nativism, according to which our "speech organ" must have originated within a relatively short period of time (e.g., in evolutionary psychology "certainly" during the Pleistocene) or possibly even through a single supermutation (Chomsky 1965). If language now has to lose its special status as a knowledge-transmitting ability of the species *Homo sapiens* within a consistent evolutionary approach, and instead is even revealed as going hand in hand with our genetic evolution—Foley (1991) has proved that there is a strong relationship between human genetic variation and linguistic groups, and, in addition, Cavalli-Sforza, Menozzi and Piazza (1994) found "important correlations between the genetic tree and what is understood of the linguistic evolutionary tree" (p. 380)—one may ask what then can be the actual function of verbal behavior. An answer to this is relatively easy to give since it has been available for quite a long time already by a series of new research disciplines, such as evolutionary game theory, sociobiology, and behavioral ecology (Trivers 1985; Hamilton 1964; Axelrod 1984; Maynard Smith 1982; Krebs and Davies 1991; Wilson 1975). From the perspective of these disciplines, what counts is alone the result of a certain behavior evoked during the interaction with the partner (or opponent). The aim of any behavior is to achieve a relative maximization of the biological total fitness of the user by any sort of conscious or unconscious cost-profit comparison (= *inclusive fitness*: direct fitness [offspring] + indirect [offspring of close relatives] fitness; Hamilton 1964;). Usually, this so-called Hamilton's rule is applied to single genes, that is alleles [cf. Grafen 1991], but at least theoretically it should also be possible to develop a kind of summation function for all the genes of a given genome. In this regard, Rushton's "genetic similarity theory" (GST) (Rushton 1984; Rushton, Russell and Wells 1989) represents an interesting theoretical alternative to the explicitly single gene-centered Hamiltonian approach, even though the latter is still dominating the scene by Dawkins' indefatigable drumfire for the "gene's eye view."

The qualifying adjective "relative" when speaking of maximation is used intentionally here since empirical studies of the problems of the optimization of biological characters have shown that the theoretical optimum is not achieved in every case as has been postulated because of some all too idealistic model assumptions. Intraspecific communication, i.e., communication between living organisms belonging to the same species, is only a special case among many other regular interactions between the organism and the environment which consists of living as well as nonliving components. The essential functions which take effect during intraspecific dealings are aspects of social behavior which can be described with expressions such as altruism, cooperation, synchronization, if kin is concerned versus reciprocity, competition, and finally pure attempts of manipulation, if nonkin is concerned (for a comprehensive account of the potentially adaptive functions of human language, see Fitch [in press]). Human speech can naturally be applied in various ways within the framework of these various functional areas and to discuss all these aspects in great detail lies outside the scope of this work. A good overview that deals with the various forms, disorders, and paradoxes of mainly verbal interaction is available in "Menschliche Kommunikation" (Human Communication) by Watzlawick, Beavin and Jackson, but also in Berger and Luckmann's standard work on "The social construction of reality" and this although both concepts are already completely erroneous from their point of departure. A typical example can be found in Watzlawick, Beavin and Jackson (retranslated from 1982, p. 30): "... it is exactly in this area that it is imperative to take the term *exchange of information*, i.e., communication into consideration". And in Berger and Luckmann, one can hear what in principle represents the same opinion: "The thought process which we have developed is a general systematic attempt to put the role played by (note: transferable) knowledge for society in its true light" (retranslated from 1982, p. 197).

However, it is not the aim of this endeavor to give an overview of all the existing variants of linguistic interaction but rather what we wanted to clarify was simply the question that is still controversial as to whether speech, and with it, all of human culture really represents what almost all theorists proclaim in unison, namely the great Lamarckian exception to the purely Darwinian workaday routine of evolution. An answer to this based on evolutionary theory must be a definitive no and with that this topic is closed for the time being. It is interesting, however, to see that modern behavioral biologists have not concerned themselves explicitly with this topic but in some areas nevertheless have come very close to the correct view of things. This is not so surprising since it is they who have to translate the purely theoretical concept of evolution everyday into the empiricism of their research projects. So, for example, in an influential textbook of modern behavioral research, Krebs and Davies' *Behavioural Ecology*, we find the following statement under the paradoxical heading "Information about nothing":

Finally, communication may have nothing to do with semantic information, but simply be a method by which signallers manipulate receivers (Dawkins and Krebs 1978; Krebs and Dawkins 1984) (David Harper in Krebs and Davies 1991, p. 384).

Then, a little bit later David Harper, the author of the whole chapter on communication in animals comes to the conclusion at the end of his discourse that with the conventional application of information theory to animal behavior some things cannot be absolutely correct:

Many studies of communication tacitly assume that animals classify signals in the same way that we do. The weakness of this assumption undermines the use of Shannon information theory... (David Harper in Krebs and Davies 1991, p. 397).

This citation now confirms nicely what we have deduced here from purely evolution theoretical considerations and it does so with regard to two different aspects. First Dawkins and Krebs, whose publications are quoted by Harper, speculated much earlier about an informationless-, i.e., empty-of-meaning, nature of communication although both of them always hesitated to apply this concept generally. Dawkins, for example, still continues to adhere to his idealistic concept of a so-called memesis, an evolution of ethereal memes floating around in space (Goodenough and Dawkins 1994) and going along with the alleged Lamarckian nature of cultural tradition instead of disappearing into the sinkhole of original but unfortunately misconceived hypotheses. Second, Harper's rather incidental and primarily methodologically oriented criticism of Shannon's information concept demonstrates the point at which misunderstandings usually begin. That error which makes one believe that it is possible to transfer real cognitive meanings typically originates in those trivial cases where both (or more) of the partners in the communication *already* possess the semantic information which is apparently—to a naive observer—being exchanged. In other words, what must be viewed as preconditioned is erroneously interpreted as the result of communication. Perhaps it is this reason that many behavioral biologists, including Krebs and Dawkins, still hesitate to accept the universal validity of the nontransferability of information, although this does not have to be valid only for the specific case of purposeful "manipulation" of a receiver (whose success is always dependent on the constitutional manipulability of this receiver and therefore cannot be real, i.e., unconditioned cognitive manipulation). Rather it is also valid in all those apparently diametrically opposed cases where—let us take an extreme example of almost perfect cooperation of genetically identical partners—very closely related individuals obviously want to communicate something important with each other without any distortion of the information with ill intent. Hence, it is not surprising that we can find in the first edition of *Behavioural Ecology* (Krebs and Davies 1981) in the chapter "Signals of animals: Information or manipulation?" in conclusion the sentence:

If information is exchanged at all, it is probably incorrect information, but it is possibly better to relinquish the concept of information completely (retranslated from Richard Dawkins and John Krebs in Krebs and Davies 1981, p. 242).

This is a recommendation with which we can fully agree. Unfortunately, it is just this very promising concept that is only mentioned marginally in the more recent edition of 1991 (see the previous citation from Harper). On the other hand, some modern neuroscientists are approaching the correct way of looking at things from a completely different starting point, namely from the evolutionary view, namely from a system-internal view. Walter Freeman, for example, belongs to this group and in the conclusion to his article on "Nonlinear neurodynamics of intentionality" (1997) he accurately states:

Representations (in the present sense: "informations") are material objects such as gestures, sound, odorants, and light waves that are shaped by motor actions for the purpose of communication between brains. There are no representations in brains, only meanings. Conversely, representations have no meaning, though they are shaped in accordance with meaning in transmitting brains and can instigate the construction of meaning in receiving brains. That dictum applies to the words on this page (Walter Freeman 1997, p. 308).

Even when two genetically absolutely identical human individuals, i.e., monozygotic twins, who, in addition, have grown up in an absolutely identical environment—assuming such a thing exists—communicate with each other, the almost perfect cooperation between the two does not allude at all to an allegedly perfect exchange of semantic information but is rather solely the identity of their genetic makeup that brings about the fact that they, in contrast to the normal eternally argumentative siblings, can often cooperate with each other harmoniously throughout their lives. Or, as we heard from Chomsky earlier in this chapter, "This relates to the similarity of their inherent makeup" (for the degree of "heritability" of this cognitive agreement, see McClearn 1997, Gottesman 1997). A citation from Nancy Segal on "Cooperation, Competition, and Altruism in Human Twinships" (1988) demonstrates the close emotional bonding that can exist between identical, i.e. monozygotic twins: "To see my brother die was like seeing myself die". In spite of this, in such cases it appears to the innocent observer obvious and convincing that real meanings are transferred—one hears constantly "the two understand each other so well"—so that it is anything but easy to persuade people that just the opposite is true. However, it is nothing other than a beautiful, but nevertheless an unfortunately completely incorrect assumption which we, when we consider its recent origin, can describe as the great Lamarckian illusion, even though an illusion that certainly existed already long before Lamarck. To finally clarify this common illusion and thereby to make the logical structure of evolutionary theory more coherent and above all, universally valid is the declared intention of the present enterprise.

In the case of human culture, things begin to get more interesting, where a defender of a certain idea through his social cooperation under some circumstances can achieve considerable selective advantages over other ideological groupings. At exactly this point, in every culture, the fairy tale of the free communication society ceases insofar as competitive social interests meet head on and they in the final instance are associated with biological interests. In this way, the idea of Karl Marx of the power of the class struggle as the motor of social development is not so far away from the truth as many people would like to think. The only difference is that the working of natural selection must be viewed more discriminatingly than ever was done with Marxist ideology. Of course, it sometimes occurs that a social class eliminates another by pure physical force, but such outbreaks of raw violence, often defined with the inflammatory slogan of the (Russian, French) Revolution, that inevitably leads to a genetic change in the population are the exception rather than the rule. However, what is more difficult to observe but which in the course of time and through the generations must lead to subtle changes in the genetic landscape is the universal and permanent contest which is going on not between ethereal ideas themselves but the biological carriers of these ideas. Popper is also mistaken here when he maintains that what is special about human cognition and above all the highly successful method of modern science consists of letting erroneous hypotheses die instead of the person himself as it was supposedly so earlier (Popper 1972). Here we meet the popular cliché of the Darwinian-brutal stone age being later on replaced by the educated-peaceful new age. However, there is also a contrast between the story of the comfortable, predictable Pleistocene horde and the universal highly technical wholesale murder which is no less a misleading cliché. By using an inversion of the statement of the first modern military strategist after Niccolo Machiavelli (1469—1527), Carl von Clausewitz (1780—1831): "War is only an extension of politics by other means" (1981, p. 39) we can now establish the exact opposite of Popper's idealistic view of things: politics and, associated with it, human culture is the extension of war by other means. More about the warlike nature of human culture will come later.

13 Intelligent Sex: A Cognitivist View of Genetic Exchange Processes

Perhaps the deepest mystery in all of biology concerns the meaning of sexual reproduction.

Robert Trivers

The likely explanation is that sex generates a greater number of different genotypes, adapted to a wider range of ecological conditions.

John Maynard Smith

If the wonder of language cannot be a real wonder, perhaps sexual reproduction, as Trivers points out so poignantly for a scientist, is such a great puzzle of nature that it could be an adequate substitute for the loss of the cultural wonder of language. Trivers is correct insofar as the mechanisms and the associated functions of sexual reproduction are concerned since, until recently, they were considered to be unproblematic and requiring no special explanation. For a long time, it was thought that sex simply increases the average rate of evolution for a species and for this reason alone, asexually reproducing species must automatically be supplanted by sexually reproducing ones in the long run. In the meantime the situation has changed somewhat and for about twenty years there has been an animated debate, which is still in full flow over the real adaptive function of sexual reproduction. If we are thirsting after a real wonder of nature then we would have found, at least with the ifs and buts of sex, a surely attractive and simultaneously, as I will attempt to demonstrate, a highly qualified candidate.

It is already strange enough that we are used to seeing wonders only, when we look more closely, where there is actually nothing extraordinary to be observed. Exactly there, where every sensible person believes he knows exactly what distinguishes us humans from all other living organisms in the world, in those great achievements with something to do with human culture such as thinking, consciousness, language, tradition, exactly there, where the wonders of human existence evaporate into thin air as soon as we introduce a systematic evolutionist way of looking at things. If we now, to our great disappointment, have to admit that human speech can have nothing to do with the transmission of knowledge but rather represents just one among the many other possible forms of behavioral interaction between individuals, so perhaps it is sexual reproduction that can release us from this dilemma of cognitive isolation. Surprisingly, up till now, evolutionary epistemologists have not concerned themselves with this possibility although superficially a comparison offers itself. The reason could be that the topic appears too common if not too vulgar for a serious philosophical debate. The thesis we wish to plead for goes a good deal further in that it claims that the presentation of sexual reproduction as an exchange of information is not just a nice metaphor but that successful sex—and not just technically good, as Woody Allen is wont to say ("Sex without love is an empty experience, but as far as empty experiences go, it is one of the best")—represents, even for us intellectually endowed humans, the only possibility of communicating with another member of our species in a really cognitive manner.

Then what actually happens during what we assume is really successful sex? Looked at superficially, not a great deal, although the not so serious press occupies itself with extraordinary regularity with new facets of this ontic topic, except that two bodies (in most cases) cuddle up close to each other and with all the tricks possible try to be as close as possible together for as long as possible. Now, there may be a certain skill hidden here which not every inexperienced person can master immediately and about which there are discussions and debates and even illustrations but does any deeper significance lie behind it? "The meaning of love is unfathomable", said Friedrich von Schiller (1759—1805), and indeed, what in reality takes place during "love" borders on a miracle. If the fusing together of the two bodies of these intimate lovers was, in reality, a committed attempt at reaching a "meta"-physical goal, the case of what occurs during a successful fertilization of a female egg cell by a male sperm puts everything that went before in the shade.

Without wanting to go into great detail, just from the problems involved, it is in some way a miracle that such a thing can function at all. This makes it understandable that even the Vatican occupies itself with all the essential details of the biological origin of a new human even for the impracticable purpose of wanting to propose a definitive starting point for (new) life (Godfrey 1995). It happens unfortunately all too often for *Homo sapiens* at present and for this reason does not appear to be anything extraordinary to our everyday perception even though with the world's population growing as it is perhaps it would be better to speak of *Homo exponentialis*. In fact people think just the opposite, it is even said that anyone, meaning even the most "stupid" is capable of achieving this form of successful sex so there cannot be much to it. This everyday topic unexpectedly conforms to the reproductive abstinence of certain spiritual elites, who by their know-all teaching and selective evaluation of others (e.g., in schools) have prescribed themselves for the multiple propagation of

small portions of their *indirect fitness*. Jokes aside, the complexity of the processes in sexual reproduction are anything but trivial and for some time a
whole band of highly specialized molecular biologists and geneticists have
been occupied with this topic. What is of interest to our theme in this course
of events is that sexual reproduction cannot be considered as just a simple
transfer of static genetic information. What happens is in no way a simple physical exchange of isolated particles of information, i.e., genes, but much more
a sequence of complex internal restructuring during the ontogenesis of metazoans which can only take place after the no-less complex merging together of
two previously separately existing living systems, i.e., egg cell + sperm, to give
the transitional end result of a fertilized egg cell:

In a broad sense, all recombination events represent different types of topological reconstruction (Benjamin Lewin 1985, p. 574).

The most important processes already take place during the maturation of the
gametes where, through genetic recombination processes, a multiplicity of
new variants of the genome are produced. Essentially, two processes are important: the new combination of previously separate chromosomes from the
grandparent generation as well as the new combination of previously separate
sections of the same chromosome. With this alone, there can be a practically
infinite number of possible combinations which leads to a constant mixing of
the genetic structure from generation to generation. The brief presentation of
all these processes as a transmission of genetic information from one system
to another is therefore confusing since a genetic particle, be it a gene section,
a complete gene, or a whole chromosome, is never purely physically exchanged in the way that one of the two communicating systems simply sends it on
its way and that the other system somehow receives it. Not even in the infection of living cells by the almost non-living viruses are naked DNA particles
whizzing through the landscape, but rather the strongly reduced minimal physiology of the virus must still be involved if it applies to finding a host organism capable of being infected which mostly to its regrettable disadvantage takes over the biggest part of the physiological housework.

True cognitive and at the same time biological communication can therefore only happen by the roundabout route of a real fusion of two material systems and cannot function how it is commonly imagined, namely as a physical transmission of materially isolated particles of information. Viewed
objectively, there is no semantic information at all in an isolated hereditary
molecule since it is only through its integration into a suitable living system
that this information comes into existence, as stated at the beginning of our
considerations. Living processes can thus be rightly considered as *the* only elementary cognitive processes on our planet. However, why do such highly or-

dered parts of a living system, as DNA molecules are, lose all their cognitive characteristics as soon as they stop being associated with the system? The reason for this is again to be found in the fact that, for the part concerned, the coherence of the life maintaining processes has been interrupted. Such a DNA has been degraded de facto to a piece of lifeless material which at best could be taken up by other still living systems like any other energetically valuable substance such as food and integrated into the new system's physiology without being able to so to speak reawaken itself to life. In this connection, it is more than just a little interesting to find out that those very enzymes, mainly various kinds of kinases, which with great probability in our prokaryotic ancestors who did not even possess a nucleus, were responsible for the repair of damaged DNA, are today responsible for exactly those recombination processes which form the basis for genetic exchange processes (Keith and Schreiber 1995). This shows very clearly that what actually happens in sexual reproduction can only be very imprecisely described as physical transmission of already existing information.

After all these critical comments about the possibility and impossibility of a separate transfer of semantic information, it is nevertheless not a serious mistake when during dealing with sexual reproduction, we talk, in contrast to language, about a transmission of genetic information. This is so because in every single case, what has to be an unconditional prerequisite, i.e., the fusion of two separate biological systems, does indeed take place. In this sense, sexual reproduction does achieve what has up to now been erroneously attributed to human language, namely real communication between separately living cognitive systems, be they humans or microbes.

Upgrading sex to a first class epistemological medium will offend not a few of the many epistemologists schooled in information theory, but this does not explain to us how such a phenomenon exists at all or, more precisely, what actually could be determined by the adaptive function of this single real language of living things which the evolutionary researcher Graham Bell even spoke of a "masterpiece of nature" (Bell 1982). For that purpose, let us return to Triver's exciting reference to the biological puzzle of sexual reproduction. The first thing discovered concerns the rate of evolution mentioned at the beginning and the sweeping assumption that it must be automatically higher for sexually reproducing species than those reproducing exclusively by simple cloning or parthenogenesis, i.e., the mitotic division of germ cells (e.g., unfertilized eggs) into genetically identical daughter cells (= clones, e.g., parthenogenetic mass propagation of homozygotic females in water flea, interrupted by occurrence of pygmy males; change of generations or heterogony, Fam. *Cladocera;* small crustaceans). It is actually so that, in sexual species, in every generation, by a multitude of recombination processes, a great variety of different genotypes are produced, which, in the next generation, lead to further new types and so on. This means that as far as the change in the genetic material is concerned, sexual reproduction with its combinatorial diversity far exceeds the passing on of identical hereditary molecules when the only changes are brought about by the comparatively rare mutations (on average 1 exchange of a nucleotide in 10^9—10^{10} identical replications in both bacteria and eucaryotes; Knippers 1997, pp. 230—234). Although ultimately all genetic innovations must stem from discrete mutations, sexual reproduction is responsible for the fact that the possible number of combinations in one and the same organism can be incomparably higher than in any individual of purely asexually evolving lineages.

It was during the 1970s that it was noted that sex could also be interpreted as a very inefficient method of reproduction since the number of offspring compared to the much simpler asexual variants was rather less, to be exact exactly half (1 clone produces 1 new clone whereas 2 gametes are necessary for 1 new individual, i.e., 1 gamete shares only half or 50% of the new organism). Sexually propagating species should therefore at least compensate for this substantial quantitative disadvantage in order to remain evolutionary competitive. Failing this, under otherwise identical conditions, sexual species would never have been created or never been able to supplant already existing asexual forms. Hence the factual advantage of sexual exchange processes must be considerable in order to rule the evolutionary roost to the extent, which we can observe. Almost all the species in existence today reproduce exclusively sexually and only a very few still resort to falling back on an asexual method and if they do so then only within the framework of a cyclic succession with the obviously more diverse sexual variant. Purely vegetatively reproducing organisms have in the meantime disappeared from our planet to a great extent. There are only relatively few completely sexually abstinent species, e.g., certain plant species in extreme habitats with no competitors, most viruses, and a few other bizarre creatures on the borderline of evolution.

According to the latest investigations, what appears to be the special advantage bestowed by sexual reproduction corresponds exactly to our question about the possibility of real cognitive communication. During sexual reproduction, semantic information is indeed exchanged between to separately existing systems in the form of a fusion and it is exactly this exchangeable information, which produces variability that under certain selective conditions can be of advantage. The keyword in this connection is "adverse", or "hostile", related to the specific conditions of selection with which a living system has to be confronted to induce the changeover from asexual to sexual propagation (Trivers 1985). The first factor one thinks of here is the hostile nonliving environment, which is sometimes understandably described, as merciless. This includes the various physical influences such as temperature, radiation, pressure, material composition, etc. which exert a constant selective pressure on every living organism. The impression of mercilessness is not an exaggeration when one considers for instance what would happen if, within a few years, the mean temperature of the earth's surface increased by an average of 100°C because of some cosmic occurrence (e.g., sudden increase in the strength of sun eruptions). Such a scenario would mean an unavoidable fate for the whole biosphere since within a short period all living things would be literally cooked regardless of all evolutionary short-term attempts to escape the threat of extinction (in fact, life has already severe problems to cope with today's "normal" global warming; cf. Etterson and Shaw 2001). A more realistic and predictable end of all life on earth is linked to the expected explosion of the sun in about 5 billion years, which would very quickly be too much for the fragile aqueous chemistry of all living things. This nicely shows how the potential immortality of life is always tied up with the historical fleetingness of this phenomenon inherent in the system. Stephen Jay Gould has drawn up a whole list of such, as he says "historical contingencies". Proceeding from seven different scenarios related in each single case to an evolutionarily significant stage (emergence of eukaryotic cells, first metazoans, the Cambrian explosion, modern fauna, land vertebrates, mammals, *Homo sapiens*), he demonstrates very impressively how incredibly "unintentional" or, better: coincidental, the origin of man must have been (Gould 1991, p. 348—363).

When they are correlated with each other, as a rule, the physical conditions only change relatively slowly over long periods of time (e.g., climatic fluctuations), so that, in principle, living organisms would have enough time available to adapt continuously to the changing conditions. In such a situation, according to model calculations, sexually reproducing species would not necessarily have an advantage over parthenogenetic ones since variability in this case is only achieved at the cost of a lower reproduction rate. Furthermore, in long-term adaptive processes, sexual propagation has a further disadvantage, namely that recombination processes can also upset the already existing "good", i.e., selection tested gene combinations.

Adverse also means that the environmental conditions more or less change *permanently* to the disadvantage of the organisms concerned in the sense that, to whatever a species has adapted itself successfully, in the next moment or at least the next generation, it may have become outdated. Hence it seems as if only under apparently extreme conditions can sexual propagation bestow a clear advantage over its chaste competitor. If the physical environment is too sluggish for producing such challenges, as a rule, the animate environment of an organism, i.e., all the existing other organisms in the same biotope, manages this with ease. This is not as merciless in the sense of how the incontrovertible physical parameters which, if it had to be, would let us all burn up without pity, but all the more purposely malicious than all the brute forces of inanimate nature taken together.

The normal competition between two species for a natural resource be it room, food, hiding places, or any other important factor still sounds rather harmless in comparison although, from the theory alone, it can happen that, in extreme cases of a 100% dependence on a particular factor, one of the species is condemned to extinction. Admittedly, it does not happen often that a species is dependent on one single resource—one famous example is the giant panda with its incorrigible preference for bamboo leaves and associated and sometimes dangerous fluctuations of the population (for a new approach to the problem of extinction, see *Nature* 394, 409). Rather, the rule is that animals know how to exploit a whole spectrum of various resources so that in the case of competition with another species with similar habits there are always a certain number of ecological alternatives. If such an alternative expresses itself finally as a change in the morphological structure of the species concerned then we talk of character displacement. The most famous example of this are the 14 varieties Darwin's finches of the sub-family *Geospiza*), living on the Galapagos islands, which with great probability are descended from one single species and after a fanning out caused by competition the resulting species are very similar to each other but having important differentiating characters. By specialization of the shape of their beaks and corresponding changes in behavior, they were able to take over very varied ecological niches occupied on the nearest mainland (Central South America) by other species.

If the universal competition between two somewhat similar species for their vital requirements and interests is still relatively harmless, then other kinds of biotic relationships between living organisms show the malicious nature spoken of earlier perfectly. Predator/prey and parasite/host relationships, from the view of the first in each case, an exploitation of the second are usually characterized by the fact that any adaptation of either one must be to the detriment of the other. This means that there is inevitably a permanent kind of arms race in which each species that was previously running on the spot so to say evolutionarily speaking runs the danger of being thrown out of the race. Naturally in this race, there are mutually positive dependencies in the way that a too great disadvantage for one side (prey/host) can also become a disadvantage for the other (predator/parasite) and vice versa, especially when the relationship between the two is isolated from all the others. However, this seldom occurs because most species interact with many other very varied species in a complex way.

That a predator is ultimately endangered by its prey being overexploited and being reduced in numbers is understandable but that the prey experiences a disadvantage from the disappearance of its main enemy less so. So can and could some potential victims be able to prosper without their predators, exactly like most host could lead a much fitter life, ontogenetically and phylogenetically, without their pests—whether they are relatively harmless intestinal parasites (tapeworms, nematodes, etc.) or dangerous viruses. A human life without viruses appears well worth living and it would hardly impair our continued evolutionary existence in any way. As an example of the absence of predators, one should think of the animals living on isolated islands such as the Galapagos who, for this reason, stupidly show no fear of the new human invaders. However, many of our hunted animals would, if we ever wanted to ask their opinion, not ask for a substitute for their former predators in the form of a human hunter as the protector of an idealized "ecological balance". Finally, we humans likewise do not really miss our old enemies or should we introduce lions and tigers into our gardens and parks so that "ecological balance" for our species could be reestablished? The social legitimization of the technical highly armed hunt of modern times using the camouflage jacket of ecology dispenses with any scientific basis. An honest philosophy of hunting, if any such thing can be appropriate in our time, can only fall back on the sheer desire, using highly technical modern weapons, to shoot defenseless and anything but really wild animals giving them next to no chance of escape. The transition from here to normal mass rearing of animals is a fluid one.

John Endler has produced a condensed overview of the involved evolutionary relationships between predator and prey and compared them with the much closer and more specific relationships between parasite and host and he comes to the conclusion that there are many factors that bring about that only in very rare cases, as far as predator and prey are concerned does it come to a real coevolution (Endler 1991). Predators usually prey on many different victims and it makes no sense for them to concentrate too much on one species. They normally exploit the most common of their prey which of course has the effect of reducing the numbers of these animals and this, in turn, forces the predator to look for another species of animal to feed on. This again makes specialization difficult. Interestingly, a special defense strategy of most victims is usually shown only towards the increasingly dangerous end of a sequence of being pursued so that the predator has to be more adapted to a general hunting strategy than to an overspecialized kind of behavior he can only rarely use. The life/dinner principle (Dawkins and Krebs 1979) goes even further and says that an unsuccessful attack by a predator means the survival of the prey but from the other side is simply a ruined mealtime for the hunter. The selection pressure on the prey is therefore much stronger than that from obtaining a single meal by the use of a highly specialized hunting strategy. Eventually, the very different population densities of predator and prey as well as the varying lengths of time between generations gives rise to very varying evolutionary scenarios. Whereas the first by relative rarity and long life has to struggle with lowered genetic variation, as it is for example extreme in some species of cats (Burke et al. 1992, O'Brien 1994; the special case of the cheetahs is described in May 1995; for the rather limited morphological variability of the domestic cat, see Stephens and Yamazaki 2000), and, consequently, a slower speed of evolution than the second (e.g., rodents) in the normal meaning of the word as well as evolutionarily speaking, is always running a little in front. This explains much better why, in the case of parasites where a mostly minute villain with a breathtaking number of generations —think of a bout of influenza which develops from a few tiny cells in a matter of hours into a potentially life threatening population of millions—makes the life of a specific host very difficult, it must be just the opposite, namely a serious problem for the much bigger organism.

Now, in the case of the parasite/host as opposed to the predator/prey relationship, we can rightly talk about a common evolutionary fate, as if the two species were tied together with an invisible string called coevolution. In the public discussion, people somewhat hastily over generalized and derived from it the fiction of the ecological balance (see above), which has to be maintained if the threatening collapse of the earth's ecological system is to be halted. If there ever were anything like ecological balance as a natural biological law, then it could not be endangered by any species including our own and would not need to be artificially maintained. It is not a seldom occurrence for a species to go into extinction but without two others or even more having to automatically disappear from the face of the earth. Our own species provides the best and most up-to-date example. We have been going from exploiting one animal to the next for a very long time now and, in the meantime, have destroyed many species of large animals. In the last 50,000 years, for example, 73% of the large mammals of North America no longer exist, 80% in South America, and not less than 86% of those who lived in Australia (Diamond 1992). We would like to see any other species do as well.

From a purely evolutionary theoretical point of view, in real coevolution such as in the case of parasite and host we are dealing with the result of a permanent destabilizing selection, which under some circumstances can cause both species to be trapped in a principally open and unlimited evolutionary race. These simple cases, however, as already said, are not the rule since we mostly find a network of involved relationships among a great number of species so that the selective demands on each individual species can be many times higher.

Hence to change permanently at any price is the motto enforced by selection under difficult conditions of survival and under this motto 99% of all the existing species have come into being after they have devoted themselves completely to sexual reproduction. The few exceptions that still exist today prove the rule: only under conditions of a strongly reduced biotic pressure, i.e., a reduced pressure to compete with other individuals of the own or other species, do such species change over to asexual propagation but if the situation changes immediately return to sexual reproduction again. The perfect example are the many species of water flea (*Cladocera*; see above) that in Spring in a time of unhindered expansion in temperate latitudes, when the water begins to warm up again, go over to the asexual method of reproduction. However, when things start getting crowded and the environment is saturated with individuals, they finally return to indulge in the more convivial sex (overview in Bell 1982). Likewise many plant species which reproduce purely vegetatively for a time in order to colonize as quickly as possible bare areas and other still unoccupied habitats (e.g., mountains, deserts) where there are few other species are also a confirmation for this relationship.

A general competition hypothesis is therefore the currently most discussed and doubtless the most plausible explanation of the wide distribution of sexual reproduction. Van Valen (1973) provided this, in his opinion, "new evolutionary law" with the somewhat mysterious designation "red queen hypothesis" with reference to the Red Queen in Lewis Carroll's famous *Alice Through the Looking Glass*, who at a certain point speaks to Alice with almost prophetic intent the words: "in this place it takes all the running you can do, to keep in the same place". Sexual exchange processes have not so much the simple function of promoting better adaptation to certain ecological demands caused by directed selection but primarily are there to just keep competitive species in constant evolutionary motion since each standstill, even momentary, could bring with it an inestimable risk. That the ubiquitous parasites must play a special role in this scenario is brought about alone from the fact that pests, as a rule, have a much shorter life span than most of their hosts. That is even valid for viruses like bacteriophages infecting different kinds of bacteria. So, right from the beginning, they have a head start on their infested hosts who can only compensate for this with a variety of tricks. One of them is undoubtedly the permanent phylogenetic zigzag course of the host species through the obligatory practicing of sexual reproduction. Therefore it is not astonishing that no animal with a somewhat longer individual life span, hence virtually all vertebrates and most of the higher invertebrates, never reverts back to the much more primitive and, at first sight, less expensive variant, i.e., asexual reproduction. Only species with a relatively modest life cycle, as far as time is concerned, can afford, as an interim measure for the purpose of a rapid increase in numbers to change over to reproducing by a parthenogenetic cloning of their gene equipment.

Parasites are simultaneously the curse and blessing of evolution because it is they alone, partly independent of physical changes, which obviously make those adaptation processes on the part of the host organisms necessary—the processes that are kept permanently in motion by the genetic roulette of sexual recombination. Although this permanent, thorough mixing of the genes does not always bring anything really new in fact, without doubt, it can and does occasionally come to a radical change with unforeseeable consequences. In simple gene-for-gene models, the partners, parasite and host alike, go in cycles in the form of their alleles (e.g., H1, H2 stand for the resistance types of the host H), P1, P2 for the virulence types of the parasite P: H1 fi P1 fi H2 fi P2 fi H1 fi …). From this, they can only escape by a correspondingly novel mutation (e.g., H3) (see Clarke 1976). This picture of a multitude of populations interacting with each other and for this reason changing permanently but, however, just now and then undergoing a fundamental change fits in well with the persistent discussion of the relative importance of a so-called punctuated equilibrium in evolution (Gould and Eldredge 1977, 1993; for a more recent empirical verification, see Elena, Cooper and Lenski 1996). Detailed comparisons between larger taxonomic units have shown more and more often that a continuous progressive evolution of characters is rather the exception than the rule, which means that phases of relative quietness alternate with phases of more or less stormy evolutionary changes. Sexual reproduction alone is certainly not an adequate explanation of this, that may be true (cf. Stenseth and Maynard Smith 1984), but perhaps it provides the necessary *élan vital* or "vital impetus" which, still for the French natural philosopher Henri Bergson (1859—1941), represented something inconceivably mysterious. It does indeed seem true that species with sexual reproduction have become specialized at waiting for the right moment in two ways, first, as an individual in order to secure the fittest reproductive partner and second, as a species and population in order to grasp, now and then, the rare offer of chances to develop.

The parasitism hypothesis, as the competitive factor is also called because of the special significance of parasites, goes back to an early work by Haldane (1949), who was the first to point out the evolutionary significance of a continuous adaptation of a host species to its coevolving antagonist. This hypothesis is in fact a very plausible one but, as recently demonstrated, alone it is insufficient to explain the extraordinary stable occurrence of sexual reproduction. To maintain this in the face of its cheaper competitor, asexual reproduction, there must be a complementary effect known as *Muller's ratchet* or the mutation accumulation model (Muller 1932). The no longer completely new idea behind it says approximately that, in a population without sexual recombination an unavoidable tendency towards an accumulation of deleterious mutations should automatically arise (Haigh 1978). Like a pawl for a cog, one deleterious mutation after the other is retained in the genome without any possibility of getting rid of it. A thinkable way out of this trap which, as if under the effect of a real ratchet tightens tighter and tighter, appears to be sexual reproduction which by recombination processes may be capable of directly reinstalling the original gene distribution (cf. Ridley's "hatchet"-mechanism; Ridley 2001; for a mathematical model on "mutation and sex in a competitive world", see Peck and Waxman 2000). However, only when we combine Muller's mutation accumulation hypothesis with the Haldane-Van Halen competition hypothesis do we obtain a model calculation of a clear and evolutionary stable advantage for sexually reproducing individuals over asexually reproducing clones (Howard and Lively 1994). Although the actual effect of Muller's ratchet has remained controversial (compare Chao and Tran 1997 "The advantage of sex in the RNA virus phi6" with Morell 1997 "Sex Frees Viruses from Genetic Ratchet")—on one hand, the high mutation rates are required to be applicable, and on the other hand, they are compensated for to a certain degree in the short term (Wagner and Gabriel 1990) —, the empirical data speak very strongly for the validity of at least the Parasitism hypothesis. Even humans provide indications of a strong influence of parasites on the diversity of a species, whereby, the most convincing example originates from Africa. It has been discovered recently that the genes for the so-called major histocompatibility complex, which are responsible for the coding of important transport molecules in the immune system of the human body, show a substantial degree of genetic polymorphism, which, as biological variability, is associated directly with the occurrence of the malarial pathogenic organism *Plasmodium falciparum*. Interestingly, in addition, the polymorphism pattern differs in the population of West Africa to the pattern in the East African population. This indicates that in this case, an interaction with other pathogenic organisms may play a role (from *Infection, Polymorphism and Evolution*, The Royal Society, London, 25— 26.5.1994; *Adaptations for Disease Resistance and Virulence*, Ciba Foundation, London, 27.5.1994). A critical question which naturally arises is the significance of such special polymorphism for other areas of evolution, not just that of the human species. Was polymorphism in other such areas only the chance recombinatorial consequence of the primary genetic variation caused by parasites or have variations occurred completely independently of them and been brought about selectively by other demands of the environment? In any case, the variability associated with malaria demonstrates very nicely the enormous influence exerted by pathogenic organisms on our evolution and in this way is an important indication of the parasitism hypothesis being partly correct in the matter of sexual reproduction. "Partly" because, in the meantime, some species have become known in which sex in the absence of parasites appears to be stable (Lyons 1997). In detail, things are always very awkward, although, in principle, Weismann's early idea that it is only the "creation of individual heritable characters for natural selection to work on" (1889) is still valid and that the simple Darwinian competition makes all this outlay is worthwhile if we look at the incredible diversity of nature.

In conclusion, what does our thesis of real cognitive communication, in which a real transfer of semantic information can only result from a material fusion process between two separate systems, contribute to a better understanding of evolution? It is true that this thesis does not allow us to substantiate the adaptive necessity of sexual reproduction—here we have the parasitism model against or combined with other hypotheses—but it does give us the possibility, at least, to formulate a supporting argument for the special form of sexual exchange processes. If a real information exchange were possible without the strong biological restriction of sexual reproduction, there would have been no necessity for such a kind of reproduction to develop in the first place independent of the selective advantages and disadvantages. However, since we can now be certain that semantic information can only be generated within living systems, we can with every justification set forth the claim that the recombination occurring during sexual reproduction represent the only possibility on earth, up till now, of realizing a real cognitive transfer between organisms. The specific system dependence of this evolutionary unique mechanism of cognitive exchange is demonstrated even further by other peculiarities of sexual reproduction which—if one were still to proceed from the false supposition of a particular transmission of material information, perhaps even measured in bits and bytes—are not comprehensible. One example would be the apparently trivial fact that many living organisms on earth can no longer communicate on that manner with each other. This does not even surprise the biology students who learn at the beginning of their studies that so many thousands of various "natural" species exist and that without exception these species, according to Darwin's important discovery, can all be related to each other by diverse intricate paths of common descent. Kinship, as we all know very well, has its limits. So nobody is seriously disturbed that, to name but one of the thousands of possibilities, a lady field mouse wants nothing to do with a bull elephant. Neither does anyone wonder that fleas, whether male or female, hide themselves unnoticed in our comfortable clothes.

This is not meant as a tasteless joke but there would be nothing to say against a flea having fun with people, a dog with a cat, and a whale with a real fish. Why not, what has happened here that many, many, actually all species do not really like others or not infrequently even hate each other? As is said by so many advisers on relationships: no proper love without real understanding! One does not realize that this piece of conventional wisdom is exactly what the successful communication between the species lacks, namely a little understanding or, if we are honest, a great deal of understanding. Cognitive communication between different species does not function any longer because the accumulated knowledge of flea and human, mouse and elephant, etc. has, in the course of time, become so different that intelligible communication is no longer possible. They have, so to speak, drifted apart a little over the years or, under some circumstances, quite considerably and now, when they come across each other, they are speechless and, viewed intellectually, do not want anything to do with each other. It could be that, in some areas and under some circumstances, we have important things to say if only the other could understand them in some way. It would sometimes be an advantage for a mouse in danger to be able to trumpet like a small elephant in order to scare away the nasty cat.

What, in all these disruptions of real communication, leads to insurmountable difficulties is given by the biological species as the cognitive boundary of a population of individuals formerly still capable of communicating. This natural phenomenon, free of all philosophical hair-splitting about the very "essence" of natural kinds, is a biological reality just like the complex relationships between populations in the process of separating (see Maynard Smith 1989: "Macroevolution" p. 273—302). The variable boundaries between these populations show us 1) that the evolutionary position of such a boundary is dependent on the kind of knowledge involved in each case, i.e., that species is not always species in the same sense—Pleijel and Rouse (2000) have therefore recently proposed that the classically defined term "species" be replaced by LITU (*least-inclusive taxonomic unit*), a kind of minimal systematic unit, that primarily takes into account the molecular-phylogenetic taxonomy after Queiroz and Gauthier (1992)—and 2) that the exchange of knowledge is indeed system dependent and tends to be holistic. If this were not so, we would have no reason to exclude the possibility that also very distant- i.e., not directly related relatives at least in that partial area where they still possess relatively similar genetic knowledge about the world could successfully communicate with each other.

Even though Dawkins' attractive metaphor of the selfishly isolated gene invites us to expect limitless possibilities of evolution, the iron curtain of non-communication between species speaks strictly against it. Communication can very obviously only take place between those individuals who possess a basic genetic structure as similar as possible where only the individual positions may come more under discussion but not the structure itself. As soon as the extraordinarily stable structure is touched on by mutation, free communication becomes difficult because then the overstepping of the boundary of the species comes into question. If the necessary unanimity within a population is lacking, an inevitable social splitting must occur and this heralds the (mostly) irreversible history of (at least) two new species. It is probably these times of phylogenetic uproar or, to use Gould's term, punctuation in which a phase of relative quiet is often replaced by several successive phases of radical change. Unfortunately, empirical details about the reason why such a lasting disturbance of the intraspecific balance of a species was created is extremely sparse and thus we only have very indirect indications from paleontology or gene comparisons of recent species which allow us to make more detailed hypotheses about the actual historical course of events in the development of species. In this respect, the breeding of domestic species could have been a much more promising way to bring, in vivo so to speak, the process of the true Darwinian origin of species before our very eyes. However, the results of all the efforts made up to now by breeders all over the world have been disappointing but at the same time also very interesting. Although humans have been trying intensively for several thousand years with certain races of domestic animals, we have not yet succeeded in the apparently simple task of producing a single new species. In fact, basically all of our domestic animals, from laboratory mouse up to circus elephants, in spite of all our intuitive methods of artificial selection, have remained that wild species with which we started with despite the greatest breeding goals.

Our oldest domestic animal, the domestic dog, which has lived for at least 12,000 years (Clutton-Brock 1995)—the wolf at least "accompanied" us for more than 300,000 years (Olsen 1977)—together with man at the same camping places, our true companion, is a good example. In spite of the great difference in their outer appearances, every dog is still a real dog (*Canis lupus familiaris*) and as such has still remained the big bad wolf (*Canis lupus*) since they are genetically compatible, i.e., they can interbreed. The superficial differences between a miniature poodle and a giant bullmastiff are so impressive that one could understand an inexperienced observer mistaking them for being two different species. However, it is clear that all these various races, estimated to be around 400, up to the present day, have remained, in principle, nothing other than a wolf with strongly varying characters. There are very large dogs (e.g., Irish wolfhound), tiny ones (e.g., Yorkshire terrier), unbelievable thin ones (e.g., greyhound), extremely fat ones (e.g., bulldog), short-haired (e.g., Jack Russel), shaggy dogs (Old English sheepdog), smooth skinned (dachshund), wrinkle-browed (boxer), quiet (St. Bernhard), and excitable (poodle). Nevertheless, the basic principle, which lies behind the extraordinarily variable dog model, is still nothing more than the morphological-functional blueprint of the wild population of the wolf (for pictures illustrating the considerable morphological variability in the domestic dog, see Yamazaki and Kojima 1995).

That the wolf and the dog belong to a single common species has been shown by comprehensive studies by—nomen est omen—Wolf Herre at the Institute for the Study of Domestic Animals in Kiel (northern Germany) (Herre and Röhrs 1973, Herre 1982; see also Zimen 1988). Herre crossed poodles (a very unwolflike race) with wolves and, as expected, the offspring were poodle-wolves a colorful mixture of characteristics from both races. In the course of further breeding over a relatively long period, the offspring regained the dominant characteristics of the wild form—the shy and careful wolf who is nothing like the cliche, which says, he is aggressive and wicked. It is also very interesting that bitches when they are allowed a free choice of mates or when they run away at the "right" time, i.e., when in heat, select partners that appear more like the wild form, e.g., German shepherd dogs and the like. This exactly is the reason that the results of unwanted crossing experiments that occasionally take place in our city parks are mostly a German shepherd mongrel. Obviously the female still has at her disposal the ideal picture of the original male wolf when she can freely choose her suitor.

That it is relatively easy to breed new races of dogs but that it is a very different thing to create a new dog species should give food for thought to us as biologists since it shows us that the matter of Darwin's natural selection process cannot be so simple as we professional adherents to the evolutionary theory often imagine. Anyone who has something to do with breeding animals quickly realizes that our self-invented methods of artificial selection are not exactly squeamish, in fact, just the opposite. When we examine the so-called harshness of natural selection in the wild we find it difficult to find such cases as were once usual, e.g., litters of newborn whelps were closely examined after birth and the death sentence carried out immediately if one of the young did not pass the test. In addition to the extreme harshness of the purely physical selection comes another selection pressure that probably seldom or never takes place in the wild. In modern breeding programs of any domestic species, it is quite normal that only animals selected by very strict criteria are allowed to breed so that over a number of generations, in a very purposeful way, the realization of an often very special combination of characters is worked at. Naturally, characters already present have to be utilized which in the course of time can show a new development. Examples of this are characters such as obesity (fattening pigs), long hair (angora forms), but also types of behavior change such as sluggishness (milk cattle), tameness (domestic cat), or even increased aggressiveness (fighting dogs).

The direction of the selection pressure by the domestication of the wolf is in the direction of more and more humanization of the dog even in races that are not primarily bred for a sweet nature and human expression. One often hears people saying that a dog and its owner grow to be more like each other the longer they live together. However one would like to interpret such stories, there are indications that people choose their oldest companions according to criteria that have been developed more with regard to the (subconscious) image of desirable human characters. This would be, in any case, a credible biological explanation of the similarity, if it does indeed exist, between dog and human owner rather than their physically growing to be more like each other. What is ontogenetically impossible can be realized phylogenetically very well by a corresponding selection pressure. As far as the shape of the dog's head is concerned, it does indeed look as if a part humanization has taken place since not a few races of dogs exhibit clearly pseudo-human features in their morphology. It is noticeable, for example, that in comparison to the long head of the wolf, the muzzle of the dog is much shorter and broader. In addition, the original, relatively straight profile of the wolf's head has, in many dogs, a well-defined fold exactly like that in humans between the nose and the forehead. All these anthropomorphic changes are most obvious in races like boxer and bull-dog, chow chow and King Charles spaniel. Sometimes one has the impression that perhaps we are dealing with members of our own species, especially when one sees fun postcards with a pair of dogs dressed up in costumes, maybe as clown and ballerina. The humanization of the dog is not confined to just the external appearance but has also taken control over the various forms of be-havior of *Canis lupus*. The most well known is the faithfulness of the dog which stems from the strong social tendency of its ancestors and represents a hyper-trophied result of extreme selection. It appears that the human influence on the dog has worked in a similar way as in our own species, namely in the direction of an elongation of youthful development, an intensified neoteny in which the permanent submissiveness but also more pronounced curiosity (for the rela-tionship to intelligence, see chapter on consciousness) of the dog is in sharp contrast to the wolf. The unique excited yapping and barking of a supposedly adult dog is probably the best example. It comes from young wolves who are still being looked after by their parents who, when their world was still in or-der, had to move silently through the woods like robber barons to exact their toll.

We are therefore not surprised when a self-confident and egoistic represen-tative of the not so distant group of *Felidae*, a tomcat called Garfield describes his evolutionary partner in misfortune Odie as a permanently driveling, thoroughly dumb, and penetratingly submissive pain in the neck. To rescue the honor of our domestic dog, we must say at this point that he can do nothing about it since he was helplessly subjected to at least 12,000 dog generations of human selective pressure. However, what we have still not achieved in spite of all our efforts is the creation of a new species although the preconditions, as we shall see in the following citation, could not have been better:

In practice, the events leading to a new species may often occur in small peripheral popu-lations, because such populations are more likely to be exposed to an atypical environment and hence no directional selection, and to be sufficiently isolated spatially that genetic changes can occur without being swamped by gene flow from outside (Maynard Smith 1989, p. 280).

The necessary conditions for the successful creation of a really new species *Canis domesticus*—the usual name *Canis familiaris* is confusing since it im-plies the status of an independent species (Linnaeus regarded wolf, jackal, and dog as separate species and, even worse, Darwin thought that "our domestic dogs have descended from several wild species" [Darwin 1859, chap. 1])—must have been present, however, in 20,000 years, not even a beginning has been made to cross the genetic border of the species despite intensive external sel-ection. Although, rather to our disgrace, it does not take much for us to force an already existing species systematically into extinction, it appears that there has to be a very different set of circumstances in order that a new species can be created. Our lovable domestic dog shows us not only that the big bad wolf of our fairy tales can be extraordinarily willing and faithful but also that the system-dependent constraints of species creation can sustain a very strong external pressure. Defining the natural boundary of a biological species as being identical with the border of genetic recombination, as was done by the spiritus rector of morphological systematics Ernst Mayr in 1963, is the first logical step but is still half of the truth. Certainly, at the point where two different genomes can no longer tolerate each other, when the reciprocal interchange of information is interrupted, a turning point in evolution has been reached insofar as a separation is introduced which with all probability will continue to exist forever. Such irrevocable separation lines can be tested experimentally quite simply in that one carries out interbreeding experiments and observes if they succeed or not. However, we still need to clarify the not unimportant question of whether the actual given limits and restrictions of sexual reproduction have been adequately covered, i.e., if panmixia as the total mixing of the gene pool of a species is a realistic view or if, in real life, other limitations come into play.

As we can, in view of these considerations, already guess in advance, the latter is just not occasionally the case but instead the rule with only few exceptions. Animals, and apparently people as well, are far more demanding than one was previously prepared to accept. This is especially true where sexuality is concerned. The effective limits of genetic exchange processes lie a lot closer together as one expects from classical population genetics which, even today, views a biological species as to a certain extent as a catchment area of isolated single genes, the co-called gene pool, which in principle is accessible to each individual like in a kind of unlimited molecular lottery (cf. "the gene pool approach" in Maynard Smith 1989, p. 32). Instead of an absolutely free choice of partner up to the definitive limits of genetic compatibility, as a rule, a preferential selection of the partner predominates to some degree with a highly specific outline of requirements. A kind of "same-type" pairing (*assortative mating*; in humans: Thiessen and Gregg 1980; Buss and Barnes 1986) causes individuals who are similar, and therefore probably close genetically, to find each other. There are, however, effective incest barriers, which prevent individuals who have a too strong similarity or are too closely related from pairing which keeps the disadvantages of certain genetic burdens as low as possible (in spite of this ca. 16% of all human pairs remain infertile). The saying "opposites attract" sounds attractive but is not really valid in this form for us choosy humans (Burgess and Wallin 1949). Because many disadvantageous gene variants or alleles only develop their full phenotypical efficiency and therefore harmfulness when the contributions to the new genome by both parents are homozygous, e.g., hemophilia, sickle-cell anemia, etc., the probability of such a coming together of individuals too closely related must be avoided. For most organisms, there is a suitable strategy of reproduction, which strikes a balance between panmictic confusion (outbreeding) and a bias toward isolationist inbreeding.

All this ultimately concerns the stability of the respective individual gene combinations which, if phylogenetically successful, consists of a certain number of genes, which are not—as a reductionist information theory could again suggest—isolated particles of (selfish) information, but just the opposite, finely tuned and selectively coadapted instructional elements. What is indeed shocking about sexual recombination and simultaneously fascinating is that each time, the functional coherence and therefore the adaptivity of the already existing gene combination, is destroyed, apparently capriciously and absolutely randomly, in order to make room for a completely new gene combination. In this picture, the adaptivity of the individual gene assortment and sexual recombination processes are direct opponents and one has to ask oneself if the advantages associated with parasite defense balance out the considerable disadvantages of the constant intermixing of evolutionarily proven gene combinations. As a rough rule, thorough mixing brings about a notable reduction in fitness of all the genes concerned and the more so, the stronger the differences between the fusing genomes. This is intuitively quite reasonable if one imagines it in an anthropomorphic way, for example, the genetic command in connection with the development of the brain from the mother's side says— we are now deliberately exaggerating: "Here we want to establish a new and nimble-minded speech and communication center", while from the father's side comes the instruction: "Let us establish a new risk-prone play center at this location". A kind of Mendelian compensation between both proposals according to the scheme: from "long" (red) and "short" (white) we get "middle" length (pink) would be quite difficult if not entirely impossible. If such a compromise solution to this phenotypical problem were not lethal, the fitness of the resulting offspring would suffer. In the place of an evolutionary free space of unhindered sexual reproduction inside the more or less widely set limits of the species, we find a continuous line of graduations placed between the disadvantages of pure inbreeding, which is close to asexual reproduction, and the rapidly increasing disadvantages of completely arbitrary mixing. Somewhere between these extreme poles there must be an optimum between the advantageous constancy of single gene combinations and selectively forced changes, which would mean that there, exists a continuous gradient of fitness stretching from absolute incompatibility to optimal tolerance and fertility.

This optimum will, however, as one can easily foresee, look completely different according to the species and population concerned. As a rule, its exact position always lies rather near the boundary of pure inbreeding, which proves that living systems are in some way "unwilling" to put their well-tried individuality at risk. Sexual reproduction is therefore more a compulsory measure imposed from outside by a corresponding selection pressure since organisms, as shown by the universal phenomenon of assortative mating, clearly try to keep genetic changes as small as possible. On the other hand, it is easy to understand that a new selective advantage is more likely to be obtained with new combinations of genes if not too differently structured genomes are fused together. The most valuable information, when viewed in isolation, is useless if it cannot be correctly "understood", that means harmoniously integrated into the existing cognitive organization of another living system. This also explains why the very impressive muscle power of a bull elephant is of no particular advantage to a little mouse even if such a combination were at least theoretically imaginable. Sexual reproduction, from this perspective, is above all a problem of the miscibility of knowledge:

Fertilization does not mean the rejuvenation or renewing of life, it would be entirely unnecessary for the continuance of life; it is only an attribute for making the mixing of two different hereditary tendencies possible (August Weismann 1892b, p. 304).

A revealing example of what significant role is played by the specific kind of information that is to be exchanged or newly combined is to be found in the basic difference in the organization of animals and plants. Most animals are highly centralized moving units in which a rapid and perfect coordination of all the bodily processes is a prerequisite for a successful evolutionary carrier. Plants as usually immobile creatures do not have to comply with such demands since the coordination between their various organs or tissue systems does not have to be so perfect or so centralized. They are therefore endowed to a certain extent with a significantly greater degree of freedom of their internal organization than animals. This fundamental organizational difference is reflected in the corresponding degree of freedom in sexual reproduction. While in animals, extreme outbreeding can be associated with a great loss in fitness, in plants, interbreeding of two different species, i.e., bastardization, is not only genetically possible but often a biological advantage as well (Arnold 1997; Rieseberg and Soltis 1991; Coyne 1996). In contrast, real bastardization has only been proven between animal species for a few species of gulls (Hoffmann et al. 1978), frogs (Remmert 1980), and fish. In the last named, occasionally there has been observed a proper "speciation in reverse" which, looked at more closely, is a kind of re-fusion of previously separate species (e.g., *Cyprinodon pecosensis* + *C. variegatus*; Rosenfield, cited in Norris 2000).

Sometimes, in so-called ring species, that are species in which a series of interconnected populations either encircle a barrier (mountain ranges, lakes, seas, climatic zones) or move along a limited range of latitudes around the globe (many bird species), it happens that in an overlapping area of previously separately living subspecies, the newly created mixed race is stable and in rare cases its distribution can even extend at the cost of the two original forms. More often, however, the two separate forms do not interbreed and we then have the paradox of two new "good" species despite a continuous chain of interbreeding populations (Irwin, Bensch and Price 2001). Examples of the origin of mixed forms are zones of interbreeding between the hooded crow (*Corvus corone cornix*) and the carrion crow (*Corvus corone corone*) and the interbreeding zone for the western and eastern forms of the long-tailed tit (*Aegithalos caudatus*). Both zones stretch across Central Europe, however, only the hybrid long-tailed tit which inhabits a relatively large area (Central Germany) appears to have suffered no great reduction in fitness. In *Homo sapiens*, there are also stable hybrid forms from previously long-separated populations, e.g., mestizos (European/American Indian), mulattos (European/ black African), zambos (black African/Indian). Generally, hybrid forms only have a chance of becoming established for any length of time when the interbreeding takes place in large numbers in a relatively short period of time so as to produce a novel intraspecific selection pressure that under some circumstances can compensate for the original selective advantage of the original forms (see also J. Weiner [1995]: Evolution Made Visible. *Science* 267: 30—33).

Also in this vein, is the observation that rejection reactions between tissues in plants are characteristically not so specific nor as strong as those in animals where cells, which are often genetically similar, are not easily compatible with each other. Herein lies one of the main problems of organ transplantation with the so-called oversensitivity reaction of the delayed type, which leads to the transplanted organ being rejected after a longer period of time (Schmidt and Thews 1997). In contrast, it is comparatively easy in plants to graft different varieties of the same species, foreign species, even species from a different genus onto the same stock. This is regularly carried out by modern wine and fruit growers and even by gardeners with their roses but the transplantation of organs between animals of one and the same species presents very great difficulties. The difference in how strict the coherence of the morphological and physiological system of plants and animals must be causes a corresponding difference in the miscibility of their genetic information.

Nevertheless, both types of life form, the sessile form of the plants and the mobile form of the animals, demonstrate the phenomenon of assortative mating, which proves that what makes a biological species can no longer be considered to be as easy to define as was previously thought. In this regard, an interesting thesis comes from Josef Reichholf, director of the State Zoological Collection in Munich. His thesis says that the boundary of a species represents the genetic result of an evolutionary stable strategy of individuals (ESS; cf. Hamilton 1964), whose aim is the maximization of fitness of a concrete set of coadapted genes (Reichholf 1979). As the almost universal distribution of assortative mating shows, Reichholf with this thesis comes nearer the truth than most of the purely nominalist or typological assessments which allowed the species to be debased to a purely theoretical construction. At this point, the circle of evolutionary theory can be closed, and we meet again the important question of the real unit of selection in evolution (for a recent debate, see Wilson and Sober 1994). The single gene, viewed as a completely isolated functional unit, cannot be the answer to the puzzle since it must be unimportant to it in which genome it is at the moment as long as it is in the next round of the reproductive game.

That the selfish gene has evidently been robbed of its perhaps earlier active evolutionary egoism and has been subordinated functionally as well as evolutionarily to the genomic structure, which, in the meantime, has become the main carrier of all the characters of the whole organism and therefore—if it does not want to disappear with it—has to serve the genome, shows, among other things, some of the functional details of the sexual recombination processes. So, for example, it is only through the structure of the whole genome that which gene is located in which position is determined in order that it, within the framework of a recombination process, can wander from one system to another. In addition, within a species, the absolutely unchangeable combination of certain genes to larger packets, called chromosomes, limits the possibilities of exchange to a great extent and, furthermore, the specific probabilities of exchange within the chromosome by *crossing over* processes is not completely random but must clearly experience the effect of natural selection (Trivers 1985). All of these limitations have the effect that the gene simply cannot move freely through all the genomes of this world but are selectively bound to a comparatively narrow and precise set of coevolving partners.

Let us turn the picture round for a moment and consider it from another perspective and see exactly what indeed would happen if the single gene really was, as Dawkins claims, the unit of selection. Then each gene would only be interested in its own advantage and therefore the genome should tend to begin to break down bit by bit. At least, with a certain probability, the secessionist tendencies, which arise in some genes, will, with time, be able to succeed since they alone would be the important replicating units. Since this is in no way the case and in fact just the opposite has occurred and all the genomes of this world are becoming larger and larger, the only explanation must be that single genes are unable to survive without the unity of the genome and therefore are automatically no longer capable of evolving on their own. It is most likely that free-living viruses, which are specialized at surviving at the cost of their hosts, are the only true examples of Dawkins' selfish genes. An interesting transitional case could be so-called *onco-* or *cancer genes* (e.g., *c-Sis* in primates, *c-Abl* in mice, *c-Src* in chickens) which preferentially become "active"—they simply mutate and thereby promote proliferation of their original cell and hence themselves—only after the end of the reproductive and parental care phase of their "host", i.e., the associated multicellular body. However, as soon as they succeed in some way—usually with the help of retroviruses being able to "recruit" new genes from their hosts (Abelson and Rabstein 1970)—in leaving the body, they mutate to really dangerous foes (i.e., *v-sis*, *v-abl*, *v-src*) which, under the condition that more than one accessible host to be infected exists, no longer have to worry about the extinction of their former "ancestor" ("dangerous" means here at least in an evolutionary sense where successful long-term reproduction and not a "nice" end of ontogeny is of relevance). It could even be that, perhaps viruses in general originated in a similar way by "escaping" by some means from an already more complex organism. However, the apparent lack of original homologous RNA- or DNA-sequences between eukaryotic hosts and viruses rather speaks against such a hypothesis. In fact, the opposite seems to be true since meanwhile a respectable fraction of the human genome has been revealed to derive from prokaryotic origins, as for instance Bacteria (Salzberg, White, Peterson and Eisen 2001; Andersson, Doolittle and Nesbø 2001).

Genes, which have the fate of being found permanently within a stable genome, however, have to behave differently if they are not to be thrown out of the evolutionary race at some time. A good example of this are the recently discovered meiotic drive genes, which can in fact egoistically spread "at the cost of the individual" in a sexually reproducing population. The initially threatening effect of their spreading comes to a standstill, however, as soon as there are fewer and fewer individuals still capable of surviving and transporting the "bad" genes. Apart from that, they die out apparently on their own. The upshot is—in contrast to the really malignant viruses—apparently selfish genes are already inseparably tied to the survival of a very specific host organism and can therefore only evolve *together* with all its other genes as a constituent of the genome, as for instance in the case of the *segregation-distorter* gene in the fruit fly (Hurst 1996). In the meantime it has been established that the effect of other genes, namely those which play an important role in partner choice, also put an effective stop to the attempted egoism of such meiotic drive genes. Experiments with flies with eyes on noticeably long stalks (*Cyrtodiopis*) have shown that the length of these extravagant characteristics in the males is an indicator of the presence of a Y chromosome that suppresses meiotic drive genes, and the fascinating thing about it is that for this reason such males are preferred by the females as mates (Burghardt and de la Motte 1988; Wilkinson, Presgraves and Crymes 1998).

One of the main reasons for the somewhat still great disagreements in the question of the unit of evolution is to be found in the apparently identity-disrupting activity of sexual reproduction. If there were only asexually reproducing organisms on this planet, we would be spared a great deal of the heated debate over this topic. Every double-stranded DNA molecule and every single-stranded RNA molecule, large or small, in pieces or as a whole, would double itself perfectly identically in every reproductive step and so ultimately bring forth a wonderfully homogenous world of obligatorily homozygous organisms where, even for the convinced molecular reductionist, who sees the world populated alone by selfish genes, there would be no more doubt over the real unit of selection in evolution:

> If there were sex but no crossing-over, each chromosome would be a replicator, and we should speak of adaptations as being for the good of the chromosome. If there is no sex we can, by the same token, treat the entire genome of an asexual organism as a replicator. But the organism itself is *not* a replicator (Richard Dawkins 1982, p. 95).

If we now throw overboard the absurd separation of replicator and vehicle, which, in theory and in practice, is not capable of survival, the last bulwark of a too one-sided gene-centered theory of evolution will have gone. It will be obvious that, what Dawkins says cannot possibly be true, since the whole organism, if counterbalanced for its extreme structural complexity, is at least an equally perfect replicator as the isolated DNA-string. A fictitious friend and his fictitious monozygotic twin brother, as we all know, would be so similar as to be easily mistaken for each other and if for the sake of producing a guaranteed identical environment (which, for technical reasons, is not easy), they were mutually able to slip into each other, they would be indeed perfect replicators since one would no longer be able to decide which one was which. All the more severe conceptual problems begin with the troublesome theme of sexual reproduction, and because of all the confusing sexual games of nature, one could be forgiven for thinking that this function was not so much for keeping a few steps in front of parasites and carnivores by an about-face of recombination but rather for using these recombination tricks to put one over the correct understanding of its real function. It is therefore understandable that the main objection to Dawkin's gene selection theory had to come from this quarter:

> It has often been suggested to me that a fatal objection to replicator selectionism is the existence of within-cistron crossing-over (Richard Dawkins 1982, p. 90).

This is without question a fatal objection since within-cistron crossing over, by occurring within the coding regions (= genes) of the genome (Watson 1976), cuts through the gene. One could now go a step further and claim that Dawkins' quite precise characterization of sexual recombination as a merciless genetic grinding machine which leaves behind ground genome ready for fast food (are there MacGenes?) ultimately also makes his own cherished selfish genes to dissolve into thin air or rather single bases:

It is changes at the single nucleotide level that are responsible for evolutionarily significant phenotypic changes, although of course the unvarying remainder of the genome is necessary to produce a phenotype at all. Have we, then, arrived at an absurdly reductionistic *reductio ad absurdum?* (Richard Dawkins 1982, p. 90)

We have, in fact, and Dawkins is not so shortsighted as not to perceive the severe problems that recombination entails for his somewhat lost replicator machine. Already at the beginning of the next paragraph he recognizes that:

At the very least, this is not a helpful way to express what is going on. It becomes downright misleading if it suggests to the student that adenine at one locus is, in some sense, allied with adenine at other loci, pulling together for an adenine team. If there is any sense in which purines and pyrimidines compete with each other for heterozygous loci, the struggle at each locus is insulated from the struggle at other loci. (Richard Dawkins 1982, p. 90/91)

But what is even more interesting he then chooses an argument, which again goes back to a more holistic approach to genetic information:

An active replicator is a chunk of genome that, when compared to its alleles, exerts phenotypic power over its world, such that its frequency increases or decreases relative to that of its alleles. While it is undoubtedly meaningful to speak of a single nucleotide as exerting power in this sense, it is much more useful, since the nucleotide only exerts a given type of power when embedded in a larger unit, to treat the larger unit as exerting power and hence altering the frequency of its copies. (Richard Dawkins 1982, p. 91)

Nevertheless, this surprising admission is not convincing enough to persuade Dawkins to take over a more genomic view, but instead—and that is a very tricky procedure—he relies on the same argument as advanced before by critics of his gene-centered view:

It might be thought that the same argument could be used to justify treating an even larger unit, such as the whole genome, as the unit that exerts power. This is not so, at least for sexual genomes. We reject the whole sexual genome as a candidate replicator, because of its high risk of being fragmented at meiosis. (Richard Dawkins 1982, p. 91)

In other words, if a meiotic incision strikes the whole genome it is fatal for its candidature for being the unit of selection, but if it concerns a gene it is not! And then, if there is no sexual recombination at all, we suddenly have equivalence of all levels again? (Cf. first citation by Dawkins in this chapter) The solution to this unnecessary conceptual confusion is much simpler than one would expect. For that purpose it is sufficient to separate the concrete content or functional meaning of the genetic material from the purely physical process of natural selection: if an individual with a certain genome is negatively selected (death without previous reproduction) all its associated genes automatically have to share that same irreversible fate, but if for some reason a particular gene variant disappears from the genome the larger unit must not necessarily dissolve (Heschl 1994a).

Let us still stick to the hard facts for a while and examine even closer what an unexplainable drama sexual reproduction has to offer to the gene-centered approach. The worst that could theoretically happen to us, the uninhibited panmixia of all existing living organisms on earth, has fortunately never been the case so that we can confidently cross it off the list. Sex has well-defined limits of decency and the majority of those well-behaved organisms endowed with the Darwinian honorary title of species keep within them. How does it look within the species? Here, yet again, we do not find what one would at first logically assume, namely unlimited promiscuity of all the individuals in a whole population capable of reproducing. In addition to the problems posed by geographical distance, which of course strongly limits panmixia even in populations linked together, most animals demonstrate unexpectedly exacting standards in their choice of reproductive partners.

However, exceptions prove the rule. In some cases, where the expenditure involved in courting, investment, and the risks are comparatively low, many individuals allow themselves to be carried away by indiscriminate sex. Characteristically, in mammals it is mainly the male who possesses the invaluable advantage over the female of being able, after donating a small unobtrusive germ cell, to disappear without a word. As a rule, however, the female understands very well the use of diverse countermeasures to compensate for this unpleasant physiological disadvantage. Among the best and not exactly unknown methods are a general coyness combined with an abundance of the most varied tests with regard to the willingness, generosity, and last but not least, the faithfulness of the potential partner. This last instance changes the disadvantage into an advantage since the disproportionate relationship of a few female germ cells to a whole army of recombination-addicted sperm cells has the effect that most females in a population, with great probability, can be sure of reproducing (this does not rule out reproductive competition between females as recently shown for chimpanzees; cf. Pusey, Williams and Goodall 1997) whereas the males can never be sure if their subjectively valuable hereditary information can be brought to some female or other. This circumstance allows the female to lean back so to speak and relax to a certain extent and wait for things to happen, whereas the males on this planet, however, remain subjected to a more continual stress in order to find at least one, but also perhaps many willing sexual partners. The sociobiological term for this fundamental asymmetry in the sexual self-interest of the two sexes is appropriately called *female choice*. Interestingly, this is exactly what occurs in dancing lessons and at balls, which are nothing other than occasions for comparing and meeting possible partners, although very refined, where whenever "ladies' choice" is announced, those ladies have a free choice of partner for the next dance as if this would be the exceptional case in most men-dominated cultures where it is mostly the gentlemen who are faced with the problem of making a good choice.

In reality, things are not so simple since, from an evolutionary point of view, a strategy of one sex must always result in a corresponding counter strategy from the other. This results in the fact that, in contrast to all the other essentially harder interspecific competition (e.g., competitors for food, predator/prey and parasite/host relationships), through the wide-spread obligation to reproduce sexually, both partners are damned to love each other in the both senses of the word if they are not to suffer the fate of dropping out completely from the race. Every female egg cell requires a male germ cell, and vice versa, every male sperm cell absolutely requires a ripe and fertilizable egg so that, in consequence, what is dramatically known as the "battle of the sexes" (for entertaining descriptions of this battle without foreseeable end, see Pease 2001; Schweder and Riedl 1997; for a comprehensive sociobiological treatment, see Barash and Lipton 1997) must end in a draw as a matter of principle. In the words of Loriot, a famous German satirist, "man and woman are not suited to each other but one can interbreed them quite successfully." On the other hand, with regard to the often heated controversy about some supposedly significant gender differences in intelligence, Herrnstein and Murray get the facts clear:

When it comes to gender, the consistent story has been that men and women have identical mean IQs but that men have a broader distribution. ... The larger variation among men means that there are more men than women at either extreme of the IQ distribution (Richard Herrnstein and Charles Murray 1996, p. 275).

That of course does not mean that conflicts between the sexes could not play an important role in evolution. In fact, just the opposite is true, recently it has become clearer and clearer how strong the influence of a permanent sexual conflict between the sexes can be on the speed of producing new species in very different groups of animals (Gavrilets 2000).

A good example of this on the behavioral level is again the selection of a partner, where it has been found in the mean time that both sexes can be very choosy in a comparable way, above all if the personal stake is a high one. Since the males are often coerced into make a greater investment than just contributing a few cheaply produced sperms by certain behavioral strategies of the females, they should also try to counter with a proper strategy of mate selection, then called *male choice* (the "official" procedure on Western dance floors) and meaning at the same time that the definitive area of free choice of partner is even more strongly reduced than before. It was Darwin who first recognized the great influence of this choosy behavior, especially that of the female, on the evolution of sexually reproducing species and even gave it its own name. In contrast to natural selection, which collects together all the other environmental influences on an organism into one expression, Darwin's term "sexual selection" is only related to this, in particular female, selective behavior in reproduction determined by the advantage of the egg cell:

This (kind of selection) depends, not on a struggle for existence, but on a struggle between the males for possession of the females; the result is not death to the unsuccessful competitor, but few or no offspring. Sexual selection is, therefore, less rigorous than natural selection. Generally, the most vigorous males, those which are best fitted for their places in nature, will leave most progeny. But in many cases, victory will depend not on general vigour, but on having special weapons, confined to the male sex. A hornless stag or spurless cock would have a poor chance of leaving offspring. Sexual selection by always allowing the victor to breed might surely give indomitable courage, length to the spur, and strength to the wing to strike in the spurred leg, as well as the brutal cock-fighter, who knows well that he can improve his breed by careful selection of the best cocks (Charles Darwin 1859, chap. 4).

The results of this very ingenious female selective breeding, in which the female waits patiently on the sidelines for the result of the tournament to reward the winner with affection, have been closely investigated by many behavioral researchers and have contributed, not without cause, to the macho image of the male sex. Such a cliché is, however, for this reason not completely without a certain justification since the greatest number of all the kinds males in the world love getting dressed up, if not permanently, at least for as much of the time as possible, in special costumes for display to show themselves or to advertise themselves as conspicuously as possible to any females present. The last is especially so if, for example, a special morphological outfit is not exactly opportune for other selective reasons such as increased predator pressure. However, there are a few cases where the usual relationship of female to male is exactly the reverse, e.g., in the species red-necked Phalarope (*Phalaropus lobatus*) and dotterel (*Eudromius morinellus*). Here, brightly colored females compete for unobtrusive, camouflage-colored males who alone have the full responsibility for both incubating the eggs and looking after the brood. The reason for this unusual switching of roles may be found in the higher arctic zones where the summer is very short but luxuriant with respect to food supply („airplankton": insects in high density). Under these extreme conditions, only one brood per year with four eggs per nest is usual (Maclean 1972) which means that to achieve a greater reproductive rate, the female has the only alternative to compete with other females for males to hatch her additional eggs (Erckmann 1983). And then, there are some rare species of birds where the male and female share the fledgling young between them in an "exemplary" manner and raise them separately (simultaneous brood division: Hoi 1987; Horsfall 1984; Price and Gibbs 1987).

Among the most famous peacocks of evolution are to be found in the impressive male variants of, to name but a few illustrious examples, domestic fowl, pheasant, capercaillie, red deer, baboon, lion—all species in which the competition between the males is exacerbated by polygynous relationships. What must occur, however, if, as often implicitly assumed with reference to Darwin, only this form of sexual selection played a role in sexual reproduction? If indeed unchangeable criteria of an optimal species-specific partner selection should be put into practice, the final consequence, within a relatively short time, must be a stable population of perfectly adapted types, in fact completely independent of whether the influence of sexual selection is limited by other selective factors or not. The latter would only put off the final result in the direction of an evolutionary optimal compromise without being able to prevent a certain extreme type being produced (*runaway process*: Fisher 1930, Kirkpatrick 1982). If we were to transpose this situation to the relationships of humans, where man and woman, as in any other species, can take up various positions in the battle of the sexes, it would mean that logically every male lineage, if successful in the competition, would end in producing a perfect Adonis (e.g., Mr. Universe) in quantity and every female lineage a Venus (e.g., Mrs. Amazon). Since this is in no way the case in our and most other species and, as a rule, there is a great deal of variability of both sexes even in our secondary sexual characters, some additional and perhaps completely different factors must play a role in the selection of our partners. Darwin himself pointed out such a possibility:

Those who have closely attended to birds in confinement well know that they often take individual preferences and dislikes: thus Sir R. Heron has described how one pied peacock was eminently attractive to all his hen birds. It may appear childish to attribute any effect to such apparently weak means: I cannot here enter on the details necessary to support this view; ... (Charles Darwin 1859, chap. 4).

The actual key to a better understanding of sexual selection is not so much to be sought in the strongly idealized partner selection mechanism of classical ethology, still based on the group (= species) selection paradigm (cf. Eibl-Eibesfeldt 1989) where visual "releasers" or "sign stimuli" (Lorenz 1935, 1943) alone are considered effective, but rather in the already mentioned assortative mating procedure which represents an evolutionary compromise between sexual reproduction caused by pressure from parasites and other unpleasant enemies and the otherwise optimal unsullied, i.e., purely vegetative reproduction. That would really mean, however, that the individual must be primarily concerned about genetic similarity and only secondarily looking out for certain other general criteria of possibly fitness-increasing characters such as sexual attraction (Perret, May and Yoshikawa 1994; Enquist and Arak 1994; Johnstone 1994), physical strength, faithfulness, stamina, good health, social interest, considerateness, etc. Only such differentiated behavior can explain the simultaneous presence of particular quantitative trends (e.g., more reliable, faster, quieter, more economical, etc.) and the maintenance of differences independent of these, i.e., a lasting variability of various genetic types, from Type T_1 to T_n within the population.

The optimum between becoming different (outbreeding) and remaining the same (inbreeding) should, in most cases, lie as near to the inbreeding boundary as possible to reduce an automatic loss of inclusive fitness caused by an increase in genetic mixing. The greater the genetic difference between two different partners, the greater the loss of inclusive fitness for both sides, since the sexual partner can also be seen as a social companion which is altruistically helped in its attempt to reproduce. The loss would be zero only in the case of perfect inbreeding between identical, i.e., monozygotic partners that, in the end, would equal again the purely vegetative reproduction with its twofold quantitative advantage (see above). This demonstrates yet again that, anthropomorphically speaking, living organisms essentially do not really want to reproduce sexually but instead have been forced to by certain selection pressures such as Muller's ratchet and, above all, by a permanent pressure from the side of coevolving parasites. The latter becomes particularly obvious if one looks at those genetic factors, which, in mammals, are responsible for the defense of the multicellular body against all smaller enemies such as viruses and microbes. Independent studies have shown that both in humans (Wedekind et al. 1995) and mice (Egid and Brown 1989; Potts, Manning and Wakeland 1991; Yamazaki et al. 1976) potential mating partners are preferred who differ significantly in their body odor as a reliable indicator of the respective type of immune system (up to 128 expressed gene loci [MHC sequencing consortium 1999] constitute the *major histocompatibility complex* or MHC which shows the highest degree of variability among all human coding regions; main variants: HLA-A, HLA-B, HLA-DR). Interestingly, this result even translates to our preference for certain artificially produced perfumes where similar MHC types tend to use the same category of fragrances on themselves whereas they are attracted to other people preferentially to other, i.e., non-self fragrances (Milinski and Wedekind 2001).

In contrast, concerning general morphological features, the situation is rather different. So, for example, Patrick Bateson, a specialist in experimental research into assortative mating behavior (Bateson 1978, 1983) was able to demonstrate in the Japanese quail that the grade of a cousin apparently plays a very special role. These birds showed in simple choice experiments that they preferred a partner that outwardly resembled the siblings of the animal being tested, i.e., the birds with which they had been reared (Bateson 1979). Their first preference was logically for first cousins (ca. 35% of the time was spent in front of them) and then third cousins (ca. 25%), followed by their own but unknown siblings (ca. 15%) with whom they have had no previous contact (Bateson

1982). The least interest (ca. 10%) was shown for their own familiar siblings. One does not have to go as far as to claim that the dainty quail with its preference for close relatives is directly comparable to us clumsy humans but it is at least noticeable that also in our species, the topic of cousin and cousin appears to be a fascinating one, and, as we will presently see, in every respect from real life to art and culture:

A last sound is to be reported, a sound from a form so delicate and intricate that it could have disappeared in all the rumor of the universe. But the sound remained and did not disappear. It came from Eschberg. It was the soft heartbeat of an unborn child, a fetus, a female human being. What Elias had heard and saw he forgot, but the sound of the unborn heart no more. Since it was the heartbeat of the person that was predestined for him from eternity. ... And Elsbeth's heart lay on Elias' heart, and Elsbeth's heartbeat entered Elias' heartbeat. Then Johannes Elias Alder roared so terribly, so pitifully as if he, with clear understanding, must die. The cry instantaneously shattered the consciousness of the girl and she sank helplessly into the body of the young lover. The revelation was fulfilled that the five-year old had once listened to in the streambed of the Emmer, when he heard the heartbeat of an unborn child. In this night of universal horror, Johannes Elias Alder fell in love with his cousin Elsbeth Alder (translated from Robert Schneider 1995, p. 35/78).

What Robert Schneider described with such moving words as an emotional stirring of a young person is nothing more than that what has obviously moved us humans very deeply again and again since ancient times. The evolutionary voice of the heart appears to tell us deliberately that, later in life, we should find that stranger interesting and attractive who is not a really close relative (incest taboo), but instead resembles our own closest relatives the most. This people that fit this description best are often cousins, i.e., the children of the brothers and sisters of one's parents and it is therefore not astonishing to learn that in many human societies the marriage of first cousins does represent a widespread phenomenon (Fox 1967). In this respect, it is interesting to see that the cultural and partly the judicial barriers against incest fall in exactly this area. For instance, in the German Criminal Code, only incest between members of the nuclear family constitutes a crime punishable by imprisonment. Thus the selection of a cousin as a sexual partner seems to represent a widely distributed optimal compromise between inbreeding and outbreeding (van den Berghe 1983), at least, this strategy, in most of the species closely investigated so far, appears to be the dominant one. The use of the term "optimal" here does not exclude that a completely different decision will often be or must be made under natural conditions if, for example, the theoretically best variant is not achievable. Assortative mating describes only the general tendency of many animals including *Homo sapiens* to pair with partners who are genetically similar but not too similar and this with regard to sexual reproduction as well as in relation to practically all social affairs (cf. Bischof 1985). One could, generally speaking, refer to a kind of *assortative association*, for which a particularly charming description of the phenomenon comes from Astrid Lindgren:

If two small boys meet each for the first time and they are two of a kind, then they only need to look at each other to know at once whether they will get on (Astrid Lindgren; from *Michel from Lönneberga*).

In this respect, humans do not only not represent a supposed exception to biology, but exactly the opposite, they are, to all intents and purposes, the paradigmatic textbook example of sociobiology. Their whole social behavior beginning with "Calculation of Parental Love" (Voland 1995) through the complete set of primary emotional reactions (sympathy/antipathy) to strangers to the family up to the most strongly institutionalized areas of living together (legislation, social systems, political systems, etc.) presents the genetic relationship as being the most important structuring factors of social behavior. So, for example, many investigations show that the—not always completely rational—choice of the sexual partner in the matter of practically all characteristics (physical attraction, intelligence, interests, education, social status, etc.) is mainly based on diverse criteria of similarity (Thiessen and Gregg 1980; Bouchard and McGue 1981; Rushton 1989; Mascie-Taylor and Boyce 1988; Mascie-Taylor 1987, 1989). Even supposedly completely rational decisions in complicated cooperation experiments of the prisoner's dilemma kind—this consists of having to choose a safe strategy of deception, although a cooperative tactic would give a better result for both partners (formal details in Maynard Smith 1991; Sigmund 1993)—were decisively influenced by a short opportunity, lasting only 30 minutes, of getting to know the game partner in the sense of an intuitive appraisal. This was always in the direction of the precise prediction of the behavior of the future partner, in the case of cooperation, or enemy, in the case of rejection, made during this appraisal (Frank, Gilovich and Regan 1993). Even in fields such as politics, where one would not readily imagine that biological determinants come into play, phenotypical characters, and therefore also ultimately genetic characters, do play an unexpectedly important role. Among others, they do that when they as physiognomic similarities between political allies or, as in the investigated case of the Russian Perestroika, with the comrades who wanted reform influenced the amount of cooperation between them (Heschl 1993a).

From all the preceding, we arrive at the informative situation that during the sexual recombination process not many, as one could have originally suspected, of the genes but rather only a small percentage of the total gene material is actually randomly mixed and exchanged, whether by a new distribution of the parental chromosomes or by crossing-over processes within individual chromosomes. This means nothing other than that in all the vast regions of DNA, in which, in any case, perfect genetic identity between the mating partners prevails, despite extensive meiotic rearrangements no actual change is brought about to the sequence of nucleotides. Our thesis now goes on to claim that the actual function of sexual reproduction primarily consists of taking over only that foreign, but evolutionarily promising (e.g., Miller 2000; Welch, Semlitsch and Gerhardt 1998; Wilkinson, Presgraves and Crymes 1998) genetic material which, in principle, could have already been successfully integrated into the own genome by independent mutations. In other words, sexual exchange processes essentially allow living organisms to make evolutionary possible, i.e., with greater probability successful steps much *earlier* than would have been the case if they had had to wait step by step for the suitable mutation.

The complete mixing of all the genes by the random distribution of the chromosomes to the new germ cells as well as the crossing over to give new combinations of genes within the chromosomes now appears, in this more cognitive way of looking at sexual reproduction, more as a technical necessity rather than a somehow especially adaptive process that is at work. The question remains what other conceivable possibilities are there in order to, starting from the phylogenic precursor stage of an identical DNA replication, exchange only very specific and limited regions of the extremely long nucleotide strand? For this purpose, let us make an attempt as genetic engineers and, as an example, to try to transfer a very specific single gene, which for the receiver is assumed to be an extremely valuable one, from the transmitter genome G1 to the receiver genome G2. How can we best go about it? To be more explicit: is it at all possible, to transfer deliberately only a certain fraction of DNA without any complications and completely isolated from the rest of the genome? The answer to this for natural organisms must be a system-determined no since, without information about the whole genome structure, we would have no possibility of defining the location and therefore, at the same time, the molecular access to the particular gene, whatever amount of DNA is occupied by it. The only solution, starting from the already proven and therefore most proficient method of identical replication, which comes into question as far as biological systems are concerned, evidently consists of, after a successful duplication, risking a more or less arbitrary mixing of the whole gene material (for the somewhat different processes of conjugation in Bacteria, see Willets and Skurray 1987). Meiosis as the main source of sexual recombination represents nothing other than a somewhat altered sequence of two normal mitotic cell cycles and this cannot be a coincidence. The essential difference is that during the second maturation division of the chromosomes, not two diploid (2n; n = 23 chromosomes) body cells are produced, but the once duplicated, hence $2 \_ 2 = 4$ chromosomes are distributed among four then haploid (n) germ cells with each one having a simple set of chromosomes instead of a double. As far as what could be interpreted as a purposeful exchange is concerned, not even the isolated transfer of single chromosomes has been established. Of course, it can and does occasionally happen that, in spite of the indiscriminate mixing, in end effect only a single chromosome or none at all is exchanged but this only occurs coincidentally, as a rule, and is not related to the type of genome concerned. The whole procedure is comparable with shuffling two packs of playing cards (e.g., blue = father, pink = mother) so long until there is a perfect separation of the two colors again. How does it look within the chromosomes? Crossing over, which can take place in the first maturation division by a cross over of the previously still separate paternal and maternal chromosome halves (chromatids), is here responsible for what at the level of the genetic super packet, the chromosomes, is brought about by the cellular division apparatus, the complicated filigree patterned and efficiently working spindle, and also causes—partly with different mechanisms in different organisms (McKim et al. 1998)—a random mixing of all the genes of each individual chromosome. Whereby, the function of the spindle itself is controlled by the chromosomes themselves, another example of the self-organizing, cyclic-causal nature of living systems (McKim and Hawley 1995; Nagele, Freeman, McMorrow and Lee 1995; for the basic principles underlying the genetic self-control of DNA replication and cell cycle in eucaryotes, see Kornberg and Baker 1992; Murray and Hunt 1993; Hartwell and Kastan 1994; Nurse 1997). Only because, as a rule, the similarity of the paired genomes is very great and assortative mating still additionally enhances this basic similarity, can such a technique, which at first sight gives a chaotic impression, succeed in evolution. What finally remains after fertilization is a genome, which exists in a unique schizophrenic form (Trivers 1997): as under circumstances an optimally adapted double self from two subordinate selves, very similar to each other but also slightly different from each other, from the parental sets of chromosomes:

Perhaps the most interesting thing to come out of the realization of possible conflict within the genome is a philosophical one. We see that we are not even in principle the wholes that some schools of philosophy would have us to be. Perhaps this is some comfort when we face agonizing decisions, when we cannot "make sense" of the decisions we do make, when the bitterness of a civil war seems to be breaking out in our innermost heart. (W. D. Hamilton 1987)

So it is no wonder that in the meantime the neo-Darwinian theory of evolution is beginning to occupy itself with the potential internal conflicts of a "divided self" (Haig, 1999) or between "two selves" (Sigmund and Nowak 2000) with highly interesting consequences for a better understanding of some peculiarities of our indeed very peculiar social behavior. While Hamilton's classical kinship theory from 1964 proceeds from the fact that the average degree of kinship between individuals can explain certain variants of social behavior including altruism, David Haig now goes a great deal further and examines the exact social distribution of paternal and maternal genes coming together in a particular genome (Haig 1992, 1997). He sees there the conflict between father and mother carried on to a certain degree in the children and, depending on the respective origin, with completely different influences on the behavior of the individual in question. In this way, relationships between full brothers and sisters, father to daughter, and mother to son are characterized by a symmetrical proportion of paternal and maternal genes whereas, in the case of half sisters and brothers, daughter to father, and son to mother the same genetic relationships are asymmetrical. This alone appears to cause, that under the assumption of certain losses of genetic fitness, occurrences of incest in certain combinations take place more often and furthermore are experienced by the participants in quite different ways. This seems to be clearest in the case of father/daughter incest, which from the viewpoint of the female victim is often described, as especially traumatic (Finkelhor 1979; Russel 1986). Apart from the influence of uncertain fatherhood, the share of paternal and maternal genes of the father are to be found to 50% in the daughter and her children whereas from the standpoint of the daughter, the situation is clearly different: her paternal genes can be found with the same probability in her own gametes as in the gametes from her father. The maternal genes are naturally lacking in the genome of the father which leads to the striking "disinterest" in incest (Haig 1999). The mother/son relationship is, in principle, the same only in this case, the "stealing" of paternal genes in the father would accrue an additional disadvantage which causes mother/son incest to appear less seldom (Finkelhor 1979).

To all these complications comes the additional genetic asymmetry from the different distribution of the sex chromosomes (X, Y) which, in contrast to the other chromosomes, the autosomes, are transmitted differently according to sex. For example, the paternal relationship to a sister with reference to the Y chromosome is equal to zero, since a brother always inherits the Y chromosome and his sister always the X chromosome from their father, and this could lead to an evolutionary conflict between the genes on the sex chromosomes and those on the autosomes (Haig 1999). A further logical step in this direction would be to examine more closely the concurrence of certain individual gene combinations within the whole genome, whether from the father or the mother (for the phenomenon of genetic imprinting where one allele is silenced according to its parental origin and its intricate relationships to behavior, see Isles and Wilkinson 2000; for the phenomenon of "directed" or at least not fully random mechanisms of genetic recombination, see Gruss and Michel 2001), and to test in which way well proven gene combinations can be maintained over generations of sexual reproduction. If, in this way, the laborious, in principle random and therefore always risky way of cognitive gain is additionally complicated by the ingenious strategy of sexual reproduction (for a recent revival of the discussion, see Lenski 2001; Rice and Chippindale 2001), it is also manifestly true that sex is still by far the most sophisticated kind of entertainment between individuals since the discovery of life. Whether one likes it or not, and for whatever reason, it is here to stay for the next few million years.

14 How to Explain Consciousness

We don't know how it works and we
need to try all kinds of different
ideas.

John Searle

Oh, he is right, that popular philosopher when he says that "to be" is only
an expression with marked electro-
magnetic-galvanoplastic moments.

Johann Nestroy

However complicated the so-called consciousness problem still appears in the
eyes of many philosophers, its scientific treatment is comparatively simple.
The decisive thing about it is the capability to overcome two wide-spread prejudices, namely an anthropomorphic one (Ryle 1949) and a dualistic one
(Popper and Eccles 1976). The first says that man alone, as a social organism
with a gift for language and hence verbal communication, can be capable of
being conscious, whereas the latter proceeds from the Cartesian idea that consciousness can never be reduced to "simple" material processes in the brain.
Both views are equally wrong and disprove themselves immediately, since first,
it cannot be strictly proved that other people also have consciousness and second, the fact that we, in the course of our lives and without any difficulty, develop the ability (Bischof-Kohler 1988) to attribute purely material processes
in our environment—e.g., the appearance of members of our own species—to
consciousness processes (cf. Bühler's "You-evidence" 1922), that without the
acceptance of the thesis: "consciousness is linked to material phenomena"
(Patricia Churchland: "Consciousness is not based on brain processes, it is
identical with them", translated from *Psychologie heute*, July 1999) we can save
ourselves the trouble of any further meaningful discussion of consciousness.
The subject is very similar to the reality problem dealt with previously or the
"direct" perception of animacy (cf. Scholl and Tremoulet 2000) which both are
supposedly just as intricate, however, it is not half so difficult to deal with as
Searle would have us think. If it is allowed to attribute consciousness to our fellow humans without any really compulsive grounds for doing so, but which we
do everyday and every moment, consequently it must also be possible in principle to do so for our somewhat more distant relatives, which, of course, ani-

mals are. That we are dealing with material processes here is then almost a triviality. One can quite simply turn this explanatory perspective around. The precise statement of the problem, as succinctly put by the Austrian satirist Alfred Polgar (1873—1955) as long ago as a century, with regard to the animal/human comparison is also valid in the same way for the relationships within our species whereby instead of street maps and cities we can simply substitute suitable biological terms:

In the darkness of the animal soul, man shines the light that his inborn individual knowledge of the human soul has ignited. This, incidentally, as if he wanted, with the aid of a street map of Paris, to find his way around in London (Alfred Polgar).

If we apply this then to within our species:

In the darkness of the psyche of his fellow man, man shines the light that his inborn individual knowledge of the human psyche has ignited. This, incidentally, is as if he wanted, with the aid of his own genome, to find his way around in another.

With this equating of animal and human, or the emancipation of the former in relation to latter, which, incidentally, can only be a real problem for people who do not know animals, we can now speculate about and discuss the consciousness of animals in exactly the same way as we do about the no less hypothetical consciousness of our fellow humans. The arrogant and contemptuous criticism of an unjustified anthropomorphization of animals can therefore be laid to rest as a further myth (Fisher 1996). We then find ourselves again in a situation in which every individual, whether man or beast, observer or observed, is in a fascinating world of flealike , mouselike, or apelike behavioral researchers who, imprisoned together for all eternity in a kind of oversized laboratory, have to scrape a living, observing each other permanently, and secretly drawing their own conclusions and, in spite of all that, never really knowing what may be going on in any of the others. That our conclusions and the words linked to them can be so strikingly similar or occasionally even almost identical has nothing to do with that it may somehow be possible to get a glance directly into the interior of another person. It is much more the factor that Chomsky, already some time ago, expressed in unusually precise words:

It is a fact of nature that the cognitive structures developed by people in similar circumstances, within cognitive capacity, are similar, by virtue of their similar innate constitution (Noam Chomsky 1977, p. 71).

If we strongly feel the need to bestow upon each other a certain degree of consciousness, this does not mean that we humans can talk objectively over subjective inner conditions but it is only, with great probability, due to our not unique ability to attribute hypothetically to our fellows such a condition and the like from examining external criteria. With regard to humans this innate procedure is usually, that means at least since the early Greeks (to cite just the most important ones: Socrates, Plato, Aristotle, Epicurus), called epistemology or theory of cognition, whereas for nonhuman animals for a few years we have been using the term "theory of mind" or, in short, ToM (Premack and Woodruff 1978). And such good hereditary "consciousness researchers" are expected to be present, as we will presently understand better, everywhere where a most narrow mutual coordination of the behavior in space and time is required.

Let us say it the other way round: The consciousness problem in its three separate though closely intertwined domains of the world- (that is our main topic here), self- (one could say "theory of self", ToS) and other-consciousness (ToM) is not, as most dualist or even trialist (see Popper) philosophers and some incorrigible behavioristic psychologists (e.g., Heyes 1998) would have us believe, a basically disputable if not insoluble one, as long as we are concerned with animals, and in contrast, an objective and—because, finally, we can speak to each other!—therefore basically "solved" topic if it concerns us as humans. Just the opposite is true, if taken as a simple phylogenetically evolved and hence, with great probability, adaptive behavioral mechanism of how to attribute meaning to the world or like-mindedness to one's self and others, the character consciousness can, if one is really interested in its distribution in the animal kingdom and its evolutionary origin, be used similarly to any other biological character for taxonomic description of animal groups. Or, to speak with the words of David Hume with regard to "world-consciousness":

It seems evident, that animals as well as men learn many things from experience, and infer, that the same events will always follow from the same causes (David Hume 1777/1975, p.105).

Our strategy will simply consist of starting out from our own understanding and experience with consciousness (where else) and look at the animal kingdom to see if there are any candidates that show any indication and characterization in this direction, we will call this the consciousness phene. Perhaps sometime in the future, we will discover an especially long consciousness gene sequence, or more likely a complete series of them, like the recently discovered gene for *emotionality*—by which is meant anxiety—in mice (Flint 1995) and has been traced in humans in the meantime (Lesch et al. 1996). Allowing ourselves to be inspired by the saying from Descartes, *Cogito, ergo sum*, for our purpose we can modify it somewhat to give *Cogito, ergo sum consciens* or *Sentio* (I perceive), *ergo sum consciens*. Armed with this saying as an experimental testing device, we will attempt to scan the biological field in the hope of finding something. In this way, we can deliberately save ourselves having to explain the question whether, with a mere 6 billion concepts of personal consciousness, it is at all possible to develop a generally acceptable concept of human consciousness—a statistically significant mean value of all the concepts of consciousness of the human population. My feelings tell me that, in principle, it must be just as possible in the same way as defining the nature of mammals by the presence of hair or milk glands. We should satisfy ourselves with introducing a broadly directed scientific proposal with at least a realistic chance, as an empirically applicable concept, of making a small contribution to the present debate.

Viewed from a biological perspective, there are only two really interesting conceptual possibilities of defining consciousness. The first, more general, is closely connected to the second, more specific and fundamentally more exacting one, and was advanced in the stimulating book by Griffin *Animal Thinking* (1984) at least in broad outline, although the author did not manage to give a precise definition of what he really wanted to understand about this phenomenon. As an argument, or excuse, for this shortcoming, Griffin asks us to consider that it is nonsense to attempt to restrict this phenomenon artificially with a too exact, or a pseudo exact definition of an area that is already so difficult when we consider it in humans. As a reason, he quotes a statement by Woodfield (1976), which sounds plausible:

If the term to be defined is vague and indistinctly prescribed, a definition can only be appropriate if it is itself vague and indistinctly prescribed" (retranslated from Donald Griffin 1984).

Nevertheless, a little later, he ventures making a material approach to mental experience and states categorically: "I assume that it is certain that behavior and consciousness originate exclusively from the processes which take place in the central nervous system." Although one could be tempted to say "so what?", the question about the central nervous system, CNS, does have a deeper meaning. It is exactly this remarkable morphological structure, the CNS in animals, on which we would like to hang a precise definition of consciousness. This primitive form or archetype of consciousness could be described as *primary consciousness* or *elementary consciousness*, is then exactly given when a living system has acquired the ability to control the behavior of the *whole* organism by a locally circumscribed center, as for instance the filigree network of the CNS. This is evident for the first time in those living systems consisting of many individual cells, who, in the course of evolution have become specialized, not at passing their time just lazing around and assimilating sunlight but as a unified and coordinated creature able to move around with goals in sight. All multicellular animals must solve the difficult problem of how one can bring about that a multitude of genetically identical cells which are basically prepared to cooperate do not just refrain from eating each other but progress beyond an uncoordinated and occasionally undisciplined package to acting jointly together in time and space and in perfect harmony. This explains why even Darwin attributed a kind of intellect, let us call it primary consciousness to be on the safe side, to even the simplest animals:

A little dose of judgment or reason, often comes into play, even in animals very low in the scale of nature (Charles Darwin 1859).

As we all know in the meantime, and not just the neurobiologists, the evolutionary solution to the problem of behavioral coordination consisted of a selected group of reliable and not too-short-lived cells taking over the supreme command of the undertaking, i.e., animal. The so-called nerve cells were entrusted, on one hand, with establishing a central control unit (e.g., cerebral ganglia, brain) in a location as safe as possible, that means away from the unprotected surface and/or enclosed by a robust protective structure (e.g., vertebral column, skull) and, on the other, to set up a network of not-easily broken connections to this exclusive center from the many important parts of the organism. In order to perform this function, these selected cells had to undergo corresponding changes in their appearance and their function. First they had to stretch themselves as far as possible and still be able to tolerate this physiologically and second, with the aid of a special apparatus, called a synapse, they had to ensure direct contact with the neighboring cells as well as at the end terminals of the conducting fibers to the corresponding target cells (e.g., muscle, gland, or special sensory cells). As a novel function, they took over the task of transmitting information on the cellular level of the organism, and this in such a way that they are now specialized in nothing but influencing what other cells, be they nerve- or any other body cells, should do. Transmission of information in this special sense has nothing to do with stochastic information but is nothing other than purposeful influencing of other cells, not in the comparatively slow and therefore primarily chemically mediated developmental processes in embryology, but rather as an often split-second synchronization of a body operating as a whole.

The consequences for the organisms involved in this development are remarkable insofar as here a kind of central control unit has come into existence. This, for the first time, is in the position of being able to "know" what is happening overall in the body from the innermost parts to the outermost periphery. This can occur in relation to exterior influences which have an effect on the body through specialized gateways, the so-called sense organs but also in relation to internal changes which the organism itself, using effectors such as muscles or glands, can undertake. The organizing center of such organisms must possess the ability of being more or less informed in detail about the current situation of the whole body as much as possible. It is this alone that gives us the right, when talking about organisms possessing a CNS, to speak of the existence of consciousness—an awareness of its situation. The etymology of the word "consciousness" comes from the Latin *com-* or *con-* (with) and *scire* (to know). This form of consciousness is so obvious to us as humans that it is rarely, if at all, considered a worthwhile subject for researchers, be they psychologists, behavioral researchers, or even philosophers since they are primarily used to dealing with only the highest levels of our existence. Whereby, the whole business is much less inconsequential than it appears at first sight if one examines it more closely. For this purpose, let us imagine a completely simple, everyday perception, e.g., the sense of pressure when we shake some one's hand. What could be so extraordinary about that?

Well, we perceive the pressure and feel the characteristic contact on our hand, however, in reality, as far as our neurons are concerned, we do not feel it on our hand but completely in our head. We do not sense this pressure only in our heads, but rather we sense the *location* of this pressure completely in our heads! In spite of this, we have the unmistakable feel and the *certain knowledge* that some one has grasped our hand and has not grasped our head and shaken it. I would like to recommend to consciously make a practical test to exercise this unique relationship between the periphery and the organizing center. Hold an object in your hand and concentrate on the perception of this object, ideally with both eyes shut. The impression of holding something in your hand is such a strong one, no so much better, so much more a real one that even with a great effort one does not succeed in simply *imagining* that the sensation does indeed take place in one's head and not outside at the periphery of the body. The neurobiologists know, in the meantime, in fact, have known for some time that all this conscious perception and sensation come into being first in the control center of the brain and not, as it seems, in the sense organs.

In this sense, we do not see with our eyes, as often erroneously claimed, we do not hear with our ears, we do not feel with our skin, or smell with our nose, or even taste with our tongue. Not even pain, the most unpleasant of all the different categories of stimuli, is paradoxically perceived most accurately at the location where it hurts so much. This is demonstrated by the reaction that we and all animals as well show in such situations. We do not shake or wildly jerk our heads as a rule—except when we have a real headache—in order to get away from the pain we *perceive in our heads*, but rather we react with an abundance of fantastically complex safety reactions to remove that part of our body in danger out of harm's way.

This somewhat emotionally charged excursion into the paradoxes of conscious perception, which are naturally adaptive in character—the opposite would indeed be an unimaginable paradox—may not seem so important since it has been described and elaborated sufficiently in the relevant physiology textbooks (Schmidt, Thews and Lang 2000) and especially in the books on neuro- and perception physiology (Schmidt 1998). Examined in detail, things are a bit more complicated since in reactions to pain or emergencies (e.g., blinking reflex, avoidance reactions, movement reflexes), whose function is to prevent or ameliorate any acute injury as swiftly as possible, the spinal chord and the brain stem coordinate some important bodily functions autonomously. For example, we often react so quickly to a strong pain stimulus (the classic case of touching hot metal) that we have reacted before we become aware of the pain. In all these cases, we could say that our spinal chord already knows what is going on at the periphery, i.e., the rest of the body and what reaction it has to initiate as quickly as possible. We have known since the rather repulsive experiments by Sherrington on domestic cats with their brains removed, i.e., clinically dead cats, that a whole series of complex pain reactions can occur without any participation of the cerebrum (Woodworth and Sherrington 1904). Any impartial observer would have thought that the animals were completely conscious and feeling great pain. In any case, they exhibited a full variety of typical cat-like reactions to being subjected to pain—pitiful meowing, wriggling, snarling, baring their teeth, increased blood pressure and heartbeat, etc. From this experiment and others like it, the categorical conclusion that "behavioral reactions to painful stimuli observed in humans and animals do not tell us anything about if or how intensively the pain is consciously experienced" (Schmidt and Struppler 1983) is, however, rather premature since we can readily assume that, at the level of the spinal chord or brain stem, conscious processes occur even if the dimensions and duration are probably limited. In any case, the perception in the spinal chord of events at the periphery is of a completely different nature to the local perception process itself. Expressed anthropomorphically one could say that the spinal chord, or brain stem, knows exactly what it has to do if, e.g., a strong pain stimulus reaches it from an extremity. That this spinal consciousness is naturally subordinate to the cerebral consciousness and without any direct possibility of connection to our speech centers is a completely different thing. As far as its function is concerned the best comparison is the ladder-like nerve system of the earthworm. When such a creature has been skewered by the beak of a blackbird and wriggles back and forth, we can suppose that this poor worm feels at least a grain of despair without, of course, any intellectual frills about the significance of life but certainly a little local consciousness of what is happening to it. The often cited argument that conscious feelings in humans, in contrast to animals, can only be proved through spoken communication (MacPhail 1998) also misses the mark completely since the speech for the observer is nothing other than external behavior whose interpretation is totally dependent on him.

Let us have a closer look at the distribution of this kind of consciousness. Most animals—revealingly, the exceptions are some plant-like animals such as sponges and corals—are the chief candidates for such a primitive form of consciousness that is alone dependent on the existence of a CNS. The theoretical minimal model of such a CNS would consist of at least a sensory and a motor nerve branch, both of which would be connected to each other via a central switching location, let us call it the minimal brain. Consciousness would then come into existence through the realization of a simple response code (amplification/weakening) whereby, what is unusual is to be searched in the special kind of centralized network. The behavior of such a creature could even be investigated by the behaviorists since the sense channel, the perception of a stimulus and the movement apparatus through the (hopefully) intelligent regulation by the central control unit could bring about a movement. The behaviorist model of human behavior does indeed still looks like this. This kind of basic consciousness can reach various levels of complexity in different species so that it is a rough simplification to talk about a unified basic consciousness in all animals possessing a CNS. The minds of earthworms and other primitive groups of animals must be based on a much more modest range of perception of the world than the complex crossmodal or intersensory perception processing of vertebrates where the neuronal integration of several previously separate sensory channels facilitates a more coherent view of things (Nicolelis et al. 1995; Schlechter 1996; Sjoelander 1995; for the fascinating human ability of lipreading, a crossmodal process at a quite high cognitive level, see Vroomen and de Gelder 2000; Massaro 1999; Massaro et al. 1998; Campbell, De Gelder and De Haan 1996). While, in earthworms, only isolated, very local sensory impressions (e.g., light/dark, damp/dry, painless/painful) are transmitted to the central pharyngeal ganglia, in the brains of higher vertebrates, real objects are created which simultaneously can be defined olfactorally, tactilely, visually, acoustically, and even gustatorily. One of the first overviews of this fascinating topic is available in a book by Barry Stein and Alex Meredith with the telling title "The Merging of the Senses" (1993).

Who are then the big losers in the evolution of consciousness and, above all, for what reasons exactly must they be the ones? All plants and fungi must be counted as examples because they lack any internal structure that could in any way control a coordination of behavior for the whole body. Plants are the real masters of a decentralized and therefore perfectly democratic life form in which decisions are made simultaneously in many, more or less independent locations. The disadvantage of such a lifestyle is obvious: it rules out any joint action of the whole organism. Therefore, we should not wonder that we seldom see trees running round the countryside whereas a deer living in the forest does so going about its business with pleasure. However, the life as an unconscious plant does also have an invaluable advantage, however. It is comparatively difficult to put an end to the life of a plant but can often be done to animals with a well-aimed blow to the center of consciousness, i.e., the head. One can bite off pieces of a plant, gnaw its roots, eat the leaves one after the other from the branches, nibble the flowers, or steal the nectar. It would, in comparison to an animal allows this to happen with stoical composure. With the exception of certain sensitive developmental stages, mostly at the compact dissemination stages e.g., seeds or pollen, it is rather difficult to kill a plant with one stroke.

The great difference that in an absolutely "purposeful" way consciousness brings either with it or not, is to be found nothing more than in the ecological conditions, which as a special selective pressure made some of the multicellular organisms into plants and others into animals. Animals as heterotrophs need ready prepared food and are obliged to move around as compact bodies whereas plants can concentrate completely on a relatively continuous development of a widely spread out productive body of biomass where it is of no importance for the toes, using a brain, to know what the ends of the hair are doing at the moment. Consciousness—and viewed evolutionarily, there is nothing more to it—is simply the adaptive consequence of an existence as a professional vagabond who has to wander endlessly from place to searching for uncertain resources. Plotkin stated the facts of the case briefly as follows:

This requirement for movement, and the associated consequences of that movement, have been fundamental selection forces in the evolution of some kind of organ for the co-ordination of movement (the brain and nervous system), for the evolution of organs that produce the movements and for the evolution of sense organs that allow for better-directed movement. As we shall see, it has also been the fundamental selection force in the evolution of learning and intelligence (Henry Plotkin 1994, p. 103).

For this reason alone, we find, in the smallest, single-minded worm to the largest vertebrate racing through the world, that unique inconceivable phenomenon which to the present day causes so many conscious headaches. It is then clear, owing to the existence of a CNS, we can rightly conclude, as Griffin does, that there is consciousness even if it is rather a modest one. In order to justify, as far as evolution is concerned, the biological necessity of such a central control unit, we need to know the general selective conditions of life as a multicellular animal.

Hence while one can describe the previously discussed form of consciousness as *primary central representation*, in that the current situation at the periphery, e.g., sensory impressions of the external and internal world, is relayed in a more or less detailed manner to a variously complex neuronal central control unit and then back to the periphery to take action, we now switch over to dealing with a *secondary* kind of consciousness in a cognitive-organizational extension with completely new dimensions. We are talking about human consciousness in a narrow sense—here let us clarify our definition right from the start—which can be characterized by the spatiotemporal dependence of the primary consciousness of continuous current perceptions being exceeded as far as it enters an extended area of *purely fictive* experience and behavior, in other words: in imagination and thinking to achieve completely novel possibilities of adaptive behavior. This going beyond direct perception is for us the really interesting form of consciousness, since the primary central representation which precedes it and exists in practically all mobile animals, is naturally no longer very extraordinary (that "direct" visual perception also has its limits and biases is nicely demonstrated by the fact that the light coming from the sun needs time, namely about 8 minutes to reach the earth which makes us see the position of the sun always shifted by an angle of exactly 2 degrees or, roughly, 4 sun diameters; consequently, in the morning we cannot see the sun even though it has already moved above the geometrical horizon whereas, in the evening when we are convinced to see an impressive sunset, the sun has already disappeared below the geometrical horizon; Ganzberger 1996). Certainly, concerning the extent and complexity there are also enormous differences in the primary representation of the individual classes of animals but the qualitatively especially interesting differences all begin at the next higher level. It is therefore no coincidence that Griffin himself based his own definition of consciousness not on the primary but on this further developed secondary level of an internal neuronal representation "of something", be it something external (i.e., the outer world) or internal (i.e., other parts of one's own inner world). Under the rubric "A Materialistic Approach to Mental Experience" he wrote:

> For our discussion, the significant aspect of consciousness is the ability to think about objects and events, whether they belong to the present situation or not. ... The content of conscious thought may consist of current perceptions, of memories of incidents in the past, or an anticipation of the future (retranslated from Donald Griffin 1991, p. 24).

In spite of this clear definition, serious differences appear over the assessment of some species and this is impressively shown by the controversial example of the so-called "language of bees", of which, Griffin thought here we have a "communication system", "that is truly symbolic and corresponds to the principle of 'transposition', i.e., of transferred meaning" (retranslated from Griffin 1991, p. 225). Without any doubt, bees possess a central nervous representation of their behavior in the primary form like all animals having a CNS. From now on, however, we want to speak only about the existence of a real consciousness if the criterion is fulfilled that the behavior of an animal, if not strictly provable, at least empirically *verifiable*, goes far beyond that what is dependent on direct perception (Heschl 1992b; Schleidt 1992). In this respect Francis Crick also hit the nail on the head: "It is hardly conceivable that somebody has consciousness without being able to remember what has just happened even if it is only for a short time" (Crick and Koch 1990). That is exactly the point at which the consciousness *of something* emerges. In this regard, we must eliminate the well-loved bees from the competition because there is no longer any reason for answering the question posed by Gould as long ago as 1979: "Do honeybees know what they are doing?" with yes. Gould had been able to demonstrate convincingly that there was nothing wrong with the classical experiments by von Frisch as far as the method was concerned (Gould 1975)—in fact, for that time they were by far the most ingenious experiments in the whole of ethology—however, he had by no means proved the existence of bees acting consciously. In principle, Frisch himself was right with his own theory of the language of bees but to understand this one has to take the trouble to read more carefully what he had really understood about it. Perhaps the complicated puzzle's simple solution is best explained in his standard work from 1965 in the chapter "Modifications of the Language of Bees" and above all in "Phylogeny and Symbolism of the 'Language' of Bees". One can already see from the title of this chapter that Frisch *always* set the controversial term "language" within quotation marks. The essence of his thinking on the origins of the dancing of the bees was summarized with the following words:

The phylogenetic step from directly showing the way to transposition was made easier by the fact that this ability is primarily present in many insects without having any biological significance for them. If they, during their movement over a horizontal surface, keep to a certain angle to a light source (light compass movement) and one sets the surface so that it is vertical and simultaneously makes it dark, they move so as maintain almost the same angle to gravity as they were previously doing to the light. Some primary positional centers steer the animal at a specific angle to the source of the stimulus, whether it is light or gravity. The biologically significant linking of positive phototaxis and negative geotaxis can have contributed to the unmistakable symbolization in the bee dance (Karl v. Frisch 1965, p. 330).

Only one more thing need be added to finally make the nonexistence of true consciousness clear. Frisch's "unmistakable symbolization" is namely an unmistakable phylogenetic, i.e., in the course of the bee's evolution, symbolization origin brought about step by step, that means mutation by mutation, correctly formulated ethologically as a phylogenetic ritualization (Tinbergen 1939; for the "evolutionary origins of bee dances", see Dornhaus and Chittka 1999), which, as a rule, consists of a rhythmic accentuation of intention movements from the most varied functional fields (to give some examples: walk away ⇨ lift leg rhythmically, scrape the ground; fly away ⇨ clap the wings, lift the head; attack ⇨ duck down; evade ⇨ turn away; mate ⇨ rhythmic flexing of the pelvis muscles; eat ⇨ chewing movements; social grooming ⇨ lousing; touching ⇨ tickling; biting out of fear ⇨ gnashing of teeth; surrender ⇨ throw oneself on the floor). Such things must be strictly differentiated from the conscious symbolization in the course of the cognitive ontogeny of a metazoan. That Frisch was only using a metaphor when he talked about the "sign language" of his beloved bees emerges from the last citation that we will look at for this reason:

The waggle-dance shows, in relation to the sun, the direction of the goal and can be considered as an intention movement which stimulates its comrades into flying off in this direction; the waggle duration indicates (note: phylogenetically) symbolically the distance to the
goal (Karl v. Frisch 1965, p. 330).

With this characterization, the instinctlike "language of bees" (for the importance of a kind of real flight simulation through rhythmic air movements produced by the dancer's vibrating wings, see Kirchner and Towne 1994) takes its
place among the rest of the already known "animal languages", from the "language of fishes" as first described by Tinbergen (Pelkwijk and Tinbergen 1937),
via the discovery of the "language of birds" by Lorenz (1935, 1982), to the phylogenetic "body language" of people as systematically investigated by Eibl-
Eibesfeldt (1973) for the first time (the true, i.e., both artistic and scientific ex-
pert in all questions of body language, however, is Samy Molcho). All these
"languages" have in common the lack of real symbolization, since their meaning always expresses itself directly in the phylogenetic "symbolization" or rather as "ritualization". So when people really cry for help, there is no conscious
symbolic indication of any impending danger, it is simply an automatic absolutely correct and unconscious reaction in our bodies. If the word "help" is added to the cry, we find ourselves involved with an interesting mixture of a purely instinctive reaction with real symbolically communicated meaning
typical of many naturally arising human sign languages (Eibl-Eibesfeldt and
Sütterlin 1992; Morris et al. 1994).
    The discussion of the supposed language of bees shows, on the other hand,
just how difficult it really is to formulate a clear definition of consciousness for
concrete empirical research, i.e., in scientific practice. What is the situation
now as far as empirical evidence of consciousness in animals if we apply the
criterion of going beyond the narrowly bound area of direct perception? Not
only the whole of cognitive ethology, where there have never been doubts about the existence of animal consciousness, is full of it, but in the meantime,
what is all the more astonishing, also the whole of modern research into learning which was originally based on purely behaviorist principles. That the
word "consciousness" is used rather reservedly for ideological reasons no longer plays an important role since hardly any scientists confuses the associated
scientific-empirical problems with the outdated body/soul debate of philosophers. What is more decisive is that a kind of consensus has been reached
among most behavioral scientists which proceeds from the conviction that
"consciousness", "cognition", "thinking", or whatever one wants to call this ability can be most precisely characterized by this going beyond the limitations
of the momentary (visual) perception, whereby behavioral phenomena such
as "memory", "remembering", or, more generally, the "power of imagination"
in space and time become possible (Bekoff and Jamieson 1996; Gallistel 1990;

Terrace 1984; for the neuropsychological transition from seeing to imagining, see Farah 1995; Goldenberg, Podreka and Steiner 1990; for the exceptional capability of olfactory recalling, see Sacks 1985, chapter 18).

It is therefore no wonder that more and more hard evidence has recently come to light for behavior in animals that does go beyond direct perception, be it in memory, orienteering, or any other problem-solving experiments. From these experiments, the list of species that today with some certainty are candidates for a certificate in consciousness or thinking has already become so long that it is here sufficient to recommend the currently best overview of this whole field, namely the book by Sara Shettleworth, "Cognition, Evolution, and Behavior" from 1998. There, one can find the step by step attempt to make a methodical separation of purely associative- from increasingly active interpretive behavior for the most varied functional areas and species including man.

Perhaps we should take an actual example to show the general trend in cognitive research. Until recently only the primates were thought to be able to recognize themselves in a mirror and then essentially only chimpanzees (Gallup 1970). In the meantime, the orangutan and the bonobo—but only "partly" the gorilla (see Povinelli's "anomalous gorilla", 1987, p. 496; however, at least some gorillas seem to be capable of self-recognition if confronted with a video presentation of themselves; cf. Law and Lock 1994)—belong to this elite class since they also allow themselves, more or less, with the aid of a mirror to remove an ugly smear of color that has secretly been put on their foreheads. What gives an even greater impression of intelligence is when they give themselves an extensive beauty treatment in front of the "mirror, mirror on the wall ..." (Hemme 1988) in order to look as presentable as possible (Parker and Mitchell 1994; Povinelli and Cant 1995; Tomasello and Call 1997). However, what is really much more remarkable is that increasingly the so-called lower primates, even if rather seldom if at all (compare Hauser et al. 1995 with Anderson and Gallup 1997) with actions relating to themselves are on the way, at least with the important first stages (instrumental use of mirror, spatial orientation with the help of a mirror) along the obviously difficult evolutionary route to at least prestages real self-recognition. For example, it was recently possible to demonstrate, in the squirrel-like common marmosets (*Callithrix jacchus*), the ability to use a mirror as an instrument for tracking down hidden food (Burkart and Heschl 2000). This finding fits nicely into other recently made discoveries of some remarkable cognitive capabilities in lower monkeys such as true imitation in the common marmoset (Voelkl and Huber 2000) and even flexible, i.e., partly insightful tool use in the cotton-top tamarin (Hauser 1997; Hauser, Kralik and Botto-Mahan 1999). Such a development appears to be the perspective for the future, namely, a softening of the previously all-too rigorous setting of assumedly "definitive" cognitive limits in favor of a stronger evolutionary, i.e., more comparative and hence more realistic phylogenetic system of behavioral graduations. In addition, comes the increasing integration from previously separate areas of research which brings about an innovative impetus to the investigation of cognitive structures. A good example is classical developmental psychology, which in the meantime has become much more strongly interested in the development of animals. In a review article which examined the relevance of Piaget's development-psychological investigations to zoological behavioral research and attempted the first synthesis of ontogenetic human intelligence research and modern cognitive ethology, the new conceptual approaches are increasingly changing the scientific status of comparative research:

In conclusion, object permanence is doubtlessly one of the prerequisites for anticipatory or insightful behavior, and is therefore attained by all "intelligent" subjects. The above-mentioned species which display at least the sensorimotor object concept are characterized by ontogenetic processes leading to major changes in their cognitive strategies. These animals do not merely take up isolated items of information. They must organize information about the environment (and themselves) through processes of abstraction and of generalization of relevant relationships, the extrapolation of which in turn influences the modalities of their cognitive functioning (Ariane Etienne 1984, p. 318/319).

The listing of all the families of animals, which is here certainly not complete (see Parrots: Pepperberg, Willner and Gravitz 1997; Pigeons, starlings: Plowright, Reid and Killian 1998), shows an interesting connection. It is apparent that it is above all animals which live by hunting or by exacting searching for food (e.g., certain fruits) which are at the front of the list of cognitive rankings, whereas herbivores whose food is there in great quantities are not so involved in the natural selection to increased intelligence (perhaps this is the reason for there being a "silly cow" or "silly goose" but no "silly cat" or "silly hawk"). Probably a similar difference has already to be assumed for the different categories of dinosaurs, i.e., a confrontation of rather sluggish herbivores with incredibly quick and—perhaps—even to some degree intelligent carnivores. This difference becomes immediately understandable if one imagines which special demands a shy prey can make on a hunter even though, on the other hand, fleeing and hiding from a clever predator also makes certain qualities necessary. A good example of how large the differences in this respect can be, even within the same animal family, is the group of whales of which it is known that the plankton-filtering baleen whales (e.g., blue whale, humpback whale, fin whale) are nowhere near so good in demanding learning and cognition tests as the toothed whales (e.g., sperm whale, killer whale, dolphin) adapted to hunting large prey (it is know known that Orcas sometimes gather in organized bands to hunt the great blue whale). That sea lions (Schusterman and Krieger 1984) and dolphins (Herman, Richards and Wolz 1984), for example, as cunning fish hunters, are indeed capable, not of speaking, but of understanding simple three-word orders (e.g., name of the person spoken to, verb, object) from a person has been proved experimentally many times. This is similar to Bickerton's (1990) proto-language—here a typical example produced by children: "Daddy push car" (Horne and Lowe 1996). Furthermore, it has been shown that they are capable of producing symbolic categories beyond various sensory modalities (Schusterman 1990). Additionally, dolphins are known for their extensive and extremely creative way of playing (Marten et al. 1996). We will see presently that this is what is rather special about it.

At this point, we want to introduce an additional criterion for the existence of consciousness and it is one which optimally supplements the criterion that has been the main one so far, i.e., the provable independence from any kind of direct perception. It concerns the kind of origin of consciousness during the development of the individual. Since conscious behavior cannot be present right at the beginning in its full dimensions, as we can see in a newborn baby, the question arises as to whether there is a general mechanism that in the course of ontogeny, i.e., our individual lives, has been developed. Namely, this question, which, in Piaget's stage model of intellectual development gains a still more central importance, cannot be answered so easily with a general yes for all the animals that come into question but, nevertheless, there is at least a practical and therefore relatively good describable behavioral characteristic that seems absolutely specific for the acquisition of conscious problem-solving strategies. Paradoxically, it is a behavior that, even today, is still described as having no function at all or whose only relevant function may consist of a kind of unexplainable purpose of its own. The behavior in question is play, whose biological investigation has only just managed to eke out a shadowy existence. Because play behavior is the exact opposite of serious and therefore apparently important survival behavior, it has also never been of great interest for most mainstream Darwinists. It is therefore not surprising that, except for a few specialists, the scientific exploration of such an unbelievably wide-spread phenomenon has been so neglected. What can be the fundamental biological function of play in animals? Robert Fagen, whose study on "Animal Play behavior" (1981) is one of the few exceptions to the generally wide-spread rather low estimation of play by biologists, summarizes the most important possibilities in a few words:

Play has measurable costs and apparent benefits. Structure and context of play suggest that it functions to develop physical ability, including strength, skill, and endurance. Play may also be very important for cognitive skill development. Hypotheses about cognitive functions will definitely be worth pursuing when the biology of cognition is better understood. Because play functions in development, its adaptive effects are delayed in time. This time delay may be on the order of days or weeks, as in physical training; or, as in cognitive skill development, adaptive effects may be delayed until adulthood (Robert Fagen 1981, p. 355).

A special play theory of consciousness (cf. Heschl 1992b) states that play, possibly, could be not just an important ingredient of general cognitive development but much more than that. It must be the indispensable precondition for the ontogenetical development of any intelligent behavior strategies. The ecological background for this belief can be quickly explained. In order to obtain enough material for its consciousness memory, the developing subject, whether animal or human, is forced to concern itself intensely with as many perspectives of the environment (e.g., optically different) as possible to be able to develop coherent internal representations of it. In normal serious situations of everyday survival pressure, the current situation always blocks this toilsome accumulation of potentially important experiences because any negligence in a serious situation could mean an early death. Just through play, that characteristically appears only in such situations where obviously no direct need for action occurs, i.e., in Bally's famous "relaxed field" (Bally 1945) of so-called leisure time, do those prerequisites arise which make possible the active development of conscious strategies.

This strictly causative connection between play and consciousness can be explained best with a simple example. Let us assume that a young animal or a child is confronted with a new object for the first time. If the situation is a serious or even a possibly fatal one, the young animal will act appropriately and take every momentary change in the perception of the object seriously and therefore set aside any complicated consideration of the situation since this would simply be too risky. The corresponding behavioral attitude can be characterized as neophobia which, as a rule, is more frequently shown by either rather unintelligent or adult animals (there exists, however, a kind of "secondary neophobia" in particularly inquisitive animals such as crows, parrots, and monkeys, but it consists of inspecting a new or altered situation very carefully before taking any action and as such is different from real neophobia). In contrast, the seemingly "groundless" playing of young and, at least for a certain time, extremely neophilic animals fulfills exactly this function for which, as a rule, there is not enough time in the brutal seriousness of the Darwinian situation of everyday life. It facilitates the purposeful elaboration of conscious strategies which are often used much later in serious situations to the individual's advantage (an instructive example of the important cognitive function of play behavior in squirrel monkeys is described in Biben 1998).

The functional connection between play behavior and the capability of conscious problem-solving strategies becomes clearer if one only looks more closely at how far playful exploring can be an advantage for the organism. In order that the experience gained during play can be applied later in the life of an animal, it is very important that the individual has the ability to recall this experience more or less at any time when it is needed and therefore consciously use it in a new situation. For that purpose it does not always need a complex consideration of the current situation and all the possibly relevant experiences from earlier, it is often sufficient if, at the right time, the individual consciously conceives a practical solution. Conceptions of this kind can be observed in very many animals and we do not always need costly laboratory apparatus to be able to prove this empirically. I can assume that our cat who was inadvertently locked out and then meowed pitifully for five minutes before running round the block of houses for 50 meters with quite a few obstacles to overcome before reaching the patio door which is often left open and then pouncing on the wonderfully smelling breakfast did not immediately think of making this long detour—she probably hoped that we would open the door again—but must have occurred to her very quickly. Such an idea requires that the cat can consciously envisage the solution to this small problem and then carry it out. If she were to try to solve the problem by behavioristic trial and error we would have had no cat for some time since after the thousandth fruitless attempt she would have died of hunger.

For play behavior to occur at all and be furthered by natural selection, we need first a protected area within which it does not really matter if mistakes are made. It is therefore not surprising that real play is only found in species which have at least a minimum of social life, where at least one parent looks after the safety of the developing offspring for a time. It is not surprising and not from chance that *Homo sapiens* is an extraordinary species where the young are protected, looked after, and specifically brought up to be adult members of the human race for many years. In species where the young animals are left to their own devices from the very beginning of their lives, e.g., most fish and amphibians, it would be too great a risk to subject them to the danger of playing and for this reason, there has been no evolution of play and therefore no development of consciousness.

Play behavior as a direct expression of the development of consciousness, has, from an evolutionary perspective a very serious purpose. That the connection between play and the development of cognitive capabilities cannot be a trivial one is implied further by observations of its exact taxonomic distribution. As a rule, the more a species displays play behavior, the more intelligent it is. The best proof of this relationship is our own species in which play behavior during the development phases is awarded such a very high status (Sylva 1977; Sylva, Bruner and Genova 1976) that one could, as already once suggested, rename us as *Homo ludens*, the playing man (Huizinga 1956). However, not only conscious cognition is concerned: even the emergence of vocal language seems to be connected to an extensive phase of playful pre-speech babbling which is done by the young infant just for pure fun, often in reaction to an adult's incitation in "Motherese", but very often also in a completely spontaneous manner. However, its function is important because this phase leads over to the formation of the first so-called "protophones" which, ultimately, allow the child to produce simple syllables and its first true words (Oller 2000; for the recent discovery of a very similar kind of "babbling" in young monkeys, see Elowson, Snowdon and Lazaro-Perea 1998; for vocal play in gibbons, see Carpenter 1964, p. 254). Interestingly, young infants, if not challenged acoustically because they are being reared by deaf parents, are even able to babble with the help of a primitive sign language (Petitto et al. 2001; for older children who are able to develop their own sign language, see Helmuth 2001), an observation which again shows the unusually close relationships between manual pointing (or signing), producing distinct sound patterns ("words") and naming (and categorize) objects (cf. chapter 12). And again, play seems to be the central mediator towards higher, i.e., conscious processes of cognition.

In this connection, it is noticeable that not one great thinker, inventor, or any person famous for their wealth of ideas grew up under dangerous and therefore play-inhibiting conditions. In fact, just the opposite, most of the great geniuses in history did not have developmental conditions in their early childhood that were just not bad but rather noticeably good and provided above-average opportunities for development. Even the most famous of all the poor proletarians, Karl Marx, who, as one of the first modern social scientists, correctly recognized that—exactly like animals—the economical, or more generally the ecological conditions determines the social life of humans, came from a wealthy, middle-class family and at least during the important developmental period of his early childhood and adolescence did not have to put up with any serious restrictions:

His father was a successful lawyer, who allowed the young Marx to study what was in fashion: law. Marx went to Bonn for his law studies. He spent most of his energy (according to information from his teachers) on wine, women, and song which ended up with him fighting a duel over a lady and receiving a scar over his eye! One can hardly claim that he learned strenuously ... (retranslated from Rius 1983, p. 15).

Every biography will only be really informative in all its thousand details. Nevertheless, it is now understandable why even the allegedly most serious of all human occupations are basically nothing more than grownup games:

..., finally, I would like to express the hope that I have succeeded in making clear the idea that the productive scientist must be a traditionalist who enjoys playing a complicated game according to the rules in order to be a successful innovator who discovers new rules and new figures with which he can play (retranslated from Thomas Kuhn 1959/92, p. 322).

The appearance of play behavior by the young animals of a species, including our own one (Stone, Smith and Murphy 1973), is one of the most reliable indicators of the existence of real consciousness, since we can be convinced that play is the absolute precondition for its development. It is therefore not surprising that the list of the species that up to now have proved to be playful (see Fagen

1981) corresponds almost exactly with the list of animal species of which one assumes or, in the meantime, knows exactly that they are capable of intelligent behavior of which the capability of conscious representation is an integral part. An
excellent example of this is a recent study on "Ring Bubbles of Dolphins" (Marten et al. 1996) which corroborates the general research results into intelligence
in dolphins up to that time. What is interesting in this respect are the borderline
cases which make the developmental biological connection between play and
consciousness much clearer. There is, on the one side the large barrier from the
generally playful mammals and birds to the reptiles where, if at all, only the modern crocodiles and monitor lizards, and recently also the primitive tortoises
(Burghardt 1998), come into question as candidates for conscious intelligence.
In any case, large crocodiles exhibit the initial stages of parental care of the
young as the evolutionary prerequisite for the development of youthful play behavior which has already been mentioned. The American alligator appears to
have made use of this possibility since there exists at least one substantiated
proof of play in published form (Lazell and Spitzer 1977). There are also some
scanty indications in monitor lizards but this still needs to be verified (Hill 1946;
Fagen 1976). I personally once had the opportunity during my university studies
as a novice comparative ethologist to observe the behavior of young savannah
monitor lizards (*Varanus exanthematicus*). On their own, the results of this initially promising undertaking were less than encouraging. Not just that, for an
almost permanently active warm-blooded animal, it is rather a tall order to observe the behavior of reptiles which appears to consist of remaining absolutely
still for 99% of the time (sunbathing, resting, digesting) which was the little I
could in fact observe and not exactly a paragon of all too great creativity. Not
even a single time did I notice play as such but instead some distinct cognitive
limitations of monitor behavior became more than evident. If there were white
mice in the terrarium, the lizards were only able to follow them as long as they
were directly visible. When a mouse disappeared behind an obstacle, the chasing
lizard abruptly froze as if struck by lightning only to start moving again, if at all,
when any other perception event occurred. When a few cunning victims succeeded in surviving the first impetuous attacks by the monitors, they were even
able to turn the tables somewhat and so the hunted became the hunters. The fo-
rewarned mice quickly learned to remain well-hidden during the day and then
during the night to creep out and nibble at the tails of their armored pursuers
(there was little else suitable for the mice to eat). These geniuses at survival were
obviously, to a certain degree, conscious of the danger that threatened them during the day while the wicked predators had no idea that the white mice they had
seen were not simply imagined or not and that they were still there even though
they were not visible at that moment. It cannot be excluded that the duel between
the up-and-coming mammals and the saurians which had degenerated to mon-
strosities in the Jurassic age was not decided in exactly this unspectacular way.

In the majority of the nonplaying reptiles it is more than doubtful that they have real consciousness and the same question in regard to fish is therefore purely academic. Nevertheless, in this class of animals, two proved examples of play-like behavior has been found and this interestingly in that group of fishes which have the largest brain by far than any other. The electric fishes *Gnathonemus petersii* and *Mormyrus kannume* have been caught playing with small balls, under water naturally, supposedly in the same way as young mammals usually do (Meder 1958; Meyer-Holzapfel 1960). Both species have an average brain/body weight ratio of 1:50 in contrast to other fishes such as pike which only have a ratio of 1:1300—this even exceeds that of humans who have not more than 1:55. A great part of this comparatively gigantic brain mass, however, is there for the processing of motor and sensory signals of the electrical sense organ typical for this group of fishes. How much brain mass remains for the development of consciousness going beyond mere perception has still to be clarified.

That play, in addition to having a simple physical training function for various forms of behavior (*physical training hypothesis*: Brownlee 1954; Fagen 1976; Symons 1978), has to be seen in a close causal relationship with the cognitive development of animals and humans (*skill development hypothesis*: Bruner, Jolly and Sylva 1976) has been shown by recent investigations into some interesting molecular-biological effects at the neuronal level. It was established that, through play behavior, the post-synaptic expression of certain gene products of the genes c-*fos* and c-*jun* (Dragunow and Faull 1989) which have been shown to be involved in learning and memory processes (Kaczmarek 1993; Morgan and Curran 1991), not only strongly activate that region in the brain suspected of being a kind of subcortical "play center" (PFA, *parafascicular area of the thalamus*; Mountcastle 1984) but also simultaneously extensive parts in all the investigated cortical regions (parietal, temporal, occipital) and the whole hippocampus and, in fact, up to between 70 and 90% (Siviy 1998). Since a general increase in stimulation in large areas of the brain occurs, it is to be expected that there will be a positive effect on all the activities taking place at the time concerned which should be linked with a considerable increase in learning capacity. That this increase in performance is very probably connected with an increase in creativity is shown by a simple behavioral experiment that was carried out with children almost 30 years ago. A group of preschool children before taking a test were allowed to play with a number of objects in whatever way they wanted. Another group, in contrast, had a certain kind of manipulation prescribed for them or were allowed carry out a completely different activity. In the following test, the children who had been allowed to play freely were able to list far more novel possibilities for using the objects on offer than all the other children (Dansky and Silverman 1973). The development of creativity, which is nothing other than the conscious processing of flexible and hence generalizable problem-solving strategies, is therefore indeed subject to the play activities of the subject, a relationship which among other things is expressed by the fact that both young animals and children who play more being more resistant to social stress (Hol et al. 1994) since at the end of their development they are more flexible and confident with other members of their species (cf. Biben 1998). Many of our human imprinted domestic cats demonstrate this to us again and again: as playful kittens they grew up together with humans and, possibly, dogs, and today they are almost frightened to death by other "real" cats but are completely happy and relaxed even with the biggest human masters and dogs.

As we can see, the evidence for consciousness, in whatever terminology it is presented has increased to such a degree that a biologist must smile to himself when he is confronted with claims that the question about the consciousness of animals has not been solved at all or even never will be. To quote Wolfgang Schleidt, one of the first collaborators of Konrad Lorenz, "The question I posed at the beginning: *May, can*, or *must* we accept consciousness in non-humans as a scientific concept? I herewith answer three times with YES" (Schleidt 1992). One need say no more. How simple it often is to conclude that an animal has the ability to mentally represent a variety of things is once again shown by the case of the domestic cat (*Felis silvestris domestica*), which, as every cat owner will experience at some time, can rapidly learn to open doors without any difficulty by jumping up to pull down the handle. Most cats demonstrate this extraordinarily intelligent behavior quite spontaneously one day without any previous indication that they could. They often do so when they have inadvertently been locked in a room. If somebody is there, whether in the room they want to get out of or the room they want to get into, they know exactly how to get the message across successfully by scratching quite insistently and sometimes unbearably at the appropriate door. When they then jump for the door handle and on about the third attempt achieve success, i.e., the door does indeed open, they have never previously had the opportunity to train themselves by repeated attempts to do the correct thing. In most cases, they also have not had the opportunity of observing a person opening the door shortly before and then intuitively copying him. They just sit there, noticing that they are shut out and then they *must*, if they want to get out of the awkward situation, have thought about what they could do in the situation. Hence they must have observed and at least partially *understood* the causal relationship between the handle and the door's opening, which is no trivial thing. That does not mean that such a cat knows how a handle is connected with opening a door in any detail (how could she?) but she does know that moving it in a certain direction leads to right conclusion. In order to achieve this, the handle has to be moved down and not up which would be much easier for the cat by jumping up underneath it.

Right at the end of this chapter, we have to pose that critical question once more which determines the methodical basis of every evolutionary consideration of cognitive abilities: can it be, in conscious ways of behaving, that we are dealing with a possibility of an individual to achieve real, i.e., really new knowledge during his ontogeny? For Konrad Lorenz, for example, it was clear that all conscious processes were simultaneously connected to an increase in cognition since for him they were always concerned with things which "become conscious for the first time" and hence could not be "instinctively" preprogrammed. Or, as alluded to on the cover of a book by Griffin (1991): "Is the behavior of animals determined by instinct or are there (conscious) thinking processes which, even if they are simple, can control the activities of animals?" (retranslation) The Lamarckian Piaget would agree with this distinction between "instinct" and "insight" without hesitation and would even go so far as to claim that the potential cognitive gain by an individual by conscious insight can be infinite if were not for "unsuitable" events such as death of the person. An evolutionary consideration of this question would naturally use completely different yardsticks and principally examine whether consciousness was sufficient to meet the fundamental criterion of undirected, i.e., neither predictable nor understandable retrospectively as being simply random. As it was foreseeable, however, this is exactly not the case since cognitive mechanisms are characterized by proceeding according to well-defined rules. Hence, every larger internal physical chance event would be an inestimable intervening factor which eventually would cause the functionality of complete algorithms to break down. The whole development of intelligence follows a genetically predetermined plan in which all the necessary details of a piece of information are already laid down and it regulates in which way, during the course of a long difficult life, each problem is to be tackled. Empirical research has at least proved this fact in its essential aspects. In addition to J. S. Bruner (1964, 1968), it was, above all, the Scottish developmental psychologist T. G. R. Bower who, with numerous impressive, since simple experiments was able to prove that Piaget's concepts with regard to the development of intelligence need a comprehensive revision (Bower 1971, 1977, 1979). Bower's investigations showed beautifully that really *new* cognitive structures are not developed from the interaction with the environment, but rather, in direct opposition to Piaget, developing kinds of behavior react selectively to the environment according to a prepared plan. The manifold interaction between the initially still isolated individual reactions (in infants, e.g., grasping, hitting, touching, hearing, seeing, etc.) whose increasing coordination and integration into an intermodal conscious whole, the something special that does more than just determine human intelligence (Stein and Meredith 1993), must already be laid down ready to mature. If this were not so, the independent solving of certain problems such as finding lost objects, using purpose/means relationships, understanding symbols, introducing abstractions, and so on (see Piaget 1997) simply could not function in the way it actually does or would not be at all possible.

An elegant proof of this was produced not so long ago in an experimental work which investigated the type and the time-point of the transition from still considerably direct perception to the first real representations of an absent object (DeLoache 1987). Children from 2.5 to 3 years old were given the task of finding a real object hidden in a room by looking at a size-reduced model of the room where the exact place of the hidden object could be seen. The younger children failed to solve this apparently simple problem whereas those only a few months older had no difficulty whatsoever in looking for the hidden object in the right place. For the latter, their complete picture of the world had changed radically—from a primarily perception-centered experience to become a conscious conceptual and simultaneously symbolic understanding. This could not and *also would not* occur at an age between two and three years old if it were not intended in our constitutional cognitive equipment. Consciousness, contrary to general belief, is not a sign of cognitive progress of the individual but nothing other than the already—genetic—conscious solving of problems which, with the limited means of direct perception are no longer soluble. Consequently, we are primarily dealing here with simple orienteering problems set by natural selection and these primarily have to do with purely physical factors such as spatial obstruction (objects hidden behind each other) or optical opacity of most objects (exceptions: air, water, glass, empty space) in our complex ecological environment. The answer by some of the more mobile metazoan organisms called animals was, in contrast to direct and thus "real" perception processing, the formation of a virtual representation center in the brain, which, for the first time, made it possible to wander through the world in a purely mental manner to and from distant places, from the past via the present into the future and back again. It is amazing, however, to see that trees manage very well without such a serene capacity and nevertheless are able to develop a really complex branching of their external structure whose adaptive organization is by no means inferior to that of a strangely crenulated vertebrate brain. Even though the individual leaves of a tree plainly do not even want to know what is currently happening to all the many other leaves—why should they? to run away together?—one thing is clear: even the smallest phenotypical change which brings about a tiny expense of any kind has somehow its own, not coincidental adaptive purpose. Even if it is like life in general, only as a Darwinian end in itself.

15 The True Nature of Scientific Discoveries

For a biologist it is seducing to compare the evolution of ideas to the
evolution of the living world.

Jacques Monod

But what is the process by which a
new candidate for paradigm replaces
its predecessor? Any new interpretation of nature, whether a discovery or
a theory, emerges first in the mind of
one or a few individuals. It is they
who first learn to see science and the
world differently, ... How are they
able, what must they do, to convert
the entire profession or the relevant
professional subgroup to their way of
seeing science and the world?

Thomas Kuhn

What is valid for the biological world, i.e., the world of living organisms must
also be valid for the scientific world, i.e., the world of scientists, at least as far
as they are living beings and not mysterious ghosts. Biology and science in general do not evolve by themselves, rather, if at all, then only those unique persistent and self-made creatures which, for whatever reason, we call intelligent
humans. Hence we are well advised to follow Thomas Kuhn's path-breaking
new epistemological approach centered on the single individual and test how
far it can bring us in relation to a possibly novel biology of science. As we have
learned from purely theoretical considerations, real cognitive progress, which
is always simultaneously evolutionary progress and vice versa, can only be
achieved by absolutely blind attempts. These are the universally disliked genetic mutations and recombinations which ensure that, with foreseeable regularity, completely new types of individuals can originate. The developing individual himself must remain completely excluded from advances of this kind out
of purely selection theoretical reasons linked to the germ line/soma relationship. This means nothing other than that, in the actual context of science, scientists, contrary to their privileged public reputation, do not do what is asked of

them daily, namely bring knowledge into the light of day for this insatiably curious humanity. Astonishingly, Thomas Kuhn's considerations of what we understand by a scientific paradigm come unbelievably close to this purely evolutionary context:

> The proponents of competing paradigms are always at least slightly at cross-purposes. ... Like Proust and Berthollet arguing about the composition of chemical compounds, they are bound partly to talk through each other. Though each may hope to convert the other to his way of seeing his science and its problems, neither may hope to prove his case. The competition between paradigms is not the sort of battle that can be resolved by proofs (Thomas Kuhn 1962/96, p. 148).
> Communication across the revolutionary divide is inevitably partial. ... In a sense that I am unable to explicate further, the proponents of competing paradigms practice their trades in different worlds. ... That is why a law that cannot even be demonstrated to one group of scientists may occasionally seem intuitively obvious to another (Thomas Kuhn 1962/96, p. 150).
> Notice now that two groups, the members of which have systematically different sensations on receipt of the same stimuli, do *in some sense* live in different worlds (Thomas Kuhn 1962/96, p. 193).

If we replace here "in some sense" by "from necessity" in the rather cautious statement in the third citation from Kuhn, we have already found the direct connection to Darwin's theory of evolution: individuals with varying biologically cognitive, i.e., ultimately genetic constitution must remain separated in an incompatible way by their differing views of one and the same thing. A paradigm is, in this sense, nothing other than the view of the world varying from one individual to the next according to their genetic constitution, whereby the transition from one paradigm to another is in a double sense strictly forbidden if we now bring into play the logical extension of the Kuhnian approach to the scientific theory of evolution and not simply any self-constructed metaphorical variant of it. First, the researcher, whether he is a genius like Einstein, Darwin, and Schrödinger, or just a Nobel prizewinner, in his research activities, is just as much excluded from the realization of cognitive gain as any saleswoman, taxi driver, or any other normal person. A saleswoman and a university professor therefore —as far as evolution or epistemology are concerned—have absolutely the same cognitive status, they only live in different social biotopes to which they are varyingly well-adapted. There are successful and less successful secretaries in just the same way there are good and bad professors. That does not exclude that a successful saleswoman could have been a successful university professor or, the other way around, some academic working at a university would not have made a better saleswoman or salesman. Of course, the same can be said for professions where an unrepeatable set of circumstances leads to the fact that not every person that looks like Walter Matthau, Bill Cosby, Michail Gorbatchev, or Karl Popper—you can see them on the streets—has to automatically develop into a famous actor, comedian, politician, or philosopher. To keep to science, if there were no jobs in our society, that is no economic niches for scientists (Holden 1995), it would also be rather difficult for potential Nobel prizewinners to achieve personal success. To a certain degree, Kuhn foresaw this fundamental impossibility of a real paradigm change—what we can now equate with cognitive gain—within the life cycle of one and the same individual, even though he never attempted to argue it theoretically. So, for example, he stated, with reference to taking over a new paradigm, that is a basically new view of the world by another individual:

Just because it is a transition between incommensurables, the transition between competing paradigms cannot be made a step at a time, forced by logic and neutral experience. Like the gestalt switch, it must occur all at once (though not necessarily in an instant) or not at all. How, then, are scientists brought to make this transposition? Part of the answer is that they are very often not (Thomas Kuhn 1962/96, p. 150).

The rest of the answer which we can support from the perspective of evolutionary theory, says that the possibility of real cognitive gain must remain inaccessible for the single individual since natural selection only allows this to be achieved, from an epistemological view, by the absolutely blind path of random mutational processes within the germ line of metazoans. For this very reason, no real cognitive gain can be transmitted in any way symbolically (e.g., through speech) let alone be simply learned and Kuhn was perhaps the very first to recognize the true significance of this insight:

To translate a theory or worldview into one's own language is not to make it one's own. For that one must go native, discover that one is thinking and working in, not simply translating out of, a language that was previously foreign. That transition is not, however, one that an individual may make or refrain from making by deliberation and choice, however good his reasons for wishing to do so (Thomas Kuhn 1962/96, p. 204).

What we have here is nothing other than the principle already discovered by Chomsky of a basic skepticism with regard to the transferability of knowledge, strengthened by the emphasis on the powerlessness of the individual with regard to his either affirmative or dismissive reactions to influences foreign to him. Simultaneously, the wide-spread information based theory of cognition extremely seductive because of its apparent simplicity, where information only needs to be taken in from the environment, is also clearly rejected by Kuhn. As we saw in the chapter on human language as the supposed medium of a true cognitive transfer, communication, in whatever form, is simply a physical interaction and, as such, has nothing to do with the transmission of any semantic content. Exactly the same as the lack of perception of color for genetically color blind people (Heisenberg 1990; Neitz, Neitz and Kainz 1996; Shyue et al.

1995; Hund 1995) or seeing when it is dark for people with genetic night blindness (Rao, Cohen and Oprian 1994) cannot be experienced by nor transmitted to normally sighted people (for the fascinating subject of how to get "from genes to perception", see Gegenfurtner and Sharpe 2001). Understanding for a completely new theory t_n is beyond people who are genetically blind to t_n and it also can never be explained to them. Then what does successful scientific communication ultimately consist of? The biological theory of evolution has no difficulties in answering this apparently difficult question in a satisfactory manner: it must consist of the interaction of the most similarly structured individuals, or in Kuhn's words:

It should already be clear that the explanation must ultimately be a psychological or sociological one, that means the description of a value system or an ideology through which this system will be passed on and gain recognition (retranslated from Thomas Kuhn 1970/92, p. 381).

and further

A paradigm is that what is common to the members of a scientific group and only them (retranslated from Thomas Kuhn 1974/92, p. 390).

Kuhn's sociological investigations of the enterprise called science led him finally almost to the very door of a first consistent evolutionary explanation of scientific development. Why only almost? A closer examination shows us that his concept which, there is no doubt in the analysis of the individual processes, goes unbelievably far in the direction of evolutionary theory—Point 1: only concrete single individuals are important and not abstract theories floating around in the ether of a fictitious objective world:

Briefly, in normal science, tests often occur but they are of a very particular kind: ultimately, individual scientists are tested and not the theory (retranslated from Thomas Kuhn 1970/92, p. 361).

Point 2: a new paradigm can neither be purposely learned nor taught —, yet again, it still remains a purely metaphorical pseudonaturalization of a process which in this form plainly cannot exist just as an autonomous development of beaver's dams cannot and, for this reason, has nothing to do with a real evolutionary process. Kuhn, on this point.

The analogy that relates the evolution of organisms to the evolution of scientific ideas can easily be pushed too far. But with respect to the issues of this closing section it is very nearly perfect. ... The entire process may have occurred, as we now suppose biological evolution did, without benefit of a set goal, a permanent fixed scientific truth, of which each stage in the development of scientific knowledge is a better exemplar (Thomas Kuhn 1962/96, p. 172/173).

Exactly like Monod (see above), Kuhn warns against a too premature and merely metaphorical application of evolutionary thinking to that phenomenon which is lurking as "scientific development" through the whole of epistemological literature on science, however, only shortly afterwards to fall, together with Monod, into the same deceptive trap. And, like Chomsky, who also got very close to the logical conclusion of a necessary evolutionary exclusion of the individual from real cognitive gain, he finally still believed in the miracle of conversion, the purely mental transformation through the mere influence of instruction:

Fortunately, there is also another sort of consideration that can lead scientists to reject an old paradigm in favor of a new. These are the arguments, rarely made entirely explicit, that appeal to the individual's sense of the appropriate or the aesthetic—the new theory is said to be "neater", "more suitable", or "simpler" than the old (Thomas Kuhn 1962/96, p. 155).

If we are now really interested in looking at what goes on in the rarified atmosphere of science through the unbiased eyes of evolutionary theory—perhaps some of us do not really want to—then we must concentrate on those units which can indeed be subject to a real evolution in the course of time. If we are then prepared to do so, we will have to make the corresponding modifications to Kuhn's still too idealized definition of what a paradigm really is and finally state:

A paradigm is that total sum of cognitive abilities which, from simply seeing or not being able to see colors up to the power, or not having the power, to construct a complete scientific theory and even to understand it, is common to the members of a social group, no matter whether it is scientific or not, *alone through their identical genetic constitution.*

Strictly speaking, even this seemingly reductionist formulation is still by far too idealistic since what is denoted as a paradigm can never exist as a real common possession of a group but only in the interaction of numerous similar copies of one and the same cognitive paradigm, i.e., one and the same genetic program that allows us humans as well as animals to experience the world in a comparable way. With this explanation, we have now landed in a first real biology of science which can be nothing other than an extended behavioral biology around a comparatively new and often quite bizarre group of organisms—namely the one of scientists—a domain in which, exactly like Darwin himself would have specified, different constitutional, i.e., genetic types of individuals compete, at first, for social recognition (going to university), then for scientific success (becoming a professor, winning the Nobel prize), and ultimately for nothing other than their very own genetic success consisting of both direct (e.g., within one's own family), but also indirect (e.g., a school of convinced followers) contributions to their mutual biological fitness. Consequently, it must be allowed, or even inevitable, that the new methodological concepts in behavioral research, i.e., mainly sociobiology and behavioral ecology, be applied to this field which, up to now has been confined to idealistic-humanistic investigations, investigations that were dedicated to, but unsuccessfully, unraveling the self-constructed mystery of science, of society, and of the whole of human culture. However, before we leave the narrow field of philosophy of science to concentrate more on the proper genetic evolution of intelligence itself in the next chapter, we should briefly immerse ourselves in the depths of the philosophical chaos surrounding the superhuman creature par excellence, the knowledge-producing scientist. What is now left over of the much praised wonder of scientific discovery, that persistent myth to which a whole social caste in a world of increasing expertness owe their monopolistic position?

To illustrate the current situation, a short excursion into the history of philosophy is again necessary. Strictly speaking, the whole affair is concentrated around the so-called induction problem, which, since the days of the great skeptic David Hume, has kept moving the erratic brain currents of all the deeply furrowed brows of thinkers. This problem represents indeed the focus of misery of all the extant philosophical epistemologies up to now. The particular misery consists of simply trying again and again to establish the correct methodology for achieving true cognitive gain once and for all. The much more basic question, which in fact can be answered in advance, as to whether a method of cognitive gain can exist at all is, as already stated, ignored. As a result of this inexcusable neglect of this most crucial question, after two thousand years of philosophical musing, we now have an abundance of the most varied suggestions with regard to the supposedly definitively correct method of cognitive gain. However, with the sole exception of the Gödelian approach in logic, we still have no really serious discussion of whether it is legitimate to award the human individual the distinction of being able to achieve something like cognitive gain. What we find yet again is the already mentioned, but now very cleverly embroidered, age-old myth of humans as the wonder child of nature, as the great exception to all the other world events which follow natural laws, who can himself make the rules of his being, and concomitantly, the rules of his particular cognitive access to the world.

The methods of procedure of the supreme authority on philosophy of science of the twentieth century, Karl Popper or, in short, "Sir Karl" (cf. Kuhn 1970/92, p. 357), with his life-long crusade against everything which could have to do in some way with induction is perhaps the most vivid and best known example. Popper's cunning strategy essentially consisted of immediately arresting every suspected candidate for supposed irrationality and to disqualify it from the race before, because of science, which by definition is there to procure new knowledge, irreparable damage could occur. Popper's inquisitorial, excessively meticulous search for a universally valid, limiting criterion between supposedly incorrect, in his choice of words: intuitively illogical and the only correct, in his choice of words: formal-logical thinking ultimately led him into an area where, in fact, the most highly interesting relationship between direct perception and the mental representation derived from it are dealt with—a relationship which, in the meantime is receiving increasing attention from neurophysiologists of all kinds (Farah 1995; Kosslyn and Sussman 1995; Kosslyn et al. 1995; Miyashita 1995; Ishai and Sagi 1995; Johnson-Laird 1995; Robin and Holyoak 1995; for the existence of "imagery neurons in the human brain", see Kreiman, Koch and Fried 2000; about the short distance to schizophrenia: Silberzweig et al. 1996; Dolan et al. 1996; about the relationship to *serious multiple personality* disorder MPD: Humphrey and Dennet 1989; interestingly, Oliver Sacks [1985, part. 4] characterizes "the world of fools" as either a morass of the concrete, if imagination is deficient, or a loss of reality, if perception is deficient)—but in no way has to do with a difference between logical and illogical dealt with in the sense of being legitimate versus not legitimate. However, it is exactly this latter difference, and not the former, which in children as young as two to three years old manifests itself. As we have already mentioned children of this age were given the task of finding a real object hidden in a room by looking at a size-reduced three-dimensional model of that same room where the hidden object could be seen (DeLoach 1987). The younger children failed to solve this apparently simple problem whereas those only a few months older had no difficulty whatsoever in looking for the object in the right place. For the latter, their complete picture of the world had changed radically—from a still primarily perception-centered experience to become a conscious conceptual and simultaneously, according to Popper, *hypothetical* understanding. This could not and *also would not* occur at an age between two and three years if it were not intended in our constitutional and hence genetically determined cognitive equipment.

So why is Popper so important or at least seems to be? Popper believed he had discovered that every attempt at a purposeful inference or induction (inducere [*latin*]: "in" + "to lead") of new theoretical propositions (hypotheses, general statements, theories) from single observations represents not a logically verifiable process, but instead only a very deceitful one. Since it is a purely psychological and subjective procedure and therefore cannot be scientific, it was, in his opinion, to be excluded for all time from any serious catalog of thinkable procedural strategies for cognitive gain. On the other hand, the reverse process which would be a logically verified deduction (deducere [*latin*] "away" + "to lead") of individual perception prognoses from an already existing general hypothesis and their testing by the objectively perceivable reality is for Popper the only possible rational though not completely certain—here Sir Charles shows himself unusually modest—way to a growth in cognition. However, Popper was mistaken in many respects. His perhaps biggest mistake consisted of his looking for the way to new cognition exactly in that area in which, from the very beginning, it can be excluded with certainty, namely, in the rational area of scientific logic and mathematics. The running through of a logically coherent algorithm or mathematical calculation specifies more than any other form of human behavior in advance—and here in a very conscious manner—what is to be done. This is the exact opposite of what we should expect and what is necessary for new cognition to be gained, namely *absolute ignorance*. The latter was of no interest to Popper. So unfortunately, he looked for what any serious theory of cognition has really to be about, namely the crucial question of about the "how" of new cognition, in just that area where it definitely cannot be found.

The openly critical, but at the same time very dogmatically rationalist Popper sweepingly condemned everything that was most likely to come at all into question for cognitive gain, namely everything which potentially can be classified under the category of so-called irrational behavior. Unfortunately, in this respect, Popper made a second serious mistake since, on closer examination, it soon becomes evident that all so-called inductive, or simply intuitive, or any other irrational, crazy methods resulting from pure human fantasy are ultimately based on definite rules of behavior and thinking if not necessarily in a consciously formulated or even formalized form. Things can be explained rapidly. If a living creature, whether man or animal, taking a single perception called observation, states, for example, that "animals are emotionally controlled fools", or vice versa that "humans are evil through and through" (the examples are freely chosen), etc., then, in each case, we have an ultimately logical behavior as far as its legitimacy is concerned, otherwise, if this were not so, living organisms ought to deduce any hypothesis from any arbitrary perception. However, as far as we know, not a single animal does so. If it were so, we would not be able to carry out any successful behavioral research. Instead the latter shows us that so-called irrational behavior neither in man nor animals can ever have anything to do with real physical chance processes. Simultaneously, this is also the reason that we must deny any possibility of cognitive gain from ways of behaving by developing individuals when the ways follow given sets of well-defined rules. Here, Popper made a third mistake when he said induction was forbidden because, in his opinion, it represented a basically irrational method, which for him meant it did not follow specific rules and hence was unreliable and random. In other words, without knowing it the great Popper succeeded in producing a subtle epistemological trick, the moment he applied a completely wrong criterion, which says that only rational behavior can lead to new cognition, to an equally completely wrongly judged area, namely inductive behavior, which he saw as unreasonably irrational and unreliable. However, the following citation shows by the ideas he tried to propagate that Popper himself was ultimately an incorrigible and thus exceedingly illogical inductionist:

At the beginning of objective knowledge, there is not always an experiment, but rather an observation and the draft of a holistic picture in which all the existing knowledge and all the observations are integrated (Karl Popper 1972).

Finally, Popper made a last fatal error when he compared his special type of epistemology, which later became the famous falsification theory, with the modern theory of evolution. This is in no way alleviated by the fact that his original theory bears a more than remarkable resemblance to Einstein's very personal reflections on the question of cognitive progress. As we know today Einstein thought that:

1. The E (experiences) are given to us.
2. A are the axioms from which we draw conclusions. Psychologically, the A are based on the E. There is no logical route from the E to A, but rather an intuitive (psychological) connection which can always be withdrawn.
3. From A, single statements S are logically derived, the derivation of which can lay claim to being correct.
4. The S are brought into relation with the E (verification with the experience).

Albert Einstein 1956, p. 120 (translation from French)

Just as the evolution of organisms has had to make do with mutation and selection, according to Popper, trial and error is the only method available for achieving any cognitive progress in the proper human sphere. This may sound good and explains why the comparison has found many supporters in the meantime, however, in its essential points it is very simply wrong. There is no method of trial and error in evolution which an organism could use to reach a better adaptation to the environment—genetic mutations change the structure of creatures in a purely random way without there being any plan whatsoever or anything similar directed method—, while on the other hand, the very practical method of trial and error has very much to do with a highly purposeful human and animal activity (cf. Gould and Marler 1987) but nothing to do with things such as evolutionary and therefore real cognitive progress. It is mainly on this erroneous confusion between these two fundamentally different processes that until now most of the many varieties of so-called "naturalist" epistemologies (Richards 1987), be it rather "selectionist" or in a more general sense "evolutionary", whatever this may mean in detail, have foundered. Even worse, Popper's appealing model of trial and error, christened by Donald Campbell into "Blind variation and selective retention" (Campbell 1960), sounded so promising that not a few biologists, among them, Popper's former playmate, Konrad Lorenz, fully agreed to its basic ideas (Lorenz in a letter to

Campbell: "Dear Donald, ... I found that what I wanted to say was much more clearly expressed by you than I could have done." retranslated from Evans 1977, p. 107).

Nevertheless, the essential difference between the methods of trial and error at the behavioral level, i.e., occurring during ontogeny and the evolutionary interaction of mutation and selection at the level of phylogeny can be represented very graphically in terms of epistemology. If a living organism has had its whole functional structure changed by a single genetic mutation in one of its parental germ lines, it does not have any possibility of knowing what was going on or even understanding, that means in any directed way following or realizing it. A mutation in the genome of the germ cells produces something fundamentally new, namely a new kind of organism and this new organism of course cannot realize that a possibly important change concerning its own cognitive, i.e., genetic constitution has taken place. This potentially (as we know since Kimura there are also many phenotypically neutral genetic changes) new being with its new structure interprets, from the very first moment of its existence, the world in a new way and be it only something that all people cannot do equally well, as for instance being able to "digest" lactose (milk sugar, selection pressure: cattle rearing) or even higher mathematics (selection pressure: Cartesian school system). In any case, it represents real cognitive progress if natural selection endows the new living form with a certain minimal, in comparison with its closest competitors at least not less than average, reproductive success.

Now, the situation is completely different with respect to that behavior that we like to call learning by trial and error and that, if applied in science is not fundamentally different from an extension of childhood play with perhaps some more serious intentions. Here the question that has to be posed for a serious epistemology is how can it be at all possible to learn something from a real chance experiment. If each attempt was really blind chance, one would not be able to learn anything since one would not be able to recognize what one had actually done during complete blindness. How is it then that one can apparently learn so much from trial and error? Quite simply, because one always knows exactly what those supposedly blind attempts look like! Otherwise again one would never be able to reach the conclusion that for example, in the case of an empirical disproof of a hypothesis, one would have made a grave error compared with the case of a confirmation which warrants the assumption of no error. However, that is exactly Popper's model and that is exactly what has nothing to do with the genuine problem of cognitive gain. Two examples from very different areas may illustrate this briefly.

Let us begin with a very simple case of learning by trial and error. We will pretend that, from now on, we want to try to play a completely new sport, e.g., tennis and we begin with the first attempt to serve. We throw the ball in the air as told to by our coach, or if not available do-it-yourself book and attempt to hit it with a powerful stroke of our racquet and by at the same time springing in the air try to maximize the chances of it going over the net and landing in the prescribed service area. Our first enthusiastic attempt, we freely admit it, is a disaster, completely missing the ball, which, due to gravity, drops neatly onto our own head respectively our newly bought cap with its fashionable logo. Shall we shorten the misery? What will happen with our next attempt? We will try to make an improvement to what we carried out in a rather worse manner than we had planned. Hence we will change our behavior with relation to the previous attempt, and step by step try to get nearer to the quality of a top-ten player since we are sure that the world number one started out just as modestly as we have. How can the following attempts be an improvement on the preceding ones? Very simply, by already knowing and not forgetting, exactly what we have done previously since otherwise we could never relate to what we have already learned. Even a temporary worsening of our service can, as a rule, be regarded as something positive because eventually, if we practice long enough, we will recognize what we have been doing wrong. In other words, the *complete* knowledge necessary to learn how to play tennis must already exist in advance! If we or rather our whole bodies, did not know all the time what they are doing, or what they have just done, or what we want to do (e.g., hit the ball with the racquet and not with our heads), we would not be able to recognize either success or failure. We would have to resort to absolutely random attempts and would need an infinite eternity to reach even the standard of normal average class players and be able to compete with them. The astonishing fact that despite all these many severe "cognitive" problems not a few people manage to achieve a kind of incredible perfection after a relatively short period of time in their youth (one only needs to think of the grand slam title winner in recent years: Sampras, US Open 1995; Muster, Paris 1995; Becker, Australian Open 1996; Kafelnikov, Paris 1996; Korda, Australian Open 1998) should give us food for thought. Perhaps, it is simply the comparison of trial and error in learning and mutation and selection in evolution which is really not a good idea?

We now leave tennis and climb, although the potential earnings are far lower, some steps higher and examine more closely what, in the illustrious domain of scientific research, can be understood about the proven methods of trial an error. For this purpose, we would like, since we all see membership of the only *sapiens* species in the world as a special duty, to achieve a doctorate or if possible several of them (there are triple and even "higher" grades of doctors...) and to do this, we publicly set up a highly speculative thesis to be tested in a scientific fashion for its objective truth. In principle, the whole business functions just as in the case of learning strategies. The very first thing we have to know is what precisely we want to prove empirically and by using our hypothesis, i.e., from the ideas we are presenting, how we can deduce a prognosis for the expected observations. For that purpose, let us take the example very popular among some philosophers of science, that famous one of the swans, which we naively want at first, in common with Sir Charles, to assume are always birds with a white plumage. To be actually able to recognize these birds in the wild as such, this will not be sufficient if we concentrate, as Popper would like to have it, exclusively on the color. We must know in advance from an ornithological textbook, or have an idea from some other experience, what an animal like a swan could be.

Now we are in a dilemma similar to the one we were in at the beginning of our enthusiastic attempt to learn to play tennis the moment when we are confronted with a bird which has all the characteristics of a swan except that instead of being the universal white of swans it is almost completely black. Then, if we follow Popper's "logic of scientific discovery" (Popper 1935/59), we have to throw the hypothesis of eternally white swans overboard and establish a new hypothesis. A completely different hypothesis, as postulated by the falsification model—one that has nothing at all to do with our previous observation? Similarly to when learning to play tennis, it would take an eternity with all these wasteful random attempts till finally, probably when all swans of the world have already become extinct, we would by a random combination of all the characters of a conceivable theoretical object come across the thesis, or rather the wonder of a black swan. Even when we only want to talk about living organisms generally, we already have to have the knowledge how such a thing can be recognized reliably. As one can readily see, the feedback from seeing the black swan on our original swan concept, which is strictly forbidden by Popper, has nothing to do with a contravention of scientific logic since, if this were so, we would never have been able to perceive anything significant about the world we live in.

In order to keep everything short and to the point, the supposedly sought for information and that means both *theoretical as well as empirical information*, must already exist in advance in the person doing the seeking in order for him to be able to form any scientific theories at all. This is clearly related to the perception of the so-called "objective truth", on which, according to Popper, our hypotheses have to be constantly measured to see if we are getting at least a little bit nearer to it. If Popper's idealistic view of things were actually correct, then something would be allowed which has to be rigorously excluded for reasons supplied by evolutionary theory, namely a direct instruction of the subject or, in a more general view, the organism by the environment, which would lead us back to the error of all naive information theories of cognition. What happened to Popper was that he erroneously believed that the undoubtedly interesting relationship between perception and mental respresentation or, in short, thinking was relevant for answering the much more fundamental question for the way to real cognitive gain. What he really tried to do was to use the subjectively experienced asymmetry between supposedly real past and apparently unsure future as a reliable criterion to distinguish between merely psychological, i.e., "subjective" and truly "objective", i.e., scientific knowledge. This, however, cannot work the way Popper would have liked it to because, as we will soon see, such a fundamental difference between psychology and epistemology simply does not exist.

For example, it leads very rapidly to absurdity regarding his well-loved falsification model when one begins to criticize Popper's rather well-worn argument with reference to the factor time. The derivation can be done quickly and is easily understood. Let us assume that we have a fantastic idea in our heads and attempt as an honest scientist, to test this thesis on the basis of a prognosis derived from it. We are making a big mistake for the third time after our attempted tennis playing and our disappointing observation of swans because Mother Nature tells us in a well-controlled experiment that what we have hatched out in mind is not so. We are well-behaved Popper followers and explain in the conclusion to the public the breakdown of our prognosis in front of the, as it turns out yet again, extremely intractable since "objective" reality. Is it then compelling, from this point on and from purely theoretical reasons, to declare the therewith falsified thesis to be dead for all times? No, of course it is not since, again purely theoretically, it cannot be excluded that, in all the following experiments, exactly the same falsified, i.e., apparently finally disproved event reoccurs.

Things become even worse for the falsification theory if one looks more closely at how the apparently certain past of the scientific "facts", that is things that have been "effected", is put in order. In principle, that past is no more certain than the supposedly uncertain future. Consequently, each falsification itself from the moment of its termination on becomes itself a purely mental hypothesis, but now about the past, for which in addition a direct empirical verification must remain excluded, since past events, as we all know, are difficult to repeat (Heschl 1994c: Lindh 1993). A very similar argumentation was put forward by Alan Chalmers, who in a book with the inspiring title "What Is This Thing Called Science?" pointed out the limits of naive "inductionism" as well as Popper's "falsificationism" (Chalmers 1990). As we have seen, inconsistencies still arise even if we, like Chalmers, ignore the inevitable uncertainty in statements about observations („All observation statements are fallible", Chalmers 1990, p. 61), which can only be resolved on an evolutionary basis. As informative as the criticism of Popper by Chalmers may be the conclusion he believes must be drawn from it is rather disappointing. His supposed objectivity ultimately sounds not much different from that in the Popperian model of "objective knowledge" (1972) which he has thought to have toppled from the high horse of the philosophy of science, namely again like a sterile biology without living beings—known to be a quite futile undertaking:

Objectivism removes individuals and their judgements from a position of primacy with respect to an analysis of knowledge (Alan Chalmers 1990, p. XIX).

Homo scientificus, if he is ever achieve success within the short time period of a single lifetime, must consequently know exactly at every single moment of his highly esteemed work which at first sight seems so creative but finally is nothing other than the result of phylogeny and therefore genetically preprogrammed. In conclusion, Popper's popular philosophy of science, in all the essential points, misses the mark since, first, logical thinking has nothing to do with cognitive progress but is rather the best proof that we as subjects can have nothing to do with it and second, abstract theories cannot really, i.e., physically, be falsified but rather only living individuals who, from the beginning of their existence, try to understand the world around them with a very specific pattern of interpretation, among other things, with the help of scientific theories. Moreover, the last point can be illustrated by the fact that falsified theories, in sharp contrast to real biological falsifications, where organisms and whole ancestral lines are physically eliminated by natural selection, do not simply disappear from the world as Popper would have us believe. In fact, just the opposite, they unfortunately carry on existing as things once wrongly thought in the brain of the thinker. In principle, this is exactly what happens to all other of our so-called mistakes and errors which are not, as often presented as self-evident, the exact opposite of "correct cognition" but rather they play their own and not unimportant role in the act of cognition. The same is valid for the popular comparison of optical illusions and reason, which has to fulfill the function of correcting the illusory, hence "false" information. First, it has to be said that no illusion can be simply switched off by the influence of pure reason but, at best, can be supplemented to a different general perception (e.g., perception plus mental respresentation) and second, the supposed falsehood of many illusions only occurs by an artificial hypostatization of certain effects. No normal person is really fooled by the apparent reality of an object drawn on paper by Escher or any other impossible things, just as no normal man by the apparent reality of a pin-up girl (here fish and most reptiles and birds as more directly perceptively guided creatures are different since, as a rule, they are easily fooled into treating a well-made dummy or their own mirror image as a true opponent, a mate, etc.; for the classical case of the male stickleback fiercely attacking its own mirror image, see Pelkwijk and Tinbergen 1937, Tinbergen 1948). In a first adaptationist variant of cognitive psychology, in the meantime, it has been established that almost all what have been held so far to be "illusions" or, what is even worse for the person involved, "delusions" (Tversky and Kahneman 1983) are, in reality, simply due to differences concerning the respective presentation of the material and the solutions to the problem expected from the tested subject (Hell, Fiedler and Gigerenzer 1993). In contrast, real illusions, by definition, can only be those which we in fact can neither recognize nor prevent and for this very reason may often cost us, if not directly, our lives then at least some more subtle losses in fitness.

Again, in principle, it is no different to playing tennis: if we were to make a serious error such as holding the racquet by the wrong end, it would be engraved in our memory forever and would never ever happen again. Falsified theories and, very generally, errors in thinking also remain alive as prudently stored knowledge of how in certain problem area things do not function correctly. Trial and error, consequently means something fundamentally different when we are talking about phylogeny or ontogeny. While our many personal mistakes as well as the correct interpretation of the world can be understood without any greater problems, both actual evolutionary mistakes and successes are never recognized in any way. A scientific error is never a situation in which the environment could tell us, e.g., by the roundabout route of an experimentum cruci to resolve all the remaining unclarified points, what indeed the case is and therefore it is completely ruled out that by our wonderful capacity of logical reasoning we could surmount our own evolutionary, i.e., phylogenetically determined cognitive limits as so many philosophers and humanists would have us believe. If we could do that, we would indeed be extraordinarily wonderful creatures who would at some time be in a position to repeal the laws of evolution for our own development to finally be able to develop in an absolutely free manner. Any species of animal which could develop this capability, namely be able to obtain adaptive knowledge directly from the environment, would at once become immortal since, ultimately, its individuals would be able to get the information necessary to solve any problem they were ever faced with. The environment would always inform the threatened organism what it had to do, therefore what should it worry about? Unfortunately, the laws of evolution as well as epistemological considerations, forbid every such Lamarckian trick and this explains why an immortal species will never be discovered on earth or anywhere else in the universe. In addition, this would have nothing more to do with the process of life as it is characterized by the maintenance of dynamic structural identity in spite of permanent threats from the hostile outside world. After all, immortal life would therefore be qualitatively equivalent with nonliving and a study of it a reductio ad absurdum.

Briefly and to the point, the currently fashionable comparison of the interplay between mutation and selection in biological evolution with the mechanism of individual learning by trial and error is a lame one in many respects and can lead to a very serious mistake, namely then, if we believe that it has given us a better understanding of evolutionary processes. In future, we should prevent ourselves from equating, even in a metaphorical way, the relationship of mutation and selection with any preprogrammed behavioral strategy which

functions according to the pattern of trial and error. This serious error becomes even clearer if we attempt the opposite, i.e., if we attempt to describe biological evolution itself according to the stated pattern. This would mean that
organisms could simply try them out once on an experimental basis to find out
their effect on success. If the result were not satisfactory, the attempt would be
taken back and the whole process repeated. In other words, the organism
would already know exactly what it had to do which shows us yet again that
such systematically directed and therefore also intended progress cannot have
anything to do with real evolution. Nevertheless, there is one small consolation. Because we possess an incredibly good memory, we can remember this
exciting intellectual game, like so many of our other common mistakes, as an
attractive, although fundamentally incorrect metaphor for at least as long as
we live although falsified by our logical thinking but still extremely full of life
in our neuronally fashioned heads.

The nature of purely formal proof in mathematics and logic is ultimately the
conceivably strongest indication for the fact that after perception, learning,
and mental respresentation, now also logical abstract thinking has nothing to
do with cognitive progress. A successful formal proof is defined by its water-
tight internal consistency and coherence which shows how its spiritual father
must have proceeded in a thoroughly legitimate, i.e., logical way. At least theoretically any subsequent step can be predicted at any position within the process and at any time unless an error in thinking is made and the same is true
for the opposite direction: every previous step can also be deduced from the
subsequent one. Every purely formal system like this is perhaps the best evidence of all that human thinking, though undoubtedly by orders of magnitude
more flexible and "creative" (Piaget 1970) than any other comparable animal
kind of mental activity—in fact that's all about our supposed biological "uniqueness" —, nevertheless runs in cognitively closed pathways (undoubtedly
still the most entertaining way of realizing this is Hofstadter's "Eternal Golden
Braid" [1979/2000]). This shows that cognition does not represent a kind of unlimited "cosmopolitan" program into which any new knowledge can simply be
shoveled arbitrarily whether by sophisticated processes of physical accumulation of information or by a universal strategy of conditioned learning or
whether by convergence of our perception on objective reality as Popper
claims.

In a certain way, however, the induction problem comes quite near to the solution of the fundamental epistemological question of cognitive gain, namely
insofar as Hume and even more so Popper—this is probably the most interesting aspect of his falsification theory—cast serious doubt on the possibility
of the environment purposely instructing the subject for reasons of principle.
In this sense, Hume's considered skepticism about the influence of experience,
in which he even ends up supporting instinct, is more closely related to

Popper's passionately propagated prohibition of induction than the well-known stories told of the embittered struggle between empiricists and rationalists, nativists included, would have us believe. Chomsky recognized that here, from the different groups of adherents, an absolutely artificial and untenable conflict has been constructed. On the other hand, Kuhn provides us with an indication that the induction problem, if taken to its conclusion brings us directly in the direction of an evolutionary interpretation of human cognition:

Neither Sir Karl nor I are inductionists. We do not believe that there are rules by which one can induce correct theories from facts, or that theories, correct or incorrect, can be obtained at all by induction. Rather, we see them as statements of the power of imagination created for the sole purpose of application to nature (retranslated from Thomas Kuhn 1970/92, p. 369).

If we equate the word "induction" in this citation with "instruction by the environment"—that we are never instructed by the environment, but rather "reality" itself is only the product of our own perception apparatus can be easily demonstrated even for the convinced empiricist by gently pressing on one of our eyeballs: we can immediately see two overlapping worlds—eventually we will end up where the theory of evolution applied to the species *Homo sapiens*—and not the alleged wonder of nature—must lead us. Even Popper himself viewed this correctly as being legitimate and stated:

One can formulate the main difference between association psychology ... and learning by trial and error ... in such a way that one says that the first is ... Lamarckian (or "instructive") and the latter is Darwinian (or "selective") (retranslated from Popper 1994, p. 302).

Kuhn, who was principally interested in an evolutionary theory of cognition, could have immediately gone one decisive step further and claim:

Do let us try this once as consistent evolutionary epistemologists. From evolutionary theory, we deduce that, in principle, there can be no rule according to which new worldviews of organisms could be induced by the environment or that new worldviews whether useful, i.e., adaptive, or not useful, i.e., nonadaptive, could originate by directed changes of any kind. We must rather see in them *purely random*, i.e., mutative changes in the basic genetic structure of whole organisms caused by mutations in the germ line, which, in the course of their *directed* ontogenetic development as *new types of individuals* are exposed to the rigors of natural selection for an evolutionary evaluation.

Only having reached this point have we arrived, and this time starting out from traditional problems of pure philosophy still unsullied by science, at a real Darwinian epistemology, which, however, after that has nothing more to do with Popper's learning by trial (hypothesis) and error (falsification). Just the opposite, Popper's apparently universally valid method of the scientific refutation of hypotheses, unless understood as a feasible possibility of cognitive gain, is still Lamarckism in its purest form and as such is in complete opposition to modern evolutionary theory. In this respect, Hume was even far more up-to-date than most of his successors among the skeptics in epistemology insofar as he, as a logical extension of his critical observations on human reasoning, proposed that we should think rather along the lines of preprogrammed cognitive instincts than of acquiring something indefinable like "new knowledge".

What should a really naturalized, and therefore a first truly evolutionary theory of science, ultimately consist of? In a similar fashion to the case of Chomsky's interpretation of language it should consist of a consistent extrapolation of the at-any-rate very well developed approach by Kuhn. For this purpose, it is, in any case, just as in the anti-evolutionist Chomsky, necessary for us to remove the last Lamarckian remains in order to reach finally, in agreement with the latest research into animal behavior, an evolutionary genetics of populations but not from the wonder being science but from the subpopulation of living researchers. In such a biology of research, it is not "objective", i.e., scientific theories existing in the so-called exterior world which go through the process of selection but solely actually existing and thinking individuals who consider themselves to be scientists. Scientists, however, surprisingly cannot as individuals do what, which just from their name, they should be doing, namely providing new knowledge. Consequently, they are also *not* capable, as falsely assumed by Kuhn, of carrying out a paradigm change in the sense of real cognitive progress since individual scientific discoveries have nothing to do with such a thing. Neither from purely empirical (e.g., perceptions leading to inductive generalizations modified by falsification) nor from purely theoretical sources (e.g., theories checked for internal logical consistency) is it possible to achieve a growth in cognition since all human behavior follows a set pattern, i.e., is anything but randomly structured and, in fact, is well-defined in rules laid down in the organism itself. That means that the brain of a researcher, just like that of any other person, already knows in advance everything that will supposedly be discovered as something new. It knows exactly how it can perceive, how to arrive at purely internal pictures of the world called hypotheses, and finally also exactly the most efficient way possible of dealing with the always intricate interaction between perception and the processes of mental respresentation, i.e., thinking in every single case. Consequently, there are no cognitive gaps in this evolutionary proven system called human cognition which should have to be filled by the supernatural activity of a wonder organism, i.e., the human subject. Hence the relationship between empirical knowledge derived from direct perception and theory as secondarily deduced cognitive respresentation in the proper sense, the latter naturally also having reverse effects on the first (cf. Gigerenzer 1991), has a lot to do with the question about the kind of consciousness as virtual behavior in a mentally imagined space but nothing to do with any factual possibility of real cognitive gain.

According to Kuhn, different scientists sometimes seem to live in completely different worlds. Evolutionary considerations do not just support this view but must also promote it. If scientists, or even completely normal people (or apes) communicate verbally, they do not exchange any semantic information or even elaborated knowledge but simply interact in such a way that perhaps something like an agreement results from it. What kind of possibilities result in principle? To make things clearer, let us exaggerate, just a little, the genetic differences between the various humanlike researcher populations. To do this, we will take, in the fullest meaning of the word, an interdisciplinary group of highly motivated researchers that has decided to get together to finally solve the last great mystery of the universe. We imagine that the solitary orangutans have specialized in recognizing purely physical phenomena, the social chimpanzees, as chemists, purely chemical phenomena, and the clumsy gorillas exclusively biological phenomena. All three groups live in completely different worlds and the individuals within one group understand each other solely because they all have the same genetic makeup. Consequently, just by coincidence, we allow another group of researchers to appear and they are the bonobos. This new group has been fashioned by natural selection to recognize not only physicochemical but also biological phenomena. They are able, if only partially, to communicate with members of any of the other groups, however, there is no possibility whatsoever of them being able to communicate their more comprehensive picture of the world to any one of the other individuals.

This apparently very apelike biological picture is also valid for the world of professional human knowledge seekers called scientists and ultimately defines how something like communication can take place between researchers. A scientific theory put forward by someone cannot become accepted by the scientific community because it is objectively more correct then any other and for this reason up to now has resisted scientific falsification, but because it has been understood, that means already known, by genetically similar individuals. This also explains why a real paradigmatic change, in the sense of an evolutionary advance, can never be decided by simple proofs and afterwards transmitted by speech or writing since the proof itself is a part of the evolutionary advance which has to be realized before it can be understood. In other words, every new kind of individual who, if circumstances allowed him to have a carefree childhood and the chance to play at being a researcher, might produce a really novel scientific theory can only be understood by his fellows if they all have exactly the same evolutionary novelty already present in their genome. The colleague who would have the greatest chance of understanding the new theory would be that rare occurrence of the person's monozygotic twin where perfect communication is only a matter of having a life history as common as possible. In every other case, we have to start out from the fact that researchers, even if they are convinced they are talking about a common topic, in reality are talking at cross-purposes, some to a smaller degree and others a greater deal, without ever being able to understand the degree of discrepancy or even be able to estimate in a quantitative way how much it is. This also explains why it is so difficult, if not to a certain extent completely impossible, to prosecute something like an evolutionary theory of science. Things seem relatively simple if we talk about comparatively trivial or generally widespread theories of our perceptions, e.g., about the multitude of colors (except for the color blind), and three-dimensional shapes (except for the shape blind) of some objects or about the perception of simple positional or orientational changes of objects in space, whether self-controlled (an organism's own movement) or due to external influences (falling of inanimate objects). The more specific a comparison of a concrete cognitive capability becomes, the more difficult it becomes to judge what is in agreement and what is incompatible. It is remarkable, for example, that even the theory of evolution, which—according to still dominating idealistic views of science—nowadays is considered to be the "commonly accepted" theoretical basis of biology, is understood in unbelievably different ways by the individual biologists (compare for instance Lewontin with Wilson, or Kauffman with Maynard Smith). This is not surprising since, as a rule, we are not dealing with monozygotic twins or, at least, only seldom. Nevertheless, there is something like a basic consensus among scientists in spite of the particular difficulties of their often complex subject which makes it possible for them to cooperate to a certain degree in solving apparently the same generally understood problem. We must conclude that this consensus can only be that part of the genetically laid down worldview apparatus which is common to all the individuals within a scientific group. Hence a future theory of science will be nothing more than a special kind of behavioral research into a certain subpopulation of our own and perhaps some other chosen primate species, namely the so-called scientific one whereby, at the same time, the fundamental limits of such an undertaking are already indicated. Every philosopher of science is himself subject to the evolutionary limitations listed here—especially the impossibility of making a paradigm change within the course of his life when this is understood to be a real growth in cognition. Exactly the same limitations must equally apply to the communication of this theorist with other researchers. If among these others there is one who, by sheer chance, sees the world with his special theories a little bit different if not even better—that means more adaptive—than the first person , then there would be no possibility of the researcher—notwithstanding that linguistic communication has nothing to do with a transfer of knowledge—ever understanding something he will never be able to understand. In other words, the group of philosophers of science itself is subject to biological evolution with all its basic cognitive limitations.

Finally, there still remains the exciting question of how we should interpret the many intellectual revolutions of the last three thousand years which have been told to us by our meticulously self-made history of culture including science and technology. An evolutionary epistemology, in principle, allows us to proceed in only one way. Everywhere where something is transferable symbolically and therewith, in principle, understandable, it can be excluded with certainty that it has anything to do with real cognitive gain since this can never be transmitted by linguistic communication—it can be transmitted solely by sexual exchange processes. Whenever communication as a linguistic interaction between various individuals in the sense of a successful interaction is taking place (sample "somebody tells another person something"), one can assume that both partners have reached exactly the same evolutionary stage or cognitive level, i.e., via the phylogenetic path of their common genetic descent. As an example to illustrate this in itself trivial but simultaneously extraordinarily important relationship, let us take the discovery of the principle of special relativity by Einstein in 1905 usually described as revolutionary. Einstein's obviously new concept was not immediately accepted but, in the course of time, more and more of his contemporaries grasped it and finally adopted it in the sense of a fully accepted, positive communication. This means simply that all those people who really did understand Einstein, in principle, already carried his revolutionary concept inside them. If this were not so, then we would never have had the possibility of concerning ourselves with the topic of relativity in space and time in the way that Einstein did and through pure chance was one of the first to do so.

Another good example is the main topic of the present book, namely Darwin's discovery, or better the disclosure of the mechanisms of the origin of species since he took the step only when his closest friends and colleagues urged him to publish his ideas after it finally became clear in 1859 that, with Wallace, at least one more biologist had worked out essentially the same theoretical concept. In this same year, Wallace sent a study to Darwin, in which he presented his own independently developed version of a theory of selection. Darwin was shocked because it would, he wrote "destroy all my originality whatever dimensions it may have". We must also assume that most of his early supporters, e.g., Lyell, Huxley, Fawcett, or Mill, in principle, were already carrying a similar theory of evolution around with them to some degree in the neuronal structure of their brains, otherwise they would have had insurmountable difficulties in recognizing what was so special about Darwin's idea. In a very general sense, there were probably a large number of people who had some idea, and perhaps had had so a long time before, that a natural development concept, in contrast to all the countless god-inspired creationist theories, could be more correct even when only in a somewhat simpler form as a conceivable relationship with very similar life forms to ours such as the apes. At this level and at many other cognitive levels, Darwin's world-shaking evolutionary concept, was therefore in no way something fundamentally new that first came into the world with Darwin and definitely not only with him.

Then what kind of fate must a really unique thinker inevitably suffer when, in fact, he is the first to think up a completely new theory? Very simply, because what makes him different from all others makes it impossible for him ever to be understood. It could be that later a son or another relative will be able to understand him but he must wait until he meets a sufficiently good genetic copy of himself. How often such a case of a unique genius may occur is difficult to decide because if it does occur, there is no corresponding person in a position to give him recognition. However, we can generally assume that—starting from the fact that practically all people vary from each other in some details— the intellectual world within the species *Homo sapiens* is just as varied which means that every single one of us, with great probability, has access to areas, be they only minute differences in worldview, from which most others, if not all of them, are excluded. What, at best, remains for the successful pursuit of a complicated social undertaking such as science and culture in general can only be a symbol-assisted interaction whose functioning can only be attributed to the common cognitive denominator of all the participants. Everything beyond this is nothing other than individual peculiarities which may occasionally represent a step into virgin territory but, as a rule, will be phylogenetic, i.e., recombinations of ones already known. The situation concerning scientists is ultimately identical to that concerning beavers. They all know how to build their typical dams and lodges and are very proud of them, however, all of their sophisticated constructions differ from each other in some disputable minor detail.

Our prognosis says that an idealistically imbued philosophy of science, which up till now has been occupied with the supposedly great scientific revolutions, will be replaced one day by a biology of scientists—busy with pointing out the many and rather unobtrusively small details to which the nature of the biological evolution of living things, beginning with the first living protozoans up to the fantastic multicellular constructions of animals and man, can be attributed. It is exactly these new ideas and inventions which are known for having gone round the world like a brushfire, which must have done so solely because, as attested by their great success, they definitely were not real revolutions since if they had indeed appeared for the first and only time in a single individual, they would not have been understood by any other person. In principle, the very success of a seemingly new idea functions exactly like children learning to read and write. One gives them the material required, tearproof books and brightly colored pencils, and shows them all the wonderful things one can do with them apart from throwing them at their neighbors. The children begin to read and write as if it were the most natural thing in the world. If the children had not known beforehand how to master this fundamental skill of every modern civilization, they not only would have refused to participate in lessons but would have had to refuse. Because they are related to the notorious teachers, we even say in locus parentis, they cannot help themselves and have to learn, whether they like it or not, the highest virtues of any of our many successful cultures. In contrast, those friendly but not so close relatives like the gorillas, orangutans, chimpanzees, but also the suspiciously clever bonobos do not have the makings for the great evolutionary awakening which in the meantime is on the point of reshaping the whole biosphere for a large-scale "sapientization". Perhaps they are patiently waiting for their big chance, these (intentionally?) work-shy primates pretending to be too stupid for regular work.

Back to the esoteric world of science. Max Planck was one of the first scientists who, long before Kuhn, recognized that scientific theory has from very little to absolutely nothing, as we have just seen, to do with an ethereal, elevated development of scientific theories in the legendary realm of objective knowledge. Instead he realized very rapidly that it was primarily a matter of the evolution of correspondingly endowed epistemic subjects who can understand and, in a statement that was not meant to be witty, rather unintentionally indicated what the foundation of an evolutionary theory of science should look like:

A new scientific truth does not usually become accepted by its opponents being convinced and declaring themselves as being enlightened, but rather by the opponents gradually dying out and the coming generation being made familiar with the truth from the beginning (Max Planck 1928, p. 22).

Some philosophers of science have cited these words as evidence for the often insurmountable difficulty of converting people who think differently. Among them was Kuhn who, as long ago as 1962, did so in his book " The Structure of Scientific Revolutions". In the meantime, the idea has even been taken seriously and, in response, David Hull attempted to carry out an empirical study on the validity of this—after the discovery of h, the famous quantum of action—the second "Planck's principle" (Hull, Tessner and Diamond 1978). As was to be expected, the result of this study was negative since it is in general not very difficult to find numerous examples of new theories being accepted by their former opponents and hence to prove that a real dying out of people is not at all necessary for an orderly scientific advance to be made. In the words of Hull and his coworkers (p. 722): "Of those scientists who accepted the evolution of species before 1869, older scientists were just as fast in changing their opinion as young scientists". The results of Hull's investigation prove to us primarily that the age of the people involved only plays a minor role. What was not investigated, however, was what could be essentially interesting in such an investigation and that would be an exact comparison of the kind of researcher who very quickly accepted Darwin's theory with the other kind who, regardless of age and other secondary factors, did not do so at all. All those who were able to understand the new theory and ultimately accept it with complete conviction, should, from their cognitive constitution, be much closer in their nature to the discoverer of the theory, i.e., Darwin, than the likewise convinced opponents of the evolutionary concept. The alleged converts have not, from such an evolutionary viewpoint, accepted anything new but have simply confirmed an already existing agreement in their cognitive structures with the help of a special example. Hull's negative result, on the other hand, is symptomatic for the current view held in the traditional philosophy of science which, despite the very promising approaches of Kuhn, is still not seriously concerned with how living organisms actually come to new cognition—namely solely by pure genetic chance—and carries on considering the exact opposite, which it considers progress but, in reality, is nothing other than the most reliable sign of already existing knowledge manifested in the understandability and apparent transferability of individual insights.

Planck's allusion to the succession of generations is naturally not a really reliable criterion for talking about a real cognitive advance since how often does it happen that a new generation explicitly and defiantly throws overboard everything held dear by the previous generation in order to achieve something fundamentally new but in reality often exactly the same philosophy as that of the "stupid" old people is preserved (for an entertaining ironic portrait of this very human attitude, see Gruber 1985). Or, in science as in other domains, we very often have the influence of complicated psychological intervening variables which make it difficult for the interested sociologist to uncover the real relationships. For example, sometimes it is not to be excluded that a colleague knows and understands very well what the other means with his new theory but nevertheless "categorically" rejects it for purely personal reasons. However, the contrary can also be true, namely that a great number of people enthusiastically support a new theory but without having any true insight into its actual content and arguments (for a revealing glance into this symptomatic—we could also say completely natural—"disunity of science", see Galison and Stump 1996). This holds as well for the intricate relationships between the whole group of scientists and the so-called public where, for instance, a host of various local factors has influenced the reception of Darwin's theory at different places in the world (Numbers and Stenhouse 1999). On the other hand, the quite common idea of a linear succession of generations of people, where one generation regularly physically replaces the one before, is also an oversimplification that may have its economical justification from the government offices responsible for statistics but is very wide of the mark evolutionarily speaking when we consider the reality. Children are usually brought up by adults who are usually very closely related to them and who are therefore naturally interested in passing on their wisdom to the next generation. If acceptance of new ideas would indeed function exactly as Planck proposed, this would simultaneously mean the exact opposite of what cognitive progress means, otherwise, a successful exchange could not function. Every budding researcher must, so that he in Planck's opinion "should be made familiar with the truth from the beginning", be able to understand the new theories and concepts which are already there and which have been presented to him by model persons. Only then can the fictitious continuity of knowledge, which the philosophical theory of science has always led us to believe, be maintained. To make this scenario function, teachers and above all suitable pupils must already exist who carry the necessary cognitive equipment inside them in order to do and recognize what others, at present or even before, did and recognized. Such a Planck continuity, on the other hand, only appears to have something to do with real innovations, but in reality, the only knowledge that manifests itself must be already preprogrammed in the genomes of the individuals involved.

One can already foresee that a closer investigation of what we call in an idealistic and, at the same time, reductionistic manner, scientific progress is at least as complicated as the study of the evolution of a special subpopulation of any animal species. There a complete series of complementary methods, from comparative morphology up to physiology and behavioral research is necessary to establish if and when, and in which way, the investigated population has changed. In this field, modern genetics has developed the most important and, as far as the work involved is concerned, most efficient method for decoding evolutionary changes since it is applied directly to the most important storage structures themselves, the genomes of the organisms. An evolutionary theory of science, in future, will have to change more and more to a phylogenetic theory of knowledge acquisition if it is really interested in finding out what novel changes for instance the last two or three thousand years have brought with them for the species *Homo sapiens*. One thing can already be said in this respect: It will probably no longer be so completely fantastic, revolutionary, and wonderful as we usually like to persuade ourselves, what however will emerge will be the real biological structures of scientific revolutions. In conclusion, we can only hope that Ludwik Fleck, Kuhn's great predecessor, greatly underestimated the members of his own profession in the most exalted positions when he said:

It is an extraordinarily interesting thing how far learned men, who dedicate their lives to the task of distinguishing illusions from reality, are not in a position to distinguish their own dreams of science from the real form of the sciences (Ludwik Fleck 1946/1983, p. 128).

Or as Werner Callebaut, one of the rare convinced defenders of a truly "naturalistic turn" in philosophy of science has quite recently put it:

My impression is that philosophy of science is reaching a dead end—no *really* big things are happening right now—and that most of the work in history of science remains quite traditional (Werner Callebaut 1993, p. 478).

So up to this day, we still have comparatively little knowledge about how precisely scientists are recruited from a given population, how intelligent or creative they really are, from which particular genetic lineages they come, how they manage to survive the highly sophisticated intrigues of every day academic life (official leitmotiv: omnia pro scientia), how they set about founding their own scientific "schools" or join other rival ones, and, finally, how they, even though historically seen legitimate followers of the former celibate Christian monks and other kinds of clergymen (cf. academic titles taken from Latin like "professor" = "profess [one's faith]", "ordinarius" = "bishop", "dean" = "the oldest", "rector" = "judge" and so on), have nevertheless succeeded in maintaining through time more or less constant inclusive fitness values (direct: own children, indirect: increased promotion of—genetically closer—disciples vs. less intensive promotion of students "not gifted enough"), which has brought about that, for at least some centuries, a varying percentage of people who can call themselves "researchers", "philosophers" or even "intellectuals".

16 "Evolution Has Us in Its Grip"*

Man comes so half-baked into the
world that you can do almost anything with him without him becoming anything else, just as it has always been.

Hubert Markl

The thing about progress is that it
looks much greater than it really is.

Johann Nestroy

Let us forget scientists again for, as we have just seen, in principle, they are quite normal. Perhaps they have more education or, rather, conceit than the ordinary man in the street, but "the fact that some academics have intellectual flatulence does not mean that all mankind suffers from such flatulence" (Feyerabend 1980, p. 14). We now want to deal with the much more interesting question of how it could possibly come about that *Homo sapiens* can be defined as that special, but not unnatural primate species in which the cognitive capabilities, namely those which come under the attractive but very controversial description intelligence, have an apparently crucial importance. It is probably beneficial if we recapitulate what we have achieved with our critical considerations of the wonder of nature man and how this new picture can be applied to what we can observe everyday in representatives of *Homo sapiens*. Undoubtedly, our most important outcome is the concept that human individuals, just like all other multicellular organisms, be they animals or plants, are excluded from any form of real cognitive gain *on principle*. Each of us learns, i.e., varies and develops his behavior in manifold ways throughout his life but this has nothing to do with real learning which only occurs through evolutionary, that is undirected phylogenetic changes.

This fundamental limitation was already anticipated to some degree even by Plato who naturally at that time could not have had any idea of biological evolution. Nevertheless, in the famous dialog between Socrates and Meno, he tried to give reasons for a comparable basic limitation at the purely epistemo-

---

* This remark traces back to Peter Engelhorn, the founder of the KONRAD LORENZ INSTI-TUTE FOR EVOLUTION AND COGNITION RESEARCH in Altenberg, Austria.

logical level by repeatedly asking the following general question: "What must we presuppose in order to be able to know?". A comparable basic limitation allows us to apply Darwin's theory of evolution to our own species without any ifs and buts. This means, in other words, we must lay to rest what we always try to persuade ourselves just to be in a position to elevate ourselves as a unique exception in the universe above all of the other lower members of the animal kingdom. We must accept that learning, consciousness, cultural tradition, and other such wonderful things have nothing to do with any special position in relationship to the fundamental mechanisms of evolution. We evolve exactly like all other living organisms on this planet with the limited means of random mutation and natural selection and, furthermore, we have no possibility, with any of our many clever methods of science, culture, or technology, of placing ourselves outside this process. The basic barrier on the route to evolutionary progress, the necessity of completely random changes, cannot be circumvented by any tricks.

What then does mutation mean in relation to humans? Nothing more than the enormous variation of individual types of human who originate anew from the permanent, random processes of genetic variation occurring anew in every new generation. Here, we meet yet again the myth of the exceptional phenomenon mankind. Many educated people still believe, and nobody can blame them when it is even stated in not a few more popular accounts of human biology, that human evolution came practically to a standstill some 50,000 to 100,000 years ago and since that time—for what reason is never explained—the fantastic discovery of cultural development has taken over and outstripped what was formerly a pure matter of primitive biology. All our current knowledge of genetic variation, but also simple commonsense as well, must tell us that this fairy story may sound impressive but can hardly be correct. How could a phenomenon which has probably existed since the origin of living systems more than 4 billion years ago cease to show any effect in just one chosen species? At least since there has been genetic information storage connected to functional proteins, there has also been disturbances and changes in these storage structures, i.e., mutations, which lead to something happening in the great undertaking evolution. For our purpose, it is sufficient to take a careful look at the close—and for those who prefer the broader view—and the not so close relationships and one can see that enormous differences exist even between brothers and sisters. For most people who have brothers and sisters, it is not that we experience an exact copy of ourselves with the simple difference that we have been formed differently by the environment but rather as an unmistakable, independent individual with to some degree also very different characteristics. The great exception being the relationship between identical twins which, and this is certainly no coincidence, is distinguished by an above-average feeling of belonging together (Segal 1988).

There is no doubt that we humans mutate, i.e., we change our physical and mental constitution like all other living organisms and this at an absolutely average rate of, regardless of the fact that many opponents of evolution would like to declare discreetly that it has finished, one genetic mutation per one billion DNA replications (10^{-9}). In principle, these mutations can occur in any part of the genome even if their frequency is not evenly distributed. There are, just like in many other species, areas with a comparatively high mutation rate, so-called *hot spots* (Knippers 1997), while other areas appear to remain very stable in comparison as if they had something to do with particularly sensible information containing a dynamic "blueprint" of the body (cf. Waxman and Peck 1998) as a kind of cold mold. This means that practically all the characteristics of a person can be mutationally changeable and therefore nothing stands in the way of a continuous real evolution of *Homo sapiens*. Today we are still a long way away from a complete understanding with regard to the translation of the information stored in the genome of the fertilized egg into a developing metazoan and this ignorance must be the greater the more complex the character is we are interested in (e.g., behavior). Nevertheless, we already know in many species the sometimes highly specific way genetic mutations influence the development of the organism involved, including all its later capabilities. The culmination so far with respect to the research into this topic comes from our knowledge about the connection between gene information on one side and brain structure and behavior on the other in *Caenorhabditis elegans*, a primitive nematode worm living in the soil (Genome Maps VI. *Science* 20th October 1995). Mutation experiments have led to our knowing about 250 genes which have been proved to regulate the behavior, or as the behavioral geneticist James Thomas put it, the "mind of a worm" and this worm is nevertheless capable of learning in a simple way (Thomas 1994). Even the control of the defecation behavior, perhaps for some high-minded theorists extremely uncognitive behavior, involves, even in this primitive species, a considerable expense in genetic information in order to bring about the perfect spatiotemporal coordination of at least three different motor units. No philosopher estranged from nature may argue here that simply discharging waste and the such requires no profound wisdom.

Similar detailed research results about genetically determined instructions of behavior also exist for other organisms. As we would expect, the most investigated experimental animals are among them—fruit fly (*Drosophila*), mouse (*Mus*), and rat (*Rattus*). In the fruit fly, in the meantime, we know 6 different genes which regulate learning behavior and thus also the simple memory of this animal (Hall 1994; Levin 1992). This again reminds us that the traditional contrasting of genetically determined and learned behavior is quite an incorrect description of what is really taking place at the level of the mechanisms involved. In the two mammals most investigated up till now, mouse and rat, there are also indications that exactly that which strictly, *by definition*, should have nothing to do with genes, namely learning, is indeed widely influenced by them (Tang et al. 1999). A recently published study on the laboratory mouse, already mentioned here, shows by its title the general trend which it seems will be determining behavioral genetics more and more in the future: "A Simple Genetic Basis for a Complex Psychological Characteristic in the Laboratory Mouse" (Flint et al. 1995). In this study, it was shown for the first time that even the expression of a behavioral character as complex as fear can be almost totally controlled by a relatively small number of genes. Fear was precisely defined, or operationalized, as the relationship between curiosity and fear reactions (defecation, urination) in strange surroundings or the readiness of a mouse to dare enter the open arm of a raised labyrinth—a kind of Indiana Jones test for mice. What has not yet succeeded in investigations of humans, in contrast, was able to be proved here for the first time. It only needed three loci, i.e., gene locations on the mouse chromosomes with the numbers 1 (location D1Mit150), 12 (D12Mit147), and 15 (D15Mit28) to be able to explain to the most part the individual differences in the animals investigated which came from those differently selected strains—those with greater fear or those showing less. This does not mean to say that these gene locations are alone responsible for fearful behavior, that would be very unlikely, but the results do show that even subtle individual differences, which we have all too readily supposed as being the result of secret instructions from the environment, have something to do with the particular set of genes the organism in question possesses.

The research into this question is naturally still a very long way from a complete understanding of the regulation of learning and other forms of complex behavior by definite genes but this is more than just understandable when one is more closely involved with the material. Because in every metazoan practically everything is interconnected in the most various ways, it would be too much to expect that, as it has often been said in public discussions, we will find *the* gene for aggression, *the* gene for alcoholism, or *the* gene for an increased power of learning. Behavioral genetics gave up this all-too-simple view of the genome of multicellular organisms a long time ago since it is becoming clearer day by day that no character of any living organism can be completely independent of any other one. Hence, as a rule, many genes are responsible for the manifestation of a specific behavioral pattern (polygenic trait) on the one hand and very often, one gene simultaneously influences several different areas (pleiotropy; Hall 1994) on the other hand.

This should no longer be surprising, since—as it became clear in the discussion about the unity of evolution—without exception all genes are sitting in the same evolutionary boat, namely the many thousands of body cells of a single individual. Genes of a metazoan must, of consequence, cooperate with each other very closely so that the undertaking body can succeed at all and move about in space and time. Even if the significance of single genes for certain characteristics is recognized more and more frequently, this in no way means that further gene sequences could not play an important role in the expression of the corresponding trait. Only recently a mouse gene for spatial learning was discovered and named the *PKC*-gene since it influences the production of an enzyme (proteinkinase C) for signaling processes between nerve cells. However, no behavioral geneticist would therefore seriously claim that the actual relationship between genes and spatial learning had been fully explained. It would be rather naive to expect this to be so when alone the excretion by *Caenorhabditis* requires the instruction from at least three groups of genes. It is exactly for these reasons that the latest investigative methods are concentrated more on large areas of the genome in order to acquire more interesting and at the same time more reliable statistical evidence because of the greater possibilities of variation when comparing purely bred lines. The *quantitative trait locus* (QTL) analysis developed by Eric Lander and David Botstein (1989), and since then applied extensively and constantly refined (see *Science* 264: 1691: "A new tool for investigating multigenic traits"), is one of the important newer methods with which it is possible to determine the influence of various genes from completely different locations of the genome on the expression of an individual character in a species. In the meantime, there have been numerous investigations into the influence of genes on human behavior and more and more the signs indicate that highly complex traits such as characteristics of personality at least, as it is always so cautiously stated, possess a varyingly strong genetic component. All these statements can only be more or less true since the actual knowledge of the causal connections, in spite of impressive research successes in the last few years, are simply inadequate for making universally valid statements. The most interesting fact here is the question of whether everything that can be deduced from the theory of evolution in principle allows a prognosis to be made of how the newest findings of gene research should be interpreted.

Purely empirical gene research, from its methods, must decipher the effects of specific genes piece by piece as it were and only then be able to come to a final judgment of their influence on humans if the functional significance of all the DNA sections of the whole genome can be decoded. In contrast, evolutionary theory allows us, even at this transient stage, to make a generally valid statement on what gene research should finally bring to light. The thesis developed here must appear very provocative but it is an inevitable conclusion from the concept deduced from evolutionary theory that every multicellular individual during its ontogenetic development must remain excluded from cognitive growth since the latter can only take place by mutational changes. All the information essential for a metazoan to operate successfully must therefore already exist in the first cell, i.e., in the case of sexual reproduction, in the fertilized egg cell. This is shown alone by the fact that from cells of the various groups of animals, although very similar to each other from their external appearance, develop into very different looking animals. This may sound trivial but the complete information which allows a comparatively insignificant object such as an egg cell to develop into a burrowing mouse, a hunting cat, or a thinking human must be there in some coded form or we would not know what to expect for example from a typical mammalian egg. In other words, those in gene research will have their hands full to investigate all the information which in the case of human beings arises from the potential interaction of a considerable number of genes (for the latest estimations of the total number, see Appendix). In principle, research into these connections is possible and the crucial thing in our prognosis is that not even the smallest gap in the explanation of any character of mankind will be left over since our own species, just like all other metazoan species, is not a creature with any special kind of ontogenetic gap or freedom which could give him a special evolutionary status.

Let us call a spade a spade: the genetic constitution of an individual, the hereditary information, must determine his every single capability, without exception, down to the smallest detail—from simple physiological metabolic processes up to highly complex and conscious kinds of behavior which we usually define as intelligence. Of course, this does not mean—and this is exactly the point where most misunderstandings begin—that the environment has no significance or influence on development. In fact, just the opposite is true since practically all genetic information is related to interaction between the organism and its environment. It is therefore not surprising if behavioral geneticists, those supposedly evil materialists who want to reduce everything being a matter of genes, lay special emphasis on the constant influence of the environment, as for example Robert Plomin when he remarks: "Genetic research is the best illustration of the significance of the environment that I know" (Robert Plomin 1994).

What is claimed here, and this differentiation is very important, states nothing else than that the total information about how an organism can meet the challenges of the environment can only be found in the system itself and that the environment in this respect can contribute not a lot, not half, but nothing whatsoever. If the latter were possible, as mentioned previously, for the living organisms of this world, there would be no more problems of survival or adaptation and we would have, as far as evolution is concerned, arrived at the same tempting illusion of Lamarckism whose consequence would be immortality. Such living organisms have not yet been discovered and, for this reason, never will be because, at one stroke, the logical precondition and, in principle, the possibility of evolution would be invalidated. How could an evolution of immortal beings function if there were no more problems of survival or adaptation? Interestingly, this absurd problem poses itself when there is, if such a thing should exist and what many a believer thinks is a desirable reality, a real "heaven" in which, if we stick to the rules of play on earth before death, varying from country to country, a supernatural being rewards us with eternal life. Such a heaven would soon, if not immediately, be deprived of all imaginable rationality since everything which makes our real, i.e., biological-corporeal life worth living would automatically lose its sense with the beginning of immortality. There would be no point to show joy, fear, liking, or any other kind of feeling or thought since all that is only in relation to the real possibility of being able to die and the problems involved with it that give meaning to it all. Would that be really a heaven which could be described as a "reward" for us? Considered qualitatively, such an immortality is no more discernible from an eternal non-living state and do we not normally call the latter death?

The genetic knowledge which controls the development of a metazoan must not be understood as a special kind of technical information similar to the metaphor often incorrectly suggested, i.e., building plans for living things. Rather, one should think, as Dawkins so rightly suggested, of a kind of recipe in which are present the minimal instructions necessary for the development of an organism in the most condensed form possible (Pennisi 1996). An impressive description of this fundamental process character of genetic information comes from Harley McAdams and Lucy Shapiro who first devised "circuit simulations of genetic networks" (1995). Even better and more realistic is to imagine that an already occurring and stable living process, i.e., the survival of a cell, is simply extended to a higher evolutionary unit, i.e., the common survival of many identical cells whereby,—similar to a hologram—the information can be found in the parts as well as in the whole structure. What indeed happens is a repeated, and now more easily understood, replication of the genetic information as identical as possible to the original starting cell so that the body which is derived from it can be regarded as a system whose units communicate with themselves since, as far as the information they contain is concerned, they are equivalent. The already mentioned causal circularity, i.e., cognitive consistency of one-celled systems realized in the never-ending—except in sudden death—interaction between nucleic acids and proteins is also reflected here at a higher level.

Similar to a real building plan for a house, a 1:1 description of a whole metazoan cannot be put into the genes. The typical objection usually put forward by opponents of a biological view of things is a calculation of the statistical impossibility that such a huge amount of knowledge has enough room in our genome of limited size. How can it be at all possible that, for example, the exact place for each individual cell and perhaps also its actual specialization for specific bodily functions (digestion, immunity, movement, behavior, thinking, etc.) within the total cell number of a metazoan at each point of its development can be defined in advance? In principle, the objection is closely related to the well-loved argument that something so wonderful as human beings could never have come about by shear chance since the aggregate probability that our highly complex bodies could be attained simply by casting dice must be practically equal to zero if we take all the individual probabilities of random origin of its many characters and multiply them with each other. The random creation of a human being from a standing start as it were, i.e., from nonliving to a fully developed person, is not absolutely zero but very, very unlikely. For this reason, it is not surprising that such an event has not happened yet. However, no evolutionary biologist has ever claimed that it has because, at least since Darwin, most of them have understood that evolution and its course is not a matter of yesterday, but a matter of many millions of years. If we now look at evolution with Darwin and his successors as being a difficult and correspondingly long-drawn-out procedure of small and minute steps, the whole picture changes dramatically. Always just one small experimental change (mutation) and then, to make sure, a simultaneous testing of whether the newly created organism, even if the change is only minimal, functions adequately, i.e., survives in the first place (selection), this is something completely different to a really supernatural instant creation. The latter really would need a world moderator who would then be faced with the epistemological insoluble task of creating a world as it is and no different and then permanently supervising every change. Who can he call to help and advise him in this awkward situation?

Let us forget this idle speculation: what takes place during phylogenetic evolution of an animal phylum must do so laboriously and unwittingly, under the most difficult conditions, little by little over an almost unimaginable period of time. This is recapitulated in a more or less precise way in the already finished product at the level of the multicellular organism. This is what makes the ontogenetic development of a metazoan starting from the fertilized egg cell up to the adult living organism such an astonishing phenomenon. This supposedly miraculous development of a complex multicellular organism can, however, be easily explained if we accept the conclusion which can be deduced from evolutionary theory in this case. The total information necessary for this wonderwork is already present in the first tiny cell, i.e., precisely Weismann's germ. This is the only explanation of how it is possible within an extraordinarily short time of hours, days, or a few months for a complicated creature such as— just to name a few of the model experimental animals of molecular developmental biology—fruit flies, zebra fish, hens, mice, and, of course, we humans to come into being. This simplicity of producing such extremely complicated things such as self-functioning intelligent beings—could any computer manufacturer make the same claim?—also explains why we must assume the knowledge which forms the basis of all those intricate biological building processes as already given. If real chance still played a role here, we would have had to wait exactly the many million years that we have had to wait in order to, although we must say perhaps, have a human being of the type we have now. There are many examples to make this connection clear, at least as many as there are different metazoans on earth because every single step in the development of all these different organisms could—at least in theory—split into many numerous alternative routes of development through which, in a short time and even for some still primitive animals with just a few cells, a gigantic number of possible choices would be required to be decided by simple casting dice. Let us take the example of the tangled basic network of the human brain which, as we now know, is already more or less complete at birth. If one had to cast dice for it, it would take many thousand millions of years for something sensible to come out of it if a cumulative procedure were used as realized in biological evolution. Evolution was therefore forced to proceed step by step otherwise it could not have done much with pure chance. Thereby, however, something like the historicity of the structures of living things was established without which Darwin would not have been in a position to draw the compelling conclusion of the factual existence of evolution. We would still not be able to do this today if one creature should spontaneously appear in front of us and then again a completely different complex creature suddenly appeared at a different place as if out of nothing.

That ontogeny, i.e., the development of multicellular organisms does not have enough time available to experiment with real chance is alone an indication that the realization of adaptive evolutionary changes is out of place in this area. The great probability, that any genetic mutations (sole "exception": somatic "mutations" in the immune system, see chapter 9) that occur during ontogeny interfere with rather than promote the complicated interplay of the many cells involved which defines the actual barrier that imposes a strong selective disadvantage on each such change in development. We know, however that evolution is synonymous with the creation of new living, i.e., cognitive systems and it is for this very reason that we need the influence of chance for a real growth of knowledge and cognition. Since chance must be widely excluded as a relevant factor from ontogeny it becomes simultaneously understandable once more why ontogenetic development processes have a lot to do with complex organism—environment interactions but nothing with real cognitive gain, or any other semantic information, i.e., information important for survival.

The specific system conditions which have led to metazoans trying to restrict their receptivity for mutative and recombinant changes to the germ line, since in all other phases of development this would only destroy the existing functional coherence, confirm the essential points of the views of two major biologists of the 19th and early 20th century: Ernst Haeckel's "biogenetic law" (1866) of the recapitulation of phylogeny by ontogeny describes the purely external morphological relationships between these areas whereas August

Weismann's "doctrine" of the independence of the germ line defines more precisely the evolutionary constraints, which as selective limitations are important for the evolution of metazoans. Obviously, both laws have received a lot of criticism because they cannot be treated in exactly the same way as other famous "laws of nature" (for a discussion about their epistemological dispensability, see Callebaut 1993, p. 175, 191, 225), as for instance in physics where, under generally more stable frame conditions, universally valid laws for certain phenomena such as elementary particles are claimed. It is better to talk of simple functional regulation of metazoan evolution and to avoid using anthropomorphic expressions such as "law" or "doctrine". In addition, both "laws" are only valid for the limited field of metazoans and it makes no sense to apply them, what no biologist would do, to unicellular systems where as a rule we cannot speak of ontogenetic development even though sometimes simple single cell life cycles already exist. The content of Haeckel's idea has also come in for a lot of criticism and it is just not considered up-to-date to campaign for its correctness. Most objections concern the fact of an only very approximate repetition of the real phylogenetic ancestral sequence during the embryonic development. This apparently serious argument misses the essential point of the subject matter under consideration and that concerns the fact that the development of every multicellular animal, which means every single one of us, starts out from a unicellular stage. This results in the inevitable necessity of the whole cycle having to start from the beginning every time. If this cycle is changed by mutations in the germ line, then the change is passed on further, and on which a new change can take place and so on. Of course, through such a historical process the sequence of the separate developmental stages can change beyond recognition which means that it becomes more and more difficult to reconstruct the true line of descent alone from ontogeny. Nevertheless, the necessity of recapitulation itself is not affected since a sequence of stages A_1 fi B_1 fi C_1 fi etc. must always be followed by A_1 fi B_1 fi C_1 fi etc. , or A_1 fi B_2 fi C_1 fi etc., or A_1 fi B_1 fi C_0 (= loss) fi etc. which makes that at least a purely theoretical reconstruction should be feasible as long as all the required probabilities of assumed genetic transitions are known.

The only important point about Haeckel's idea and that of many of his predecessors (J. F. Meckel, A. Serres, K. E. v. Baer, F. Müller) about the recapitulation of ancestral history lies in the simple fact that the whole developmental cycle of each individual predecessor must be repeated by necessity. However, it is of more or less of no importance at *which* exact point of ontogeny the whole sequence is changed by mutations in the germ-cell DNA (Hanken and Richardson 1998). The usual misinterpretation of Haeckel is mostly to think that his biogenetic law implies a necessary repetition of all the adult stages of the predecessors (palingenesis), which occasionally can be pretty close to the case (e.g., in some vertebrates) but does not always have to be so (some good contrary examples can be found in Richardson et al. 1997). The only serious criticism that one has to make of Haeckel is that he used the word "law" for the recapitulation of phylogeny which, however, by this is distinguished as being a completely historical, unique process, whereas the actual lawfulness can be found only in the determinist repetition, i.e., recapitulation of ontogeny. Haeckel was therefore perfectly right in a way, but not the right way, which fits in quite well with his openly declared Lamarckian views which were based on an erroneously supposed directionality of phylogeny caused by "directed" behavior (use versus non-use). It was for this reason that Heberer correctly modified the description of the process:

During its short embryonic development, every single individual goes through the forms his ancestors went through in the long period of phylogenetic evolution as end stages. The development of the embryo is therefore a shortened repetition of the evolution gone through by all its ancestors. Naturally the phylogenetic evolution is not run through faithfully and completely with all its stages but some of the many modifications and abbreviations appear. But the basic correspondence is obvious and cannot be overlooked (Gerhard Heberer 1968, p. 517).

It is therefore not surprising that, according to the respective animal class under consideration and dependent on the in part special ecological conditions, that very different types of development can exist (for a detailed description, see Gilbert 2000). However, at the same time, this does not mean a refutation of the principle of an approximate historical recapitulation. Classical cases of modifications of the idealized series of adults are, for example, all the special organs which some vertebrate embryos and, above all, many free-living insect larvae who are strongly subjected to external influences produce at least temporarily (caenogenesis). For this reason, in vertebrates whose embryos are distinguished by being protected within the safe environment of the mother for a comparatively long period, one of the apparently most important exceptions to biogenetic recapitulation is the formation of characteristic embryonic membranes by the embryo which makes this protection possible. Another not unimportant change in the biogenetic sequence is the early formation of the germ layers (entoderm, mesoderm) as well as the first rudimentary organs which can vary even within closely related taxa. In this context, it is interesting to see that the original meaning of palingenesis is "rebirth of the soul" which in many peoples is associated with a transmigration of souls. If we push aside this latter, already Platonic possibility as mere illusory wishful thinking, then we will reach where our search for the roots of human intelligence has led us, namely, to the bygone cycles of ontogenetic development of all our ancestors. In this respect, Medicus was also mistaken with his statement that the law of biogenesis was not relevant for the ontogeny of the cognitive capabilities in humans since, as usual in this area, he too always had the misleading picture of a sequence of adult stages before his eyes (Medicus 1992). In contrast, Gopnik and Meltzoff (1997), after the first investigations by Piaget (1957, 1989) and Garcia and Piaget (1983) in this direction, recently reported an extremely interesting connection between the cognitive development of children and the history of the changes in scientific theories and this showed that Haeckel was correct with respect to the core of his idea if not in detail. The genes of our phylogenetically tested developmental program tell us all the time what we have to do since otherwise we would have no idea at all what we would have to do. In the apodictic words from Weismann:

1. Ontogeny originates from phylogeny, in fact, by pushing its stages together some of which are changed, shortened, switched off completely, or pushed apart by newly switched on stages.
2. Like each stage, each part and each organ can undergo new adaptations; very often such adaptations show the tendency to be passed onto the next newest stage (August Weismann 1913, vol. 2, p. 168).

Weismann's insight into the causal relationship between germ line and phylogeny is, in this respect, nothing other than the inspired anticipation of a concept which pertains exactly to the really interesting point in the problematic aspect of the evolution of metazoans, i.e., where real evolution does indeed take place. This limitation is valid for our own species just like any other and consequently we cannot, as we would like to persuade ourselves, occupy any privileged position among living things. This can be easily demonstrated when we examine our self-made epistemology and psychology of *Homo sapiens* rather more closely. Together with Weismann, we can now claim that anything essential in our evolution can only occur during the change over from one generation to the next but never within the ontogenetic development of single individuals. In this case, "anything essential" means, and here we return once more to our actual theme, that that thing which apparently elevates us above the rest of the animals in this world, namely our unbelievable intelligence can only have originated by going through the evolutionary bottleneck of the germ line. All of our personal knowledge comes from the genes and all our new knowledge can only result from mutational changes within the germ line and not, as has been occasionally speculated, from a not closely definable creativity due to a kind of random generator in our brains (cf. neuronal Edelmanism). The latter would only bring about a dangerous and either in the short or long term an inevitable discord in the cooperativity of the participating neurons, ultimately lethal to the whole body and in no way open a path to new cognition.

That the environment exerts a continuous, selective influence on living systems but can never have an instructive influence on them becomes immediately clear—independent of purely evolutionist considerations—if one examines more closely how different organisms react to identical external influences. As soon as we promote the analysis of any organism purposefully and keep on questioning the deeper reasons of specific behavior, as Plato did as one of the first skeptical philosophers in relation to human cognition, we inevitably come nearer and nearer to the view that only the biological make-up of the creature concerned can be the real reason for its behaving in a certain manner and no other. The few genetic investigations so far which begin to go more deeply into the behavior of an animal species show a very clear trend: the more details which come to light, the more it becomes clear how genes regulate the reactions of organisms down to the smallest details of even such flexible and seemingly nongenetic phenomena such as learning behavior. The general prognosis which can already be deduced from the modern theory of evolution and up till now has not been disproved says that the construction of not one single character of a living organism, whether unicellular or multicellular, can be realized by any kind of directed instruction through the environment. If only one single relevant character should be found which cannot be explained causally by referring to genetic information processes, a considerable revision of Darwin's theory of evolution and, in its course, a new rehabilitation of the Lamarckian edifice of ideas would be unavoidable. The everyday appearance of such things as all the kinds of problems of adaptation, from psychogenic conflicts and banal difficulties up to severe suffering, illness, and death, and, at the level of whole populations, the extinction of species speaks against this ever taking place. And if it did, it would have nothing to do with the process of life which only through the real possibility of perishing experiences a meaningful definition. In addition, as we have already seen, a kind of conceivable Lamarckian immortality arrived at by the solution of any imaginable problem made possible by the assimilation of information from the environment is qualitatively identical with the phenomenon of nonexistence, which we usually call being dead. We can therefore make the claim that, in the cosmos we are presently associated with, this should never arise. The situation, as far as scientific proof is concerned, is already extremely one-sided. Whereas day by day, new genes are discovered responsible for the expression of some complex character or other, up till now, no single case of a provably directed instruction of an organism by an environmental influence has come to light. As already stated, many great researchers, from Kammerer and Piaget (1929) to Waddington (1968b), have attempted, and there will be still many who will not give up the hope that Lamarck gets at least a few percent points as far as significance is concerned, only to have to concede finally what the principles of a comprehensive theory of evolution says from the very beginning, namely, that without a completely blind groping around, no evolutionary and, at the same time, cognitive progress can be possible.

If the total knowledge of every single one of us must be in our own genes, then this means that the model of evolution as outlined by Darwin in its basic contours is applicable without exception to all the characters of *Homo sapiens*, hence to all what we commonly refer to as human intelligence. If genetic variation in relation to intelligent ways of behavior should occur and should a selection pressure with relation to an increase in people's intelligence also be effective, then consequently an evolution must take place in exactly this direction. We can confidently assume the occurrence of genetic variation since, as already stated, there is no sign that our gene evolution came to a definitive end some thousands of years ago or perhaps a little bit more as not a few philosophers and humanists would like to see. The only question which now remains to be clarified is whether there are also signs that the world in which we humans live has wagered on something like having at least a bit more intelligence than many others being an evolutionary advantage.

17 The Survival of the Most Intelligent

The lower animals must have their bodily
structure modified in order to survive under greatly changed conditions. ... The case,
however, is widely different, as Mr. Wallace
has with justice insisted, in relation to the
intellectual and moral faculties of man.
These faculties are variable; and we have
every reason to believe that the variations
tend to be inherited. Therefore, if they were
formerly of high importance to primeval
man and to his ape-like progenitors, they
would have been perfected or advanced
through natural selection. Of the high importance of the intellectual faculties there
can be no doubt, for man mainly owes to
them his predominant position in the
world. We can see, that in the rudest state of
society, the individuals who were the most
sagacious, who invented and used the best
weapons or traps, and who were best able to
defend themselves, would rear the greatest
number of offspring. The tribes, which included the largest number of men thus endowed, would increase in number and supplant other tribes. ... It is, therefore, highly
probable that with mankind the intellectual
faculties have been mainly and gradually
perfected through natural selection;

Charles Darwin

Before we attempt to apply the idea of an evolutionary "struggle for existence"
(cf. Darwin 1859, title of chapter 3) to the development of intelligence in *Homo
sapiens*, we should stop thinking of it as being any kind of physical struggle
since the matter is a far more subtle and, above all, more imperceptible affair
than one would normally suppose. It should be emphasized that the metaphor
of there being a real physical struggle, in many respects, is even confusing. The

most common mistake is to confuse the mechanism of evolution with one of its special results. Aggressive behavior occurs in almost all animal species in intraspecific competition and has obvious evolutionarily adaptive functions for solving certain problems for individuals, e.g., acquiring territory, defense, choosing mates, defense against predators, etc. In plants and fungi, in contrast, aggressive bodily conflicts of the kind animals partake in do not make any sense since most plants cannot move from the spot and driving away their opponents by behavioral strategies such as threatening gestures, blows, and bites. However, even for peaceful beings like plants, the Darwinian metaphor of the struggle for survival still applies since this, when correctly interpreted and not understood word for word, simply states that evolution takes place by the variously successful reproduction of genetically different individuals. In this, it does not matter at all how the difference in reproductive rate comes about. This means it does not matter if it is achieved with or without aggression, by sexual or asexual reproduction, as a utilizer of sunlight or as a predator, whether cooperatively or solitarily, the only decisive factor is an actual difference in the final reproductive success and the latter is often brought about, since habitats and resources are always characterized by spatial and material limitations, by the minutest changes in the structure of the organism concerned (see: "On the manifold origins of species", *Science* 273: 1496—1502).

What makes natural selection so effective and, on the other hand, so imperceptible for those concerned, in this case, ourselves is the fact that it is operating permanently and everywhere. Because organisms, which reproduce sexually, are characterized by a comparatively greater genetic variability, individuals that have traits slightly different to others, even if the changes are minimal, are always available. This means that even the smallest change in environmental situation, under certain circumstances, can manifest itself as an evolutionarily effective selection pressure. If any description of the situation as far as the human environment is concerned can be said to be accurate, then it is above all the statement that this environment compared to the earlier epochs of vertebrate evolution has changed an extraordinary amount and continues to do so. One such change appears to be especially dramatic for our modern age of the last 3 to 5,000 years but its beginning was probably several thousand years before. We can assume that we are dealing with an epoch of approximately 4—5 (maximum 8) million years in which our already on-average large primate brain more or less suddenly—i.e., within 100,000 years —began to enlarge continuously so that our small-headed close relatives (orangutan: 400 cm³, chimpanzee: 400 cm³, gorilla: 500 cm³) would say that it has reached a completely abnormal volume of 1,500 cm³ on average (Wood 1996, 1997; Wood and Collard 1999).

Detailed morphological investigations and follow ups using computer-model calculations have shown, in the meantime, that the continuous development of the present human brain from primate preliminary stages and ultimately from simple vertebrate stages is conceivable without any problems. This means that any miraculous jumps whatsoever can be excluded. There only remains the question of what special selection pressure has compelled the evolution of the genus *Homo* to move in this direction. According to Herrnstein and Murray's controversial "bell curve" of human intelligence (1996)—and with it, we are back in the present again—at least one of these selection pressures is still effective today. The special thing about their investigation is the successful proof of such a selection pressure that unfortunately, in the extremely controversially led debate about the suspected and simultaneously feared existence of genes for intelligence, was widely neglected for the most part.

This is astonishing since in respect to a gene theory of intelligence of *Homo sapiens*, Herrnstein and Murray are extraordinarily cautious and moderate in that, nowhere in their book, do they claim that genes are the solely determining factor in intelligent behavior. However, at one place in the book they do say that "cognitive ability is substantially heritable", but immediately afterwards, also, "apparently not less than 40 percent and not more than 80 percent" (Herrnstein and Murray 1996, p. 23). Their mistake here is excusable since, as mainstream professional psychologists, they are used to trusting hereditability calculations which, however, as we have seen, can give us no information about the real origin of any adaptive knowledge but only about its varying fate under different environmental conditions. However, exactly in this respect to the existence of natural selection in humans, their investigation is in fact impressive namely when they show how differently our so modern and egalitarian-minded technical society treats individuals with different abilities. The classical intelligence quotient, as Herrnstein and Murray correctly make clear, only represents a very approximate measure of a supposed general cognitive ability g first postulated by the American psychometrician Spearman in 1904:

Socially significant individual differences include a wide range of human talents that do not fit into the classical conception of intelligence (Richard Herrnstein and Charles Murray 1996, p. 20)

but, however, significantly correlates with criteria of success in life such as school success, profession, income, married life, conduct, etc. and, finally, successful parenthood. This means that, at least in the population in the investigation, a real selection pressure in the direction of a general increase in g does exist. Such a result would have been forecast as a kind of self-fulfilling prophecy since it is clear that, from the moment that intelligence tests of a certain type determine to a great extent the distribution of chances of success in society and career, it must logically bring about effective selection in this direction. That, in principle, does not change the significance of Herrnstein's findings since, up till now, only a few other researchers have ever made the effort to do research in this direction and in comparable detail. One such exception is Volkmar Weiss (1972, 1982, 1992, 1995) who for example managed, under the most adverse conditions under the communist regime in East Germany to put together his own "psychogenetics" (the Russian term for behavioral genetics of humans; Weiss 1991). He once even published under obviously not entirely randomly chosen pseudonym H. S. Weisman (1988).

If we consider what kind of situation must there be so that—as wished for by many humanists for some reason (educability, manipulability?)—real evolution of modern-day man could no longer function, then the results obtained by Herrnstein and Murray become even more easily understood. To guarantee that really nothing more can happen on the phylogenetic level, it has to be ensured that neither any, even small genetic changes take place nor any significant changes in the environment occur. Both assumptions for our species are more than unrealistic and the same is true for the hypothesis of a definitive end of the biological evolution of humans. The only conceivable thing in this respect would be similar to an extremely strong stabilizing selection, which, in a few species—in spite of a continued genetic variability at the level of the individual—is able to maintain a certain evolutionary status quo over longer periods of time. Well-known examples are the so-called living fossils among which, according to paleontological research, belong some archetypical animals such as the sharks and surgeonfish, the latter probably being very close to the primitive form of all vertebrates (an overview is given in Hass 1979). That *Homo sapiens* could also be regarded as one of these kinds of living fossils, or is about to become one, is anything but likely.

The evolution of *Homo sapiens* is therefore nowhere near coming to a standstill. The modern human world, i.e., that technological civilization which began not too long ago by chance in Europe (Huff 1993) and which, in the meantime is on its way to —as the Bible has instructed us for several thousand years—subjugating the earth is an environment in which a selective pressure would certainly exist in the direction of increased intelligence and not one in which such a pressure would work in the opposite direction towards people staying as intelligent as they already are or even becoming more stupid. The opposite of the results of "The Bell Curve" would therefore have been a real surprise and not the terrible scandal conjured up with the hardly tenable commentary on Herrnstein and Murray's supposedly pseudoscientific handling of the data. For some considerable time in human populations, a certain selection pressure has prevailed, if not always constantly, in the direction of favoring individuals capable of intelligent behavior. There is no lack of historical examples. It can therefore be assumed that the various social elites of most earlier cultures were not always the most feeble-minded of the community (Voland

1996; Weiss 1996). It therefore also seems clear that the first medicine men and later all the other masters of ceremony, shamans, priests, politicians, and scientists of the various epochs were situated, with some probability,—in order to speak with Herrnstein and Murray—within the higher ranges of the IQ distribution. If, however, individuals with rather limited intellect ever manage by persistent and often carelessly underestimated stubbornness to reach the highest ranks, it can sometimes lead to fatal consequences for the whole group involved as sadly shown by the dark sides of the political history of the past century.

We can assume that, nowadays, the reality of such a selection pressure towards increasing intelligence is simply prolonging what has already been happening for quite a long time. One of the earlier, but perhaps still effective selection pressures that may have amplified the development of intelligent behavior is the selective favoring of increasingly larger social units which according to comparative studies of nonhuman primates (Sawaguchi and Kudo 1990)—unfortunately, there are insufficient reliable data for humans—seems to be linked to an enlargement of the neocortex. In this connection, Robin Dunbar has put forward the hypothesis of a replacement of grooming in apes, i.e., ritual delousing as social contact, by human small talk about relationships and other personal experiences, which as we meanwhile must confess to our shame determines not less than 60% of our intelligent existence as compulsory social creatures (one can imagine what a disagreeable task it would be to delouse our relatives and acquaintances day in and day out ...): "Accordingly, language may have evolved to allow individuals to learn about the behavioral characteristics of other group members more rapidly than was feasible by direct observation alone" (Dunbar 1993, p. 681). The primatologist Michael Tomasello pleads for a similar position, although without calling in a special grooming thesis:

Our contention is simply that in this context human beings evolved species-specific social-cognitive abilities to understand the psychological states of conspecific individuals in terms of their perceptions and intentions, their thoughts and beliefs, and their reflective thoughts and beliefs, which allowed them to take the perspective of others and to participate with them intersubjectively. These processes of social cognition then led humans to the species-specific ways of learning from one another that we call cultural learning, which then kicked off the evolutionary and historical processes that led to the species-specific form of social organization known as human cultures (Tomasello, Kruger, and Ratner 1993, p. 509).

This approach nicely demonstrates how species-specific human phylogeny determines what is so ethereally described by most human scientists with the term "culture". In addition, the central importance of the social aspects in hominid evolution (observation, imitation, empathy, theory of mind ToM, self-consciousness ToS, ultimately: theory of cognition, theory of society) makes it very probable that it were first higher social cognition and language which emerged conjointly (this would explain the basic anthropomorphic structure of all languages; e.g., "the sun is rising") and only later "objective" cognition and language (i.e., understanding of physical cause-effect relationships; cf. Dunbar 2000; Rumbaugh, Beran and Hillix 2000; Tomasello 2000). Now, to look for current potential intraspecific selection pressures we only need to look around our human world more closely and we will immediately recognize where we have maneuvered ourselves in the course of the years no matter if we mean 100, 1,000, or 10,000 (or more). This begins with the scientific self-image of modern man in the designation *Homo sapiens* and, because it is still insufficient, culminates in *Homo sapiens sapiens*, which at least shows that we do not want to be understood as being *Homo stupidus* although evolutionarily speaking this does not have to be a disgrace. The ability to solve problems of the most diverse kinds after an extensive, but nevertheless relatively rapid (for the close correlation between information processing speed and intelligence see Neubauer et al. 2001) phase of consideration using the conscious application of flexible general principles of association of internal representations (e.g., shapes, terms, spatiotemporal relationships, amounts, numbers, and logical relationships) must still today, if not even to a greater degree, entail a certain adaptive advantage. During ontogeny the structural basis for this ability develops, as Piaget first observed, as the spatiotemporal permanence of objects in the external world in the course of the first two years of life of every human being. In contrast to many other vertebrates whose cognitive ontogeny in fact includes the acquisition of the first steps of this ability (Etienne 1984), and in some primates even faster than in *Homo sapiens*, the subsequent development in humans can be regarded as a real cognitive explosion which without doubt is reflected in our extremely enlarged brain volume. If psychological intelligence tests or other aptitude tests of various kinds (for those who are looking for a comprehensive catalog of these: *Swets Test Services. European Test Publishers Group*, 1994; or for an amusing overview: *Can you pass these tests?* Bragdon 1996) provide the results which, in a uniquely uniform way, positively correlate with each other, then it is at least an argument for the actual existence of a general mental factor g, which seems to be identical with the expression of our special ability to draw most complicated theoretical conclusions from often surprisingly simple or "pure" (cf. Chomsky's "poverty of stimulus") perceptions (for a recent defense of *g*, see Jensen 1999).

The partly fierce argument against IQ tests of every kind is in any case psychologically understandable since who—except characteristically those who have already achieved a high IQ ranking (see below) or, since collectively humans are known to be ready to do everything, whole nations (e.g., Holland, Germany, and perhaps soon Austria) when seduced by a TV-show—would willingly allow themselves to be tested repeatedly so that they can be placed in some intellectual drawer or other and even labeled with a particular numerical value? There even exists even an international IQ society called "Mensa" which only accepts new members after they have successfully passed a special IQ test. Out of the original initiation rituals of the Stone Age where one still had to throw stones or spears a great distance (but see today's Olympic Games), in the meantime has become a very arduous procedure. Such a respect for abstract thinking was not always the rule. The German physicist Lichtenberg (1742—1799), for instance, still viewed this kind of one-sided intelligence very skeptically: "Mathematics is a wonderful science but mathematicians are often not worth hanging. The so-called mathematicians require that people think of them as deep thinkers whether at the same time there are the greatest dunderheads among them, incapable of any business that requires thinking if it cannot be done directly by an easy connection of symbols, the more the work is routine than thinking" (1941). The widespread criticism of IQ tests is, as far as the method is concerned, fully justified because with usual tests many subtle influences such as frame of mind, attitude to the test, previous experiences, etc. quite often cannot be ascertained exactly. In addition, and apart from the still unresolved question of the amount of influence of language and culture (for culture free respectively culture fair tests of intelligence, see Cattell 1966, 1968), we can state even a few more basic shortcomings with regard to the declared goals of the usual IQ-measurement procedures. For example most standardized IQ-tests do not test:

⇨ for solutions achieved more slowly but which are instead more comprehensive and abstract (cf. quiz-character of common intelligence tests)
⇨ for the recognition of cognitive problems when not brought to the subject's attention
⇨ for the recognition of a partial (or even complete) solution of a particular problem if shown incidentally to the subject (so-called "initial impulse")
⇨ the ability to solve long-term problems (example: observation and subsequent recognition of complex behavior patterns in both humans and animals)
⇨ the interaction between purely theoretical knowledge (example: theory of physical forces) and its practical application (example: constructing a functioning/working engine)
⇨ the capability of recognizing larger causal relationships, i.e., to successfully practice science
⇨ for the factual inability to solve a certain problem (this could be done by presenting different phases of the solution procedure to systematically sound how far, i.e., up to which point the subject is able to follow at all)

So in some way, the standard IQ test as applied still today is therefore comparable with a single blood test, which does indeed tell us at least something about the person at the actual moment of the test but only allows a conditional long-term prognosis to be made. Anyway, Herrnstein and Murray did not deny this shortcoming: "The individual's IQ score all by itself is a useful tool but a limited one" (p. 19). So if IQ measurements are only a kind of rather unreliable spot check, what can we do to make our measuring technique more trustworthy?

A perfect test, but as everyone would agree, somewhat difficult to implement would consist of collecting, from birth on, a kind of permanent, that is on-line measurement and observation of the varying magnitude of the intelligence level. Only such a truly Orwellian technique would be able to register something like the real intelligence development of each individual. No one would accept such a thing, either for themselves or their children. Many, if not the most people would be glad about the fact that IQ tests, except in a very few state institutions (the military) are no longer widely used and that a few countries (e.g., the USA) have forbidden their use as a selective criterion for such things as awarding positions of employment. Does that now mean that Herrnstein and Murray have concerned themselves with something that, as some humanistically minded critics would gleefully endorse, has been long outdated in a social context? Far from it, since when we examine the whole problem more closely, it can be seen that not just now and then does a test take place but exactly the opposite. An unbelievable number of tests take place all the time, whether we like it or not, from the first second of our lives till the last. Not all these tests have something strictly to do with human intelligence but we are getting a great deal closer to the essence of natural selection.

The best place to start is obviously at the beginning, i.e., at the moment of our birth. Even the life of a baby is full of tests even though to all appearances nothing of the sort is taking place. A baby, as a rule, has at least one self-sacrificing parent available to remove all the everyday worries and with loving care to ensure happiness. There is only one snag, the baby must know exactly what it has to do to make sure it obtains this optimal care from its nearest relatives otherwise the road to potential crises is already signposted. It was the psychiatrists John Bowlby (1969) and Dan Stern (1971, 1977) who, independently of each other, showed clearly how shaky and often unstable mother-child relationships can be (see also Hassenstein 1987). From very close and well-functioning relationships to extremely endangered, almost nonexistent relationships where mothers simply abandon their babies, there is every conceivable stage between. Naturally the successful course of a mother-child relationship is not totally dependent on the behavior of the child but, at the same time, it is also clear that both mother and child are being tested imperceptibly with regard to common, that means genetically shared evolutionary success. From the child's perspective, this means nothing more than that every movement, every sound, and every complicated interaction with its mother is an unavoidable test of its abilities in this respect. Whatever can this have to do with intelligence? A great deal since especially in our "wise" species it is nevertheless still necessary to fulfill the often extravagant expectations of the parents, e.g., to react and not too slowly to encouraging recitation by showing an at-least partly convincing intention to talk or a real attempt at copying what was said, in short, to present oneself as a lovable but above all unusually clever little boy or girl.

The life as a physically helpless baby is the first really serious cognitive aptitude test it has to pass after leaving the womb. In the latter, there exist other, but no less severe tests. But for the present purpose, let us skip a few years and dedicate ourselves to the unfortunately often miserably long developmental stage, which we consider to be the undisputable highlight of our cultural development so far, the time we spend at school. The compulsory attendance of school is commonly regarded as one of the most humanitarian achievements of mankind since it is the only way that children with their apparent tendency to run wild—one only has to think of the cautionary examples of the boy of Aveyron and other wolf children—can become clever and cultivated beings. The well-tried method of the school is called "teaching" and this is solely dedicated to the professional dissemination of knowledge and, what are more important, intelligent abilities such as reading, writing, and arithmetic. One would actually think that all our thinking (who still has time to teach their own children?) and all our intelligence comes from there since, as we say, you can't teach an old dog new tricks. It therefore seems to be the purposeful teaching, which turns the more or less naked ape *Homo "morrisensis"* into something very special. What speaks for this interpretation is that in no other species is there such a concentrated and well-organized instruction of the offspring. Compared to us, young apes have hardly any school stress which probably explains their enviable easy-going manner, which is diametrically opposed to our remarkably hectic and at the same time concentration-oriented way of life. However, meanwhile, we know from our considerations of a well-thought out and more comprehensive theory of evolution that we cannot learn anything really new in our short lifetimes, that means cognitive progress must be excluded as a matter of principle from this part of our metazoan existence. Then why on earth do we have to go to school at all if it gets on our nerves to such an extent if finally the only thing we can learn is that we cannot learn? The answer is again much simpler than one would suppose, since it simply says that we get tested there, not just physically but all the more meticulously for intellect and understanding. The schools set up by humans are nothing other than a huge testing factory through which, in the meantime, almost every member of our species have to go if he or she, or more likely the parents are interested in success in later life. It is nothing like an institution which can teach our children something new but rather an undertaking which—convinced humanists will not think it possible—is devoted to natural selection. What Plato demonstrated more than 2,000 years ago while instructing a young slave, namely, that one cannot teach another person anything new but only test him or her for a certain preexisting ability—in the case of Meno, it was the deduction of a simple law in geometry, the already mentioned doubling of the square using the diagonal—this is just as valid, if not more so, for our institutional as well as the private upbringing of our children. So what seems at first glance to be the great exception to the otherwise universally valid laws of biological evolution, on closer examination, proves to be only one of the million other special cases where living organisms are exposed to the rigors of natural selection.

Experienced teachers will now argue that rigorous testing and examination of pupils as carried out by previously most schools is really a thing of the (not always so "good old") past and that nowadays schools have become institutions where the most modern methods of passing on knowledge are used without conflict or authoritarian ways of showing children how to deal with the enormous profusion of knowledge of modern times. They will also mention that more and more alternatives to rigid syllabi and methods are appearing so that one can no longer talk about simply requiring the regurgitation of already present abilities as was the wont of schoolmasters in the times of our grandfathers and great grandfathers. Unfortunately, this superficially absolutely correct opinion does not change the fact that it is impossible for a person already grown up to pass on knowledge to a person growing up (and, of course, vice versa), what is changing is the type of intraspecific selection pressure varying according to the respective type of school. This selective influence, as not only Herrnstein and Murray but also all of us professed humanitarians know very well, ultimately has a very noticeable effect on the future success of our children.

Let us take as an example the very fashionable antiauthoritarian form of education that is often accepted as a teaching method especially adapted to a child's needs or at least a child-friendly one. Exaggerating just a little, one could say that, in principle, children in such schools are able to do whatever they feel like and, at the same time, great value is placed on making the lessons as less authoritarian and as near to the right level as possible. It is therefore not necessary for all the children to learn the same thing at the same time or the same pace—each child, within the framework of the school, is given the opportunity to develop as far as possible according to its own internal laws of growing up. This new kind of education was developed from a basic idea of the Italian physician Maria Montessori (1870—1952) at round about 1900. It anticipated a great deal of what Jean Piaget a little bit later believed he had discovered for the first time, namely the independence of the intellectual activity of the child, which finds its clearest expression in self-motivated play (for a comparison of Piaget and Montessori, see Elkind 1984). How can it be possible at all to talk here about Darwinian selection? An important ideological difference to the normal types of school can, of course, be seen and cannot be disputed. Compared with these, antiauthoritarian education could be best classified as the difficult method of patience, which demands a correspondingly high level of commitment from the teachers and to some extent also from the parents. At least in the Montessori method, a more or less close cooperation of the parents in the education of the children is an integral part of the method. In the meantime, most schools, be they public or private, have discovered this trick for themselves for taming all these varied small personalities called pupils. However, the only thing which consequently has changed is not the method of communication, understood as a real transfer of information between teacher and pupil, but solely the *specific* kind of applied natural selection which in no way is the same as assuming that, in the long term of course, no more selection takes place. In fact, just the opposite happens, since ultimately a concrete "result", i.e., something clearly tangible in any demonstrable form is always expected from the children, no matter in what kind of liberal way one wishes to do it. If we extend this concept a little further and construct for ourselves, just in our heads, the most perfect school system imaginable, one in which everything is done to make learning as pleasant as possible for the pupils, to encourage them, and, with all the modern aids to learning, to offer them as many interesting and creative things as we can. Would this mean that we have finally made natural selection redundant? Even in the case where we purposely undertake to expect no longer anything from the children and even to ask things from them or to check, this would also be a very special selective situation in which again differently talented children would be able to succeed to equally different degrees. It would be even more extreme to do without any kind of upbringing if it were at all possible. One would only have to provide for the child's material needs and avoid any other kind of interaction. However, even when the children are left completely to their own devices, specific advantages and disadvantages would be there for the individual child and these could have an effect on its future success in life. Understandably, up till now there have only been books or films about a similar situation since who would dare to carry out such a risky Kaspar Hauser group experiment (in the movie "Lord of the Flies", William Golding tells the pessimistic story of a group of boys marooned alone on a deserted island without any adults). On the other hand, the mostly not so rosy fate of children who run wild at an early age without any familial assistance shows us what such parental laissez-faire leads to as a rule. However, even for such extreme cases evolution has already established a kind of internal emergency program: the neglected child rapidly tries to become a "small adult" with all the known, mostly negative, concomitant symptoms (child soldiers, children in street gangs, child prostitution, etc.).

Let us return to the institution of school, permanently being revised and then reorganized by passionate reformers all over the world: whatever a teacher requires, or *not* requires, of the pupils has something to do with a concrete form of selection conditions. With respect to the IQ debate, we can thus define this process more precisely and claim that in an evolutionary perspective, the classic and in the meantime extensively standardized intelligence tests are nothing other than the attempt to practice a uniform selection pressure in the direction of promoting individuals with an increased school intelligence by designing highly concentrated, general school aptitude tests and by rigidly applying them in decision processes of all kinds that a society undertakes. It is therefore not surprising that among the most reliable prognoses, which can be inferred from most IQ tests is mainly the future school success or rather the failure of the children being tested:

Of all the social behaviors that might be linked to cognitive ability, school dropout prior to high school graduation is the most obvious. Low intelligence is one of the best predictors of school failure, and students who fail a grade or two are likely to have the least attachment to school (Richard Herrnstein and Charles Murray 1996, p. 143).

In this context, it is not just reassuring that all our close phylogenetic relatives, i.e., the other great apes have not only severe problems with every kind of a more demanding cognitive instruction offered for instance by a human experimental psychologist, but even much more so with teaching itself. Be it as may be, Herrnstein and Murray's thesis goes much further than this finding in that they claim what many critics do not want to accept, namely they show quite impressively, using extensive data, that the IQ value also predicts to some degree the success in life to be expected by the person being tested. From an evolutionary point of view, this cannot be far wrong, since otherwise we would have great problems if, for example, we want to explain why it is, in remarkable contrast to most of the other primates where the status quo has been as it is for a considerable time, that the human brain has increased in size almost continually over the last 7 million years. Among other things, the latest investigations of the brain development in mammals have also shown that the human brain is only the extrapolated increase in size of a small standard mammalian brain such as that of the hamster (Finlay and Darlington 1995). Hence any changes in local areas of the cerebral cortex (e.g., speech centers) play a much smaller role than previously assumed. Not even the part of the neocortex where the greater part of the associative functions are accommodated, is bigger than a hypothetical ape brain of comparable dimensions (Passingham 1975). However, the tripling of the neocortex in relation to the corresponding body size of an ape of the same size is an indisputable fact (Passingham 1973). The relatively low mutation rate of our species (1.2 changes per base pair in 10^9 years in comparison to 4.8 in rats) speaks for an absolutely normal evolution of our brain as well as our body (Gibbons 1995). Nevertheless, an adaptationist explanation is needed since the mutations alone provide only the essential substrate without favoring any direction. For a moment, we will disregard nonsocial environmental changes such as climate etc. which can have a selective effect under certain conditions, whose effect alone, however, should not be overestimated since many primates live in comparable environments (e.g., man and baboons). In any case, the illuminating and therefore very popular graphic description of man's transformation brought about by his standing up on his back legs and thereby freeing his hands for manipulating things is all too simple since many other species, and not just primates, would have been able to do so for a long time if the habitat of the African grassland had required it. The idea that a continuous intraspecific selection in the direction of increased intelligence served as one of the main driving forces in the evolution of *Homo sapiens*, must be complemented by the observation that many other traits of man must also have been under a strong selective pressure acting in parallel since it is only the particular combination of a whole series of special adaptations that have made our species into something remarkable (for the importance of the collective hunt in primates, compare Stanford 1999 with Lee and DeVore 1968). It is also not true that man is only a chimpanzee, which coincidently has a larger brain, no more than a chimpanzee is a kind of macaque with a somewhat larger brain. Evolution always affects entire functional units, i.e., organisms, and as a rule, the change in one character brings with it a change in a completely different character or selectively in many others. For example, the evolution of a general ability g alone without the simultaneous evolutionary changes of the social and emotional components of human behavior could hardly have functioned. Concerning social behavior, for example, there was an enormous spread and refinement of the methods of bringing up the young and this on its own shows how much work, physical and psychic, is involved in turning a newborn baby into a worthy successor to ourselves: a self-reliant person. This extremely reinforced care of our offspring was definitely one of the most important accompanying conditions for the intelligent behavior of our species to be influenced selectively and successfully. It can be shown quite easily that a playful, and therefore fantasy-stimulating and socially secure childhood is an unequivocal precondition for the development of a child's creative potential. Without such paradisiacal conditions—unfortunately, the living conditions for far too many children in the world are many miles away from this ideal (Juen 1997, Mayer 1997)—it would not have been possible to produce engineers, scientists, and philosophers since, under miserable circumstances, no inspired Einstein—who knows how many undiscovered or thwarted geniuses there have been—can flourish. Indirectly related with this general trend of favoring intelligence and artistic talent by social and economic prosperity is a continuous "civilizing process", though often interrupted by barbarian excesses which was first outlined, at least for the last few centuries, so graphically by Norbert Elias (2000). In principle, Elias, unknowingly as a nonbiologist, did nothing else but describe in more detail the intraspecific selection pressure which has probably kept our evolution going over the last few thousand years. This civilizing development can only be understood in connection with our biologically developing intelligence and this also matches with the most important factors which foster the latter: a sheltered childhood, enough free time and energy, material luxury, a safe existence, competition in culture and refinement. All these are prerequisites for the selective promotion of intelligent human beings. It is therefore not surprising that the main impetus of cultural changes in the direction of refined living came from the privileged upper classes and only spread later to the less fortunate in society. We should refrain from believing that this, as suggested in the somewhat idealistic description by Elias, was a continual process. It was just the opposite, recurring epochs of economic shortages show us how quickly other social forces can seize control. Then it has nothing to do with fine and civilized dealing with our fellow human beings and very different "qualities" are in demand: "strength" (brutality is usually meant), "virtuousness" (this can mean being against "degeneracy" and "effeminacy") "being close to nature" (meaning uncultured), finally roughness and the desire for power combined with slavish obedience. We have become all too familiar with such slogans—as a rule the more resources are limited the more likely we are to be antagonistic or uncultivated with each other (Chance 1996)—and as a trait of our species they will probably remain with us for a certain time. Very often such a barbarism is only a rebellion of the average or better: the mediocre against the over refined. By invoking a similar mediocrity the historian Hamann tries to explain the unexplainable enthusiasm of many Austrian people for national socialism before and during World War II.:

It was a broad class of otherwise unremarkable people. And it is depressing how easily people, under certain conditions, can be led and made into fanatics (Brigitte Hamann 1997, in News 5/97, p. 118).

However, even parts of the supposed intellectual elite of that time were not completely immune against such temptations (Deichmann 1995), as shown among others by the case of former NSDAP member Konrad Lorenz (his daughter, Agnes v. Cranach, today: "Like many others, I can only guess, but I am quite sure that during that time he considered himself a convinced national socialist. ... Politically, he was as unsuspecting as uninterested", Cranach 2001) who, in contrast to the unbending bee researcher Karl v. Frisch, did succumb to the at the same time both eugenic and racist Nazi ideology (for the "special" role of physicians during the Third Reich—and Lorenz' first profession was medicine—, see Gallagher 1995). From an article entitled "Taxonomy and the idea of evolution in teaching" and published by Lorenz in 1940 in the national socialist oriented journal "Der Biologe" [The Biologist], the American historian of science Robert Richards deduced the following assessment:

This is one of the most explicit Nazi oriented passage in Lorenz's work. It is one, however, that lies squarely within the Darwinian evolutionary tradition. It could have been written, mutatis mutandis, by any number of British or American evolutionary biologists in the last century or early part of this one. Darwin himself warned precisely of the comparable consequences when natural selection became disengaged in modern society (Robert Richards 1997).

And in fact, Lorenz' emphatic warnings of an impending genetic degeneration of the human race from 1940 reads very much like a bad copy of certain passages from Darwin's "Descent of Man" from 1871 (chapter 5: "On the Development of the Intellectual and Moral Faculties During Primeval and Civilised Times"), namely as exactly the same severe ignorance of the fundamental difference between science as an at least attempted impartial description (and then possibly an explanation) of natural phenomena and mere subjectively colored ideology which tries to impose its own morally judged "recommendations" to other people (hence the theoretical impossibility of an "objective" evolutionary ethics the latter being simply another version of the naturalistic fallacy; Williams 1993; cf. the discussion in Callebaut 1993, p. 78, 231, 435—441, 464):

Natural selection as affecting Civilised Nations.—... With savages, the weak in body or mind are soon eliminated; and those that survive commonly exhibit a vigorous state of health. We civilised men, on the other hand, do our utmost to check the process of elimination; we build asylums for the imbecile, the maimed, and the sick; we institute poor-laws; and our medical men exert their utmost skill to save the life of every one to the last moment. There is reason to believe that vaccination has preserved thousands, who from a weak constitution would formerly have succumbed to small-pox. Thus the weak members of civilised societies propagate their kind. No one who has attended to the breeding of domestic animals will doubt that this must be highly injurious to the race of man (Charles Darwin 1871).

In addition, this very nicely shows that both Darwin and Lorenz (as well as many other declared or hidden social Darwinists) obviously even were not capable of intellectually grasping that "survival of the fittest" simply means that all currently living creatures, no matter if "weak" or "strong" in what sense so ever, are fit in the sense of evolutionary theory, which is not a theory prescribing who "should" have the right of surviving, but nothing more than an abstract quantitative concept (i.e. differential reproduction) which alone tries to explain the factual existence of *any* kind of living being in the world and be it an extremely rare, but nevertheless stably occurring organismal variety (nonetheless, Lorenz was rather a political opportunist—at that time his academic career was blocked in Austria—rather than an active advocate of the holocaust as is repeatedly claimed by some people unless we consider that Charles Darwin was the first and certainly most influential Nazi; for a discriminating "psychogram" of Lorenz, see Bischof 1991; for the Viennese "roots of ambivalence", see Hofmann 1988, p. 1—51). In fact, we cannot even exclude that in some cases it were just the "borderliners" of a species which, by a particularly advantageous mutation—perhaps preceded by an extensive neutral phase of genetic changes, suddenly opened up an avenue to unforeseeably new evolutionary perspectives. For example, we cannot be sure that our common genetic Evas and Adams (Gibbons 2001) initially weren't some bizarre (e.g., fragile skull, small teeth, weak mandible; silly, i.e., childish and playful temperament) and hence, Darwinistically speaking, "weak" outsiders within the up to then dominating hominid lineages until they quite rapidly expanded over the whole world brought about by some unexpected, but finally highly advantageous new traits.

We will now return to "pure" intelligence, our IQ tests. Selection pressure in the direction of increased human intelligence intensified a few million years ago for some reason or other (increased size of groups, competition between groups, diet changes associated with hunting and other collective forms of food gathering, bipedal locomotion, new and more flexible mating patterns, sophistication of tool use, spread of infectious diseases, complex family structures, growing significance of spoken language, etc.; for general accounts, see Cartwright 2000; Deacon 1997; Dunbar 1993; Jones, Martin and Pilbeam 1996; Stanford 1999; for the possible influence of an abrupt climate change 2.8 million years ago, see Kerr 2001).

To be able to prove the hypothesis that it is still influencing the evolution of modern *Homo sapiens*, we need a much more wide-ranging investigation than this quite limited comparison of single tests with only a few parameters of individual success in life. To operate a real evolutionary concept on humans, one must attempt to set up a genuine fitness calculation and afterwards check which characters make their bearer the most reproductively successful in the long term within a certain population. This would also have to include a complicated estimation of the proportion of indirect fitness, i.e., fitness gains from the support of genetically close relatives and this would make it somewhat more difficult. On the other hand quite recent indications exist that at least the length of the human lifespan is clearly positively correlated with the general mental ability g, a result which, if interpreted on a long-term basis (i.e., on average better living standard, more stable social conditions, increased care for and investment in both our own and our grandchildren), could very well hint to a similar correlation of g with regard to the reproductive rate:

OBJECTIVES: To test the association between childhood IQ and mortality over the normal human lifespan. DESIGN: Longitudinal cohort study. SETTING: Aberdeen. Subjects: All 2792 children in Aberdeen born in 1921 and attending school on 1 June 1932 who sat a mental ability test as part of the Scottish mental survey 1932. MAIN OUTCOME MEASURE: Survival at 1 January 1997. RESULTS: 79.9% (2230) of the sample was traced. Childhood mental ability was positively related to survival to age 76 years in women (p<0.0001) and men (p<0.0001). A 15 point disadvantage in mental ability at age 11 conferred a relative risk of 0.79 of being alive 65 years later (95% confidence interval 0.75 to 0.84); a 30 point disadvantage reduced this to 0.63 (0.56 to 0.71). However, men who died during active service in the second world war had a relatively high IQ. Overcrowding in the school catchment area was weakly related to death. Controlling for this factor did not alter the association between mental ability and mortality. CONCLUSION: Childhood mental ability is a significant factor among the variables that predict age at death (Abstract taken from Whalley and Deary, 2001, Brit. Med. J. 322, 819).

In any case, the changes over the last two centuries testify to the fact that the selection conditions for *Homo sapiens* are beginning to change more strongly again since it would be illogical to assume that our modern civilization with its swiftly advancing technological development has no strong selective influence on both the people who are involved and on those who supposedly are not. Alone the many published family histories, which at the moment are enjoying a boom with social historians, show how unbelievably complicated and simultaneously fascinating the fate of a family can be, no matter what social class it is. Such studies show properly for the first time the extraordinarily complex dynamics of what biological evolution in humans actually means, exactly as in animal families with their own special fates. At the same time, these exciting family sagas show us how hopelessly unclear the picture of our past must remain for serious methodological reasons alone. Even if we knew the gene composition of the whole population of the world at present, it would still be a hopeless quixotic endeavor to register the exact evolutionary lines of several ancestral generations simultaneously in a study. What we can record under favorable conditions is exactly the same as in the case of other species, i.e., only miniature fragments of an unimaginably gigantic process which we, at best, could reduce to an approximate conceivability by mathematical extrapolation. Ultimately, it will not be possible to make more than a rough estimation of the gene shift tendencies over longer periods of time unless we can analyze to a great extent the hereditary material of some of our mummified ancestors such as the recently discovered and yet already famous Tyrolean Ötzi who, as it has only recently been unveiled, was killed by the arrowhead of a rather unfriendly contemporary.

Nevertheless, it would be conceivable, at least over not all-too long periods of time, to understand at least a little of the complex dynamics of human family trees, where the dying out of a side branch of the family is in the program as well as the unforeseen flourishing of a new one. The construction of a personal family tree going back several hundred years, as has always been the cu-stom of noble houses because they have always thought that they were geneti-cally better, points in this direction. This also explains the unusual fascination that such comprehensive genealogies has over us since this seemingly fixed identity of the individual among the masses gives us a safe anchorage. The great psychological need that lies behind this phenomenon of a kind of sear-ched eternity becomes immediately apparent if one sees the great lengths, sometimes for years on end, displaced people (e.g., from World War II) go to in order to find at least one of their lost relatives. The genealogies of the ari-stocracy have only one flaw in that the unavoidable diversity of the influences from the most varied genetic sources are often deliberately neglected so that only a clear and straight picture of their origins is drawn. It is therefore clear that with each marriage, a completely new family tree is added which in its turn inevitably drags its own family trees with it so that, even without civil wars and social revolutions, the highly valued blueness of noble blood must have already become a little muddy, genetically speaking. That evidence, or even suspicion of blue-bloodied ancestors which gives many people the unique feeling of being one of God's chosen few of the world would be well worth a sociological study. However, even in other areas, there are similar phenomena when it con-cerns something special such as being bodily related to Einstein or to Wittgenstein (rather less to Frankenstein) even if very distantly.

Ignoring the many limitations which the authors openly admit, Herrnstein and Murray have at least shown the actual possibility that the new age of very advanced energy and machine technology could bring with it a corresponding, serious change in the gene structure of modern societies and they also warn of the creation of an increasingly isolated cognitive elite (for their account of the relationships between the different ethnic groupings, see 1996, p. 275: "Jews—specifically, Ashkenazi Jews of European origins—test higher than any other ethnic group"). To prevent what they consider to be a fatal and hence un-desirable social development, they suggest selective countermeasures for a ge-neral raising of the level of intelligence. However, when drawn to its logical conclusions, this does not necessarily lead to a final solution to all the current social problems (Herrnstein and Murray, p. 389: " a clear, continuous, and a clear-cut rise in intelligence would circumvent many of the problems we have described"), in fact just the opposite since it would intensify further the selec-tion processes that are already underway. Then, when all said and done, an IQ value of 100, for example, if this kind of random testing procedure has any fu-ture at all, will be the mean score for the intelligence of the whole population again and all the problems that Herrnstein and Murray say an individual with low intelligence is confronted with will carry on existing. Exaggerating some-what: if a super genius who now has an IQ of 200 were transported into the fu-ture where the general population has overtaken him because of any efficient selective measures whatsoever, he would be faced with more or less similar problems a present-day person with an IQ of 70 already has. In reality, evolution must have progressed in such a manner even though not so dramatically. As the third species of chimpanzee, we could have well been content with a comfortable, idyllic existence in the forest, similar to the bonobos. However, even then there must have been individuals who thought they knew better and something was set in motion which apparently, some millennia later, still has us in its grip.

In other words, every genetic change within a population leads automatically to change in the competitive situation between the various lines of kinship but does not bring competition to an end. Only if there were a high level of genetic homogeneity, as in asexually propagated species and even then only to a certain degree, would a kind of evolutionary equalization of interests be conceivable. Our own bodies are the best example of how genetic identicalness can lead to almost perfect cooperativeness and peaceful coexistence where all the problems of a newly created super system are solved. This perfectly cooperative state can only be found within real units of evolution, i.e., within individual, physically separately existing systems, be they unicellular or multicellular. However, there is little evidence that from sexually reproducing populations, which are characterized by an often substantial genetic variation, new units of evolution could arise transcending individuals. It is of course conceivable that, at sometime, the rapid cloning of genetically identical people by modern gene technology would be possible. Those people could indeed form highly cooperative and extraordinarily peaceful hordes of monozygotic siblings fighting against each other in Darwinian competition after they have ousted completely the last romantics who still used the old sexual method. Such a fictitious scenario would somewhat resemble the situation of water fleas (seasonal change between sexual and asexual reproduction), where hordes of Einsteins, Schrödingers, and other intellectual geniuses would populate the whole world. However, this is no more than really bad science fiction and only one among many since the question of variability and thereby the ability to evolve automatically arises yet again.

Because the theory of evolution forces us to proceed from the fact that the necessary and sufficient information necessary for the development of human intelligence, like all the other abilities of a metazoan, can only be found in the fertilized egg cell, it simultaneously means that all cognitive differences between various individuals must have purely genetic causes. Since some human populations have to some extent lived separately from each other over a longer period of time, it cannot be completely excluded that, beside the already universally present individual differences, group-specific differences have also developed. Nonetheless, the genetic variation within a group can still be greater than between the groups. This does not mean in any way, what opponents of evolution in humans often erroneously believe, that for this reason there can be no serious differences between groups. If we observe different species, this misinterpretation is immediately understandable. The purely quantitative genetic variation within our own species is probably greater than that between *Homo sapiens* and any one of the present-day cats. Does that now mean that we are not very different from the cats in our nature? The essential point here consists of distinguishing a seemingly unitary measure (difference in nucleotides per genome) from the hierarchically organized semantics of the result of those measurements (e.g., a simple exchange of one single base in the DNA can either be phenotypically neutral, i.e., change nothing, or, under certain rare conditions, create a completely new character).

Completely independent of spatial isolation factors, in every population, new subpopulations are formed—no matter if we call them subspecies, races, varieties, local forms or some other term—which originate alone from the fact that the permanent and complete mixing of the whole gene pool of a species, the perfect panmixia, is only a purely theoretical possibility. Herrnstein and Murray are of the opinion that, in respect to human intelligence, such group differences between large ethnic groups exist without a doubt and, to this end, they lay out a whole series of investigations that show that the average IQ score of Asiatic, European, and Afro-American test persons did indeed deviate to a significant degree from each other in that the first attained the highest and the last the lowest mean scores. This result, which taken alone cannot be easily swept under the carpet, has led to them being accused of open racism. However, this becomes readily understandable as soon as one looks at their subsequent demands more closely: "Some federal funds now so exclusively focused on the disadvantaged should be reallocated to programs for the gifted" (p. 418). Transferred to the result of the IQ differences between the three large ethnic groups, this means that the people who should be helped are those who are already getting the biggest advantage from the current system and this is the same as intentional preferential treatment. For the others, it means lifelong permanent stress of being intentionally disadvantaged even if not always consciously. The impressive anti-racism workshops by Jane Elliot (see Verhaag 1995) have shown this very clearly, an experience which the privileged (e.g., Caucasians) or, with Elliot, "blue-eyed" people, as a rule, only have to endure for a short time if at all. An experience which alone by its cinematic presentation is experienced as extremely oppressive and is worthwhile confronting people with it, although the initiator of this highly committed project openly admits its desired effect cannot be made compulsory:

I cannot force anybody to change his behavior. People have to do this for themselves. However, I show them the consequences of their behavior and I explain, what it is like to be on the other side. ... For racism to work it is sufficient for the good people to do nothing (Jane Elliot 1995).

It is at exactly this point that the otherwise commendable work by Herrnstein and Murray has to be criticized since contrary to normal scientific practice in carrying out investigations they did not leave the social evaluation of their results to the general public or rather the reader of the book. At the end of their impressive collection of data, they discuss the conclusions from the discovered connection between IQ and success in life under the title "Restoring the Concept of the Educated Man" (p. 442). First, of course this connection still remains to be proved at the actually decisive level of biological-genetic fitness, an undertaking, which, as one can imagine, is not an easy one. This is shown by the fact that, at a time of a world-wide increase in IQ scores (Flynn 1998), an opposing hypothesis of a dysgenic trend has been put forward (Lynn 1998), and this in turn has come under fire from other critics (Preston 1998, Waldman 1998). As long as actual allelic frequencies and both local and social class-related effects cannot be included in the calculation, these results must be taken with a pinch of salt. But here it has nothing at all to do with science when Herrnstein and Murray come to the conclusion that the concept, which they admit is not completely new, of educated people supported by the no-longer popular ideals such as "wisdom" and "virtue" can solve all societal problems. Viewed scientifically, it would be possible to deduce the exact opposite of what appears to be the only conclusion following the authors of *The Bell Curve*. If one were really to see a serious danger to modern Western societies in the creation of a new cognitive IQ elite as they do, one could for instance immediately demand that all further education of intelligent people should be forbidden by disbanding all existing universities and colleges. In fact, above-average intelligence would then no longer be an unconditional advantage for personal welfare and therefore all the many chances of reproduction potentially connected with any supposed genes for intelligence would, sooner or later, be reduced to zero.

With reference to the accusation of racism, which biologists are, not completely coincidently, prone to receiving it must be emphasized that finding out that there are genetically different populations among humans is not automatically the same as valuing them subjectively. Just by neglecting these points, a great number of serious misunderstandings result since the establishment of a genetic difference per se logically cannot tell us anything about the evaluation of such a difference. The theory of evolution as a scientific theory provides us with the best evidence of this. What is of interest is solely how various types of organism with their own way of life cope with their specific environment and that without giving any kind of moral judgment. The only criterion which is available for describing biological evolution is simple survival and its associated transmission of genetic information to the succeeding generation. Considered from an evolutionary point of view, it makes no sense to talk about better or worse adapted as long as those individuals, those families, or populations, or species we are talking about are still alive. Even extinct species are not logically better or worse in an ethical sense but were only less successful than others for some reason or other. To say this briefly, one could therefore pretend that all living things are possibly variously successful but nevertheless of equal evolutionary "value" because evolution is not a person, à la Mother Nature, with any notions of value.

However, that this subtle mixing of scientific statements with social value judgments is practiced especially often in the case of questions of intelligence, throws some additional light onto our topic and shows us that our thesis of the human race still being subjected to a general selection pressure in the direction of higher intelligence cannot be totally wrong. For example, this is shown by the mere fact that being intelligent, in whatever form, i.e., having good grades at school, doing well in tests, is seen as being good and desirable whereas lack of intelligence, i.e., having poor grades at school, doing badly in tests, is generally viewed as being not so good and rarely the other way round, even though—why not?—theoretically possible. At least up till now not a single human culture is known to us which would have valued conspicuous stupidity by some of its members as being something especially positive in the same way as today each group is proud of this or that clever person believing them to be one of their group. Just the opposite, often it is the most difficult questions which arise about the "essential" or, what is typically meant, the genetic origin of some famous people which lead to the most amusing nationally biased controversies (e.g., Honolka 1976 about the Austro-Hungarian composer Ferenc Liszt: "Raised in Germany, he favored the French and announced himself, without being able to speak a word of Hungarian, to be a Hungarian, which originated from purely romantic chivalry for a repressed people." p. 395; Chazal about Louis de Funès: "A Spaniard from Courbevoie [a small town near Paris]", p. 7). Even though a first serious contest, by pure chance called the "Charles Darwin Award, an annual honor presented posthumously to persons who do the world genetic gene pool the greatest service by exterminating themselves in the most extraordinarily stupid way", has already been successfully introduced (http://www.skyhawk.org/Darwin.htm), it seems that we will still have to wait a while, hopefully, for an Olympic competition for the greatest fools to become a reality.

18 The Cultural Struggle of Genes

While culture has prompted many degrading variations in the germ, it is, on the other hand, also the source of many inheritable improvements, upgrading variations. This is a new field, ..., but how in fact did the specific talents for music, painting, mathematics, etc. originate?

August Weismann

Against all the one-sided biologistic and culturally determined explanatory accounts of human behavior, we can now set a comprehensive biological, i.e., consistent evolutionary concept that regards real cognitive gain and successful genetic adaptation by mutational changes as being identical and therefore only then becomes a uniform theory of evolution of humans. The essential difference to earlier concepts consists of, above all, recognizing the cognitive nature of all living processes and simultaneously taking note of how that quality which up to now appears to us to be special to our species relates to and at last, completely merges with biological evolution. In this somewhat more consistent kind of perspective, the apparent unlimited creativity of mankind is considered as being no different to the apparently very limited creeping ability of a snail which gives reason for justifiable doubt about the privileged position of *Homo sapiens*. The most unexpected but at the same time most important result with regard to our topic states that individual human cognition, whether primitive as in the Stone Age or highly cultured as in modern times, has nothing to do with real cognitive gain since the latter, to repeat it again, can only be achieved by the roundabout route of purely random genetic changes in the course of the Weismannian germ line.

On the other hand, such a uniform theory of evolution in the sense of, as we hope, both Darwin and Weismann gives us the chance to understand for the first time the true nature of our own personal existence as multicellular individuals and consequently its relative functional importance in the evolution of *Homo sapiens*. Hence even though we, as a multicellular phase within a complex reproductive cycle of the germ line and building of the organism do not have the possibility of writing evolution directly, we nevertheless influence it constantly by all our actions and this, in fact, independently of whether we want to or not, beginning with "homely" family life (Becker 1976a) via "harsh" economic struggle (Becker 1976b) up to "higher" primate politics (de Waal 1997, 2000). The special illusion to which we, through our own wish-fulfilling, self-concocted truism—which itself must have its own evolutionary reason—regularly succumb, mainly consists of our believing that we humans, as the sole living organisms in the world, have obtained the privilege of no longer taking part in evolution if we no longer desire to do so. With respect to our daily life this means that each single one of us can do nothing other than continuously influence the evolution of our species. Just like other metazoans, human individuals are nothing more than the most highly efficient gene manipulators who are anxious to spread their own gene combinations and, with slightly less enthusiasm to encourage those of their relatives. The reason is again quite simple: if that had not been so in the past, those of us who have the honor of living today would simply not exist. This also means, however, that each of us, through innumerable generations has been "indoctrinated" by natural selection and hence has, up to now, been a successful social Darwinist. For this purpose, it is not even necessary to belong to a great horde of followers of a power-hungry dictator, quite often even simple stratagems can be sufficient. It is enough, for example, if we, day in day out,—which it is difficult to avoid—make small decisions, since everything we do or not do very probably will have some influence on the proliferation of human genes. This prognosis is not only valid for the complicated interaction of our body and psychology with the physical environment—the fact that our ingenious methods of moving and falling as well as our many diverse sophisticated digestive and biochemical exchange processes have something to do with the long-term success of our genes seems evident—but above all for the no-less complex relationships between the innumerable members of our species.

There is no better example for illustrating this causal connection than the unquestionably highest of all the arts—the art of politics. No matter what the political inclination we have or do not wish to have, every decision we make or do not make, whether we put it into practice or not influences our lives and ultimately the success of the lives of every member of a society. Whether we now, just to cite some "important" examples, vote for or against the introduction of school milk in elementary schools, for or against cycle tracks, for or against camping in parks, for or against making people have haircuts, and lastly, for or against the dissemination of epistemological texts, evolutionary versus creationist ideas, something will be altered within the selective scope of the human conditions of existence.

Let us take the example of cycling, an exclusively human form of locomotion which, after our first crashes when we are very young, may become one of our biggest hobbies. Indeed nowadays, cycling is one of the last great adventures especially downtown when there are no specially built cycle tracks, and this completely without any high-tech frippery. It is clear that in many countries, the building or not-building of networks of bicycle pathways in the cities is influencing the survival and reproductive rate of a whole generation of cyclists. In an extreme case, the compulsorily building of such a network must lead to a corresponding increase in the number of cyclists, a following of the rule which has long been proved empirically in the case of the building of highways and the number of motorists. Joking aside, the question as to whether political decisions concerning the bicycle and its paths can lead to changes in the gene composition of a human population is not just hair-splitting but to show only how decisions of an apparently subsidiary nature can have a direct as well as a decidedly long-term influence on the expression of different human forms of existence. In the case of cycling and even more in the case of car driving, it cannot be excluded that in the course of the twentieth century some subtle displacements of gene frequencies could have been involved even if only very subtle ones.

We will leave behind us the competitive, pedal-pushing ideologists, whose hidden gene manipulations will only be in a position to be investigated by later generations of biologists, and turn to a selectively much more effective example of political traditions where the connection to our own biological evolution will be much clearer. There exists in all the races of *Homo sapiens* the good custom, or at least seen as such, of keeping one group separate from another. This practice certainly began in the first hordes of apes in the early stages of our evolution and since then reached such proportions that now there is an unbelievable complexity to our social structures. Our first freely moving and small groups of individuals have now grown to enormously large units called states which has understandably made the wandering around freely much more difficult than it was previously. Even when modern nations are no longer the small stone-age reproductive societies, most of them still attempt to achieve what in principle those primitive ape social groups so successfully practiced—in bonobos, there are nevertheless still stable groups of up to 100 members or even more all of which know each other (Savage-Rumbaugh and Levin 1995, Waal and Lanting 1997): the maintenance of most uniform selective conditions possible for the majority of their members and therefore the indirect fostering of the gene pool of this society. However, while the homogeneity of the environment of these small social units such as ape or human families can be, alone through the genetic relationships, a relatively high one, when one progresses first to a small village, a township, a county, a state, a country, and finally a community of nations there is an increasing number of different selective pressures. Nevertheless, the paramount aim of all these groups is the purposeful control of the propagation of genes and this is completely independent of the exalted aims of a political society may have given itself. Through their origins themselves, they are simultaneously and automatically a barrier to practically the rest of mankind.

This gives us a new way of defining of culture in the framework of evolutionary theory, namely insofar as this can be nothing other than the varyingly large diversity or uniformity of the sum of selective pressures which results from the attempt of every member of a social unit to influence the whole group in the sense of his or her very personal ambitions. The picture of culture we can therefore construct shows a varyingly large number of single competitors where every individual tries to change or at least noticeably influence the social conditions in his or her own sense, in other words to the individual's long-term advantage measured in biological fitness terms. One of the first predictions we could derive from such a unified biological approach to culture says for example that the smaller the size of the groups within a population the more pronounced should be the cultural diversity among the different groups since then single individuals would have a greater chance of manipulating the social life of their unit. On the other hand, the larger a group of humans is, the more limited should be the possibilities of an unrestrained particularistic development. If we roughly compare the situation of the last so-called traditional cultures (hunters, gatherer, farmers, cattlemen, shepherds, warriors) the remains of which, if not already extinct, are scattered over the whole world with the increasingly dominant great national unities, which we call countries, we quite readily notice that we are concerned here with marked differences in infrastructure. For this purpose let us take the very general field of "education through tradition" which is certainly one of the most important in our social life because it has more direct influence than any other on or even control over the fate of other individuals. To illustrate that any such closer comparison would be concerned with two basically different worlds of living, it is sufficient to take some examples from a selected number of traditional cultures. For instance, it is sufficient to move through Africa and look out for some rather "exotic" customs as described in diverse classical ethnographical monographs (compilation in Krebs 2001):

—An abortion is brought about by kicks in the abdomen. In cases of infanticide, immediately after the birth, the newborn is buried alive (SAN, "bushmen").
—At the age of 8 to 10, boys and girls have their lower incisor teeth extracted and the upper ones filed in the form of the Greek letter delta (HERERO).
—Walking begins a long time before weaning. If there are delays practical measures are taken: the knees are anointed, the child placed in an anthill (ZULU).
—If a baby is born to a chief, the magic practices of anointment and using smoke is insufficient and a human sacrifice is required (ZULU).

—It is forbidden for boys to drink beer while standing but allowed when sitting. Small girls are inculcated not to expose their breasts to thunder and lightning, otherwise their breasts will collapse (ZULU).

—As a sign of betrothal, the mother anoints the head of the bride with fat (MASSAI).

—Boys may not eat, for example, kidneys, heart, liver, and the breast of sheep. Whereas, girls are forbidden to eat the head of a male goat (CHAGGA).

—Uncircumcized men cannot be warriors, one does not like having sexual intercourse with uncircumcised women. It is forbidden for uncircumcised people to eat certain cuts of meat at feasts (KIKUYU).

—Away from the supervision of adults and outside the homestead, two or more boys indulge in masturbating competitions. Masturbation among girls is considered wrong (KIKUYU).

—All men are initiated from boyhood to manhood by a very severe operation. Their brows are cut to the bone ... in six long cuts from ear to ear (NUER).

—The direction in which one sleeps differs according to sex: the boys sleep with their heads toward the door and the girls the other way round (SCHILLUK).

—In the Tanala Ikongo, the mother is the person in authority. Only after the payment of an ox to the woman is the father allowed to bring them up and punish them (TANALA).

Independent of the exoticness of some of these rules (some of our own "modern" customs are no less exotic), what is of interest here is the extent of the existing variance between the groups when one compares more closely the different areas of life with each other. It shows that it is not chance that it is those areas most concerned with biological fitness are the ones with the greatest variability in character. Uwe Krebs, after carrying out a comprehensive worldwide comparison of a total of 38 primitive tribes (Africa: 20; North America: 2, South America: 6, Asia: 7, Australia and Oceania: 3) comes to the following conclusion:

In summary, one can state that the variance of the phenomena of upbringing in traditional cultures lies above all in the details and less in their main features. In all the traditional cultures investigated, the implicit upbringing is dominant, however, in every culture, intentional processes of upbringing reach their quantitative peak during the period of puberty. ... However, very significant is the variance between different traditional cultures in the intentional instruction within the framework of initiation or similar transitional measures. The variance extends from an almost informal transition into the adult world ... to training over many years with tests and being in action as a warrior before the status as a marriageable man is awarded (Uwe Krebs 2001, p. 578/585).

This great cultural variance finds its causal counterpart in a corresponding larger genetic variance among the traditional cultures as far as they have survived until today (for impressive pictures about the incredible diversity of both current and past cultural rites in Africa, see the "soul of Africa" by Müller, Müller and Henning 1999): Africa, for example, in spite of forced colonization and the creation of national boundaries, still has the greatest genetic variance by far among all the human populations (Jorde, Bamshad and Rogers 1998; Kaessmann et al. 1999). In fact, as Svante Pääbo succinctly put it, "the genetic variation found outside of Africa represents only a subset of that found within the African continent. From a genetic perspective, all humans are therefore Africans, either residing in Africa or in recent exile" (Pääbo 2001, p. 1219). If we cross over to the larger social units such as modern national entities, we can distinguish a comparative leveling off and homogenization of the previously prevailing cultural diversity (customs, languages, technologies) which nowadays, in the era of world-wide globalization, could be described as a kind of "end of traditions" but in many cases also as a liberation from unjustifiable prejudices. Within this seemingly unstoppable process it is quite easy to detect at least some of the main contributions coming from the natural sciences:

1) there exists no god nor any other mysterious "ghosts" in the universe (cf. Schrödinger: "I do not find God anywhere in space and time—that is what the honest naturalist tells you" (1944, p. 139)
2) both bodily and mental life ends with death
3) there are no universally valid social laws inscribed somewhere into the universe

The fundamental trend in cultural development appears to be have been more or less one and the same over a long period, i.e., an increasing departure from primarily distinctively small-group organizations in the direction of political structures constantly growing in size and based on ideological or economically unifying premises (e.g., communism, world religions, capitalism). The latter naturally requires a type of person who, in spite of their individual interests and differences, is prepared to and capable of participating successfully in a larger overall system. Here, we find ourselves again in a situation of intraspecific selection toward a higher IQ, which should make it easier for the new people of the future as well as to perceive complex profitable efficiencies of their own biological fitness under constantly changing environmental conditions. In brief, we are concerned with the tension between winners and losers in advancement, a situation we are experiencing so impressively at present causing the world from Afghanistan to America to hold its breath. Ultimately, this has everything to do with the biology of every individual, even when, understandably, in view of a globalization as peaceful as possible cannot always belong to the politically correct state of things. The development shows, however, that it is exactly these current rapid technological and civilizing changes that are in the process of forming a new type of human being. Perhaps a human being who, under the circumstances, – if one wishes to value this as being positive – is less radical not characterized by narrow outmoded tradition – and is a modern, i.e., intelligent, level-headed person who together with others may be able to build a global network of well-balanced, economic cooperation. The road to get there seems to be a long one and, in the meantime, we still find ourselves involved in unnecessary quarrels of nationalistic jealousy. However, as always, everything has its reason and our genomes are to this day the sole directors of evolution.

People who think that describing a modern nation as a political-ideological concentration of a historically grown group egoism is exaggerated should look at the legal constitution of any one of these states more closely. There one can find a great deal and in great detail about the rights and duties of the citizens but very little about foreigners as if, in fact, they even did not exist. Other nationalities are defined rather by exclusion, i.e., by a rigorous exclusion from all the advantages a state organization usually offers its citizens. The foreigner or stranger is conceived primarily as a source of problems which is shown by the fact that they may enter the country for a restricted period as a paying tourist, for example, but for any longer period are faced special hindrances or even great difficulties. It is very interesting that the possible acceptance of such an immigrant is only made easier by family reasons such as marriage and therefore genetic relationships. When one is finally accepted or becomes part of the group by obtaining official citizenship—as a paradigmatic example the many unbelievable difficulties put in the way of obtaining Swiss nationality can be admired in Emil Steinberger's wonderful film study "The Swiss Makers" —, as individual or minority, one is subject to some up to very different selection conditions within the new society (laws, customs, mores, technology, art, cuisine, etc.). This is expressed by, among other things, the valuation of the so-called "others" as against the so-called "natives" in graduations according to origin and numbers: the more ethnically different and the numbers of "foreigners" already there, the more likely one is to hear "there are too many foreigners in our country", even in modern civilizations such as the states of united Europe (Fuchs, Gerhards and Roller 1993). "Assimilation" is just another term for the resulting selection pressure which, seen in the long term affects the success in life and therefore also the genetic success of every "stranger".

The worldwide propagation of biological kinship as a generally accepted factor facilitating a possible judicial and social integration, the so-called "ethnization of law" (Bös 1993), is now interesting insofar as if it were conceivable that human societies could be formed according to any kind of abstract-rational criteria. The fact that even the otherwise impersonal and purely judicial level of the state has something to do with such a personal factor like kinship shows us one thing very clearly and that is the universal influences of real and every day social Darwinism. Concerning the future of the human species, if we should bother ourselves with this most noble of all moral questions, it is not a question of whether we have a supposedly free choice between traditional social Darwinism and any cultural development but solely a question of the *specific kind* of social Darwinism for which we decide following the precise genetic instructions of the many cells of our bodies. If some people understand by cultural developments predominantly the physical battle against or the actual destruction of all competitors, others, thank goodness, understand them to be a bloodless competition, although not the Popperian kind, but still ideas produced by our own genes. However, in one matter we must not give in to false illusions: both extremes, as well as all the levels in between the strategies of political hawks and doves, have a long-term influence on our future genetic evolution. The only remaining question is what style is preferred by each of us but to do this we must know ourselves much better than ever will be possible. Finally, in no way is this a question of an abstract or divine free will, but solely the very real question of every person's very own and completely determinate unique will which, by the way, poses the scientific question of the very diverse statements of the wishes of in the meantime more than 6,000,000,000 fascinating, idiosyncratic, unique individuals. Undoubtedly, a challenging but, at the same time, extremely exciting task for the future. And finally, the chance that Chomsky is right with his hope that it will be only the particular though most varied "character" of mankind and not any environmentally imposed "enlightenment" which eventually can give rise to at least a minimal portion of optimism in face of the many problems which will confront us:

The facts let us surmise and my trust in humanity allows me to hope that there are inborn intellectual structures. If they do not, people are only moldable and accidental organisms, then they are suitable objects for a behavior control from outside. ... I hope, naturally, it will emerge that there are internal structures which determine human needs and the fulfillment of human needs (retranslated from Noam Chomsky 1973, p. 184).

19 Requiem for a Wonder of Nature

If we are to understand evolution, we must remember that it is a process which occurs in populations, not individuals. Individual animals may dig, swim, climb or gallop, and they also develop, *but they do not evolve.* To attempt an explanation of evolution in terms of the development of individuals is to commit precisely that error of misplaced reductionism of which geneticists are sometimes accused.

                    John Maynard Smith

The outcome of our evolutionary considerations of the cleverest of all the apes can be formulated briefly and succinctly: the fictitious wonder of nature the human being is dead or, more realistically, never existed except in our imaginations. The creature *Homo sapiens*, however, with all its wonderful, and to some extent strange illusions and fictions lives on and is still evolving today exclusively according to the universal principles of the biological evolution of living things which means by the method of random genetic mutation and concurrent natural selection. The special thing about this statement above all is that a critical epistemology and the modern synthetic theory of evolution can support each other in the conceptual corroboration of this diagnosis (Heschl 1993c). The most important contribution to epistemology consists of the insight into the fact that cognitive gain by the human subject in the course of his ontogeny is impossible, since the latter, for inherent systemic reasons—as a rule somatic mutations are disadvantageous (Knippers 1997, p. 232)—must remain excluded from any adaptive changes. In other words, the personal individual development is exactly like the embryonic development or epigenesis (see special issue of *Science* 293, 1063—1105)—this latter term comes from the developmental theory of C. F. Wolff in 1759, a time when naïve minds still imagined homunculi, i.e., small manikins, in the fertilized egg cells—namely a process controlled in smallest detail by the just all-knowing intelligent genome (see Shubin, Tabin and Carrol 1997). On the other hand, this makes it necessary that human individuals must have an enormous amount of *a priori* information available and, at the same time, this fundamentally excludes that, in direct opposition to Kant and Lamarck and their numerous open or secret followers (by the way, there exists a quite special Viennese tradition of Lamarckian thinking, from famous people like Kammerer [for "useless", but detailed information about the intricacies of the Kammerer case, see www.uselessinformation.org/kammerer/index.html], Weiss, Bertalanffy, and Schrödinger: "there is a very true kernel in Lamarck's view" [1944/1996, p. 107] up to Koestler and even Lorenz) a real increase in cognition is possible *a posteriori*, i.e., in the course of individual development. Hence, to a certain degree Lamarck's superfluous idea of the inheritance of acquired characters can be equated with Kant's error of the possibility of cognitive gain *a posteriori*.

Instead of predicted 100.000 (see *Science* 272, 1098) the human genome has now been determined to comprise not much more than about 30,000 distinguishable genes, which seems a rather low number at first sight (for a short discussion, see the Appendix) and, in addition, the purely genetic difference between us and our nearest relatives, the two species of chimpanzee, amounts to, as we have already mentioned, not more than an estimated 1.6% of the whole human DNA sequence consisting of 3.15 billion base pairs. The question then heard from the innumerable opponents of any genetic explanation is "How can such a trivial difference account for the so remarkable cognitive differences?" There are two things to say about this. First, a percentage alone tells us very little about the absolute numerical proportion, which is the reason why percentages are often used to give a certain—often completely wrong—impression. In reality, if we take the absolute number of genes, this would still be about 480 genes, which, in a multitude of different combinations, would differentiate man and the closest apes, whereby, for the sake of simplicity, the total number of identifiable genes in both species can be assumed to be approximately the same. What now dramatically increases the amount of information to be stored is especially the dynamic interaction between the two kinds of nucleic acids found in the nucleus, the single-stranded RNA and the double-stranded DNA, the latter being commonly understood as the actual main store of genetic information. In simple prokaryotic organisms (bacteria, blue algae), which still do not have a separate nucleus, most of the information is translated directly from DNA into RNA and thereafter transferred on further to the ribosomes, the production plants for proteins. In the eukaryotes, however, which have a nucleus and to which all metazoans belong, one and the same DNA sequence, by a variable translation into more than one single RNA, can produce a much greater amount of different cell products according to requirements. In the meantime, complete networks of interconnected RNA sequences have been discovered, which it seems makes it legitimate to accept, alongside the genotype which produces a phenotype, a so-called "ribotype" (alternative term: "transcriptome") which makes possible a complicated but above all an essentially more flexible translation of the information stored in the DNA into diverse molecular end products and processes (Herbert and Rich 1999; Keegan, Gallo and O'Connell 2000).

These new findings could possibly make the puzzle of the so-called junk DNA somewhat more transparent—its share of the human genome can, according to some estimates, be up to 95% (see Lander and Weinberg 2000). What is meant by junk DNA is those sections of the whole DNA that have, up till now at least, no verifiable function, e.g., they do not code any protein directly, for that reason called "noncoding DNA". If indeed 95% of the human genome has nothing to say to us, we have to ask ourselves what good will decoding it have when a great part of it is genetic rubbish. Investigation of the ribotype gives us completely new perspectives, e.g., it has been proven in detail that many so-called "introns"—these are the noncoding sections within an active gene sequence consisting of the sum of the exons—can play an important role in the production of the various types of RNA, i.e., the ribotypes. All these more recent investigations are at the same time a verification of a purely linguistic analysis of noncoding DNA sequences in which, as long ago as 1994, the possibility in principle of a still to be discovered information content was predicted (Mantegna et al. 1994). It is therefore no longer to eliminate the possibility that even these noncoding sections of DNA outside gene sections—so-called "secondary DNA"—have a definite biological function and therefore contain information. However, in this area, the C-value paradox (Mirsky and Ris 1951) applies to a considerable extent up till now, which concerns the simple fact that although there has been a general increase in hereditary material in the course of evolution—from 10^8 base pairs in fungi to more than 10^9 in birds and mammals—many amphibians (e.g., salamanders, frogs) have up to 25 times more DNA than all mammals which is in direct contrast to the complexity of the latter. However, a plausible reason has been put forward for this finding, namely that the amount of secondary or non-coding DNA is primarily a matter of the size of the cells of the species concerned and indeed such an expected positive correlation has been shown between the two parameters (Bennett 1972, Shuter et al. 1983, Cavalier-Smith 1985). A proper skeleton DNA hypothesis states that a certain amount of surplus DNA is necessary to 1) increase the volume of the nucleus by the additional folding and 2) as a consequence, to keep the ratio of the size of the cell to the size of the nucleus constant during the growth phase (Cavalier-Smith 1978). The latter seems to be necessary to maintain a balance between the rate of synthesis of nuclear RNA and that of cytoplasmic proteins.

As far as our theme is concerned, we can assume that in the near future with a high degree of certainty that the rapid acceleration of functional genome research will tell us that we can no longer exclude that complex phenotypic traits such as behavior and cognition in their type of expression are completely predetermined by genetic information. In fact, there are already suggestions to link even complex cognitive functions such as consciousness and memory to those regulatory processes at the RNA level which have not been seriously investigated up till now (Schröder 2000). The very different kinetics at least speak in favor of this hypothesis: while the path to perfect three-dimensional folding and therefore functionally prepared proteins is always a comparatively complicated and lengthy process (Hartl, Hodlan and Langer 1994), RNA molecules, with their copying rate of approximately 30—60 nucleotide building blocks per second (Knippers 1997), can intervene in cell events much quicker especially if they are not too long.

Our final result here is therefore a first attempt at a really comprehensive theory of evolution that is thought to explain the development of mankind exactly like all the other living organisms on this planet by way of the uniform principles of biological evolution. Formulated more concretely, this means that expressions such as cultural evolution, social evolution, or scientific evolution exactly like the attractive idea of an "evolution" of beaver dams, anthills, birds' nests, or bowerbird's bowers is not an adequate description of what evolution really is (among other things, this explains why experts in this domain are primarily interested in the animals' skills: "Animal Architecture and Building Behaviour" [Hansell 1984], "Nest Building and Bird Behavior" [Collias and Collias 1984], namely the adaptive change of the dynamic process structures of living systems in time. All of our individual knowledge is of necessity always pre-existent knowledge, otherwise the lawful, i.e., nonrandom ontogenetic genesis of human behavior, and this includes the whole of cognitive ability, could never be explained. Otherwise we would not be able to get rid of the very annoying question which would remain of how the subject could ever know what it had to do in reaction to a challenge from the environment. Just to name a few trivial examples from everyday life: How do I react to a rapidly approaching stone that appears to be on a direct collision course with me? How do I digest all this unbelievable rubbish that we stuff into ourselves as food? In addition: how do I find the object that has unexplainably disappeared from the area of direct perception? How do I perceive causal forces in the physical world if for example an object resists its planned relocation? Finally, how do I react to emotional sounds (e.g., laughing), social gestures (e.g., winking), but also to purely symbolically meant hints and questions (e.g., "How are you?") of members of my own species? If the environment itself were to whisper unobtrusively the adequate interpretations of all those numerous problems to us, then it could immediately do so in all the other crises of our, on one side robust and on the other permanently fragile existence and we would immediately join the elite caste of angels and be immortal. Since we have already seen how terribly boring and meaningless such an immortal existence would be, whether in heaven or on earth, we should not necessarily regret this apparently desirable redemption.

The well-known philosopher, and even better behavioral researcher, David Hume, with his very skeptical view of human reasoning, recognized very quickly that this wonderfully adapted reasoning could no longer be explained by pure logic. Because he was more interested than any thinker before or after him in a natural explanation of this puzzle of nature which is generally considered divine or at least supernatural, it occurred to him as the only possible, even if not yet really evolutionary solution to the basic question of every cognitive theory. In contrast to Plato, who, although he was among the earliest cognitive skeptics (see selected parts from "Meno" and "Phaedo" in Moser and vander Nat 1987, pp. 28—34), apparently went along the wrong path of everlasting ideas for primarily religious reasons, Hume came to the conclusion that man, viewed in the light and not in a mystical twilight, is a creature endowed with understandably very complex, but nevertheless quite naturally evolved thinking instincts (Hume 1777/1992, p. 85). The totality of all his associations, no matter how daring and new to the individual they may appear, are only set in motion within a concrete kind of preset and well-defined framework.

At this point, it would be rather unfair not to honor the contribution of Kant, who, although as a declared opponent of Hume and with German thoroughness and rather too elaborately, came to a very similar conclusion. Things must remain unrecognizable since human cognition can only consist of applying predetermined thought patterns to the world. Cognition, and this is an extremely biological concept, only takes place within the cognizing subject himself who is only capable of adaptive reactions by his own cognitive interpretation of the environment. In this way Kant's a priori categories of cognition nicely confirm Hume's instincts of reason and they both lead the time-honoured tradition of epistemological skepticism, which began in early antiquity if not much earlier, increasingly nearer and nearer to the borders of biology. Nevertheless, one crucial theoretical mistake by Kant has remained to the present day. In contrast to Hume's postulate of preformed instincts, his "experiences" a posteriori are the direct results of the universal, maybe even species-specific Lamarckian illusion (paradigm: I can learn by experience) and laid the path to the later splitting of human behavior into inborn and acquired parts which still determines the public discussion of the biology of our species and considered insoluble because of the misunderstood controversy over heredity.

Only the almost simultaneous discovery of evolution by Lamarck, Darwin and Wallace caused the effigy of that wonder of nature man to waver. Darwin was no philosopher but more than any other was interested in philosophical questions. He was fundamentally skeptical of the divine origin of mankind and was the first to notice the great kinship of all living things and unlike many of his contemporaries attempted to explain it by the historical connection of an all-inclusive evolution. In principle, Carl von Linné (1707—1778), with his still used simple, but like our traditional system of personal names (forename, family name) nevertheless very efficient classification method of a binomial no-menclature (genus name followed by species name: e.g., *Homo sapiens*) for plants and animals could have also had the idea that a temporal connection is not to be excluded. A century before Darwin in the partly already extraordi-narily detailed systematic overviews of the plant and animal kingdoms no real phylogenetic tree was seen that could mirror the millions of years of evolution and slow origin of the numerous known species. Darwin's inspiration in that matter is no longer difficult to comprehend since he only noticed that new spe-cies could arise in a similar manner to how new races originate in the breeding of domestic animals by human manipulation, i.e., artificial selection, however, in this case, by natural selection. Since it was obvious that individuals of the same species could have very different characteristics, it was also very quickly understood that the phenomenon of a varyingly successful reproduction (dif-ferential reproduction), in either the short or the long term, must have an ef-fect on the chances of evolutionary survival of diverse genetic lineages. However, as often forgotten, Darwin was also a declared and convinced Lamarckian and, as such, like practically every other contemporary naturalist, just as convinced of the idea that the reasons for the differences between indi-viduals, which represent the prerequisite for evolution, where to be found in the intentional adaptation of the individual to the environment. Since the transmission of these apparently new acquisitions to the next generation was necessary to keep the game of evolution in motion, it was only logical for him to invent a proper mechanism (pangenesis) in which microscopic particles, the already mentioned gemmules, were assumed to pass on the experience of the individual to the next generation.

As we now know, Darwin wanted a great deal more than just to show that we are related physically with the apes. He was without doubt the first evolutio-nary psychologist and epistemologist embodied in one person and as such convinced that it must be possible to integrate every single one of the charac teristics of mankind, i.e., the sum of all his extraordinary intellectual capabili-ties, in the thesis of a biological descent. However, it is exactly this idea, with the subsequently invented pangenesis mechanism together with the compro-mise negotiated by later generations of biologists that only a varying percen-tage of traits was inherited, which had to remain condemned to failure up to the present day since it has allowed the human intellect, or better, its diverse famous defenders, to slip through any escape hole to avoid a natural explana-tion. The logical trick used was always the same: because our wonderful in-tellect can extract new ideas from itself, it can never be a slave to a simply ma-terialistic biology.

Only if we think out the skepticism that began purely philosophically in early Greece with respect to the possibilities of our cognitive abilities consi-stently to a conclusion can we reach the synthesis, which to a certain degree

Darwin and Wallace had in mind but were not able to take the final steps. The fundamental aim of this undertaking has remained the same to the present day. George Williams, who is known for his pithy turn of phrase, has managed to achieve this feat in packing all of it into one single sentence:

> The Darwinian process of natural selection explains all aspects of the adaptation of an organism to a particular life style in a particular environment (George Williams 1992, p. 5).

In our times, the founder of the modern philosophy of science, Karl Popper, at least recognized that a real acquisition of new cognition can never be simply induced from the environment in a Lamarckian way but instead needs a purely endogenous change of the cognitive structure of the subject to make any progress possible. For this reason Popper developed a consistent deductive epistemology in which solely processes that follow the strict rules of logic play a role. Nonetheless, this took him so far that he was already on the point of formulating the first version of a truly Darwinian theory of cognition which would have fulfilled George William's claim perfectly if it had not been for exactly 0.1% which prevented its being flawless:

> However, I claim that our knowledge is at 99%, or even 99.9% inherited; and the rest is a modification, a revolutionary upheaval of some previous knowledge, as this for its part was again a revolutionary upheaval of something preceding, and so on. However, all knowledge ultimately goes back to inherited knowledge (Karl Popper 1988, p. 36).

Consequently, Thomas Kuhn, a dissident physicist, tried to develop his own epistemology for the sciences and found that Popper's skepticism about the actual possibilities of cognitive gain was a little too orthodox, not to say paradoxical—one only has to think about his strange attempt to forbid induction. His concept of a paradigmatic change, which alludes to the central problem of scientific progress, points directly to that fundamental limitation which can only be completely understood if the theory of evolution is uncompromisingly applied not only to animals, but also to our own species, which is just as animal as all the other species. Kuhn was therefore one of the very first theorists to stop excluding that radically new things cannot be understood by us and therefore cannot be transferred by speech. Parallel to this, it occurred to Noam Chomsky that, as far as the acquisition and the use of human speech during our short lifetime is concerned, learning by trial and error would lead to, at the very most, a meaningless mumbling. This means that the structure of language as well as verbal communication itself must be something completely different than that Lamarckian wonder of nature which it is always considered to be. If Chomsky had lived in the nineteenth century, he would have been able to point out to Darwin that the organic nature of speech could give some important information for a comprehensive theory of evolution.

We can see that it was not just inconsiderate, ill-natured, and badly reductionist scientists who alone were interested in demystifying our species but also innumerous critical philosophers who have contributed to our no longer wanting to believe in humans being a unique exception of nature. All the same, even the greatest skeptics never went so far as to claim that mankind could not occupy any special position at all within the realm of living things. Plato, Hume, and Kant as well as Popper, Kuhn, and Chomsky have all openly accepted this special position. It seems that solutions to puzzles are apparently always especially difficult to accept if, on one hand, they are very simple and, on the other, convey a message that we would rather not hear. A good example of this is the gambler who is certain that the roulette ball is more likely to land on black rather than red this time because it has landed so often on red for the last few turns of the wheel. The fact that the chances of either red or black remain the same no matter what has happened before cannot convince him that he should not double his bet on the black.

We find a very similar situation when we look at the question of mankind and its relationship to biological evolution but, this time, the reverse way round. While we do not like irksome chance being in the game, we treasure everything in human nature that smacks of indeterminable freedom. So we love to talk about the limitlessness of human creativity, the unfathomable nature of individual decisions, the influence of indeterminate quantum jumps in our thoughts, about wave/particle and yin/yang dualisms in our consciousness and, of late, about the chaotic zigzag walks of our intellect. The current popularity of modern chaos research is probably explained to some degree by man's desire for apparently enigmatic and therefore unpredictable influences on our deeds and thinking over which we try to draw a cloak of inexplicability. Or, as the recently deceased Eysenck (cf. Nature 389, 794) so tellingly has expressed nearly 30 years ago:

It seems to be a need for humans to believe in an unlimited plasticity of human intellectual abilities, and every experiment that favors this belief is accepted without criticism, praised and therefore commonly known, even when no published report of its success exists! (retranslated from Hans Jürgen Eysenck 1975, p. 129).

This strong enthusiasm for the supposed freedom of chaos or any other potential indicator of basically unlimited cognitive plasticity is not dimmed at all by the fact that deterministic chaos—mathematical chaos theory is primarily what is being talked about—has little to do with real chance and indeterminacy even when, it is applied increasingly, as recently (Vandervert 1997), to online registrations of nerve impulses (Freeman 1975, 1991). In fact, not even modern brain research complies with our basic need for as much internal chaos as possible since it is becoming clearer day by day how deterministic different cortical activity patterns can be despite the accompanying random "noise" (Arieli, Sterkin, Grinvald and Aertsen 1996).

Again, looked at from an evolutionary psychological point of view, this notable dislike of everything that is predictable makes sense since who ever would like to be seen through by his more direct opponents (e.g., conspecifics, competitors, neighbors) let alone by his direct physical enemies (e.g., predators, parasites)? Surely not any of our many ancestors and, with great probability, not one of us who is still living. On the contrary, even today there is a demonstrable selection pressure towards camouflage and deception for which already in growing children a good portion of intelligence is needed (LaFreniere 1992; Perner, Baker and Hutton 1994; Peskin 1992; Sodian 1994). Two small examples may suffice: if a two-year-old toddler hides behind a bush and his view of us is obstructed, he thinks he is very clever because he is sure we cannot see him and do not know what he is up to. However, when a crafty politician regularly before an election points out the hollow speeches and empty promises his opponents are making it is much more difficult to see that he probably has not the slightest intention of keeping his own promises, even if it is "only" to put an end to state mismanagement (see Hardin's "The tragedy of the commons" from 1968).

So what is so unpleasant about wanting to apply modern evolutionary theory with all its consequences for the interpretation of human behavior if we ultimately have no choice anyway? No question about it, it tells us straight to our face the exact opposite of what we have so gladly been used to hearing. The undirected mutation, even in complicated multi-cellular animals, represents the only possibility of achieving adaptive progress and all the many fantastic ideas that we hatch out ourselves in our remarkably large brain-receptacles called heads during our lifetime have nothing to do with cognitive gain. Is this a brutal shock for philosophy? Not necessarily, because a further developed skeptic epistemology with consistent Hume-like considerations taken to its conclusions directly confirms the validity of certainly the most important statement of evolutionary theory. Since cognitive gain, by definition, can only take place through complete cognitive unawareness and lack of comprehension or, in biological terms, random indeterminacy, everything that has changed living organisms in the long course of their evolution has a very great deal to do with real cognitive gain.

Applied to mankind, this says that all our learning and thinking must be totally, i.e., not only concerning so-called "universal" or species-specific cognitive structures but also down to the smallest details, laid down in the Weismann's germ plasm. This must of course be valid for every scientific work, this one included. For those skeptics who still want to hold on to the idea of the wonder of man and therefore stand in direct opposition to the skepticism expressed here with regard to the supernaturalness of the human race, the thesis presented here can also be considered as an empirical prognosis that should be confirmed in the not too distant future by concrete molecular biological research. However, even now the trivial but simultaneously noticeably often ignored fact should be pointed out that each individual cell from the many thousands in a human body have no other possibility than to manage all the necessary complex reactions with the help of one identical copy of the genome of the germ cell. In addition, considered quantitatively, 15% of the dry weight of every individual human consists of DNA (3.7 kg from 70 kg total body weight) not to speak of the impressive theoretical total length of unfolded DNA alone in the brain. Starting from a DNA size of approximately 2 m and a number of about 100 billion neurons per brain, we end up with a length of 200 million kilometers. Who under these circumstances still believes we are only very indirectly influenced by our genetic basis—"basis" always erroneously meaning just the genome of easily forgotten fertilized egg cell since so far away from our personal experience—should remain a believer in man as a supernatural wonder of nature for ever. Furthermore, the theory of evolution also explains, why any Lamarckian form of evolution must be impossible, or better formulated in the truest sense of the word must be meaningless, since real semantic information can never be taken in from outside in any directed form whatsoever. If that were not so, we would have suddenly created a world of no longer changeable organisms since for all time they would be always perfectly adapted with the end result that such immortal angelic creatures would have been condemned to lead a completely boring existence. However, since life on earth is obviously anything but a boring affair, we can assume that the evolutionary view of things is not far wrong.

At the end of every fable there is a moral of the story and we can express this even more briefly since everyone is free to give his own opinion. The interpretation of any phenomenon is a matter for the individual, a fact which, in the meantime, thanks to the influence of one of the founders of modern constructivism, Heinz von Foerster—one only thinks of his unbeatable slogan "Objectivity is the delusion that observations can be made without an observer" (in v. Glasersfeld and Ackermann 1997, p. 43) —, is going to be taken over by an increasing number of contemporary cognitive scientists (cf. Peschl and Riegler 1997). The result of this unavoidably subjective interpretation is, as we saw in the example of language, even using symbols and other supposedly "objective" carriers of information is not transmissible. This is not completely true because fortunately there is still that age-old proven method called sex but this has something to do with morals. In conclusion, we do not need to give anyone good advice since there is no lack of evolutionarily proved givers of advice and schoolteachers as shown by the mountain of literature of innumerable supposedly progressive or conservative moral apostles. On the contrary, every one of us practices his or her evolutionary ethics and does it day by day and minute by minute. What the individual, no matter whether deliberately or unintentionally, attempts to do is to permanently influence the social environment in terms of his genes. Anyhow, from now on we all therefore have the right to insult each other as bad social Darwinists (e.g., as left-wingers, right-wingers, liberals, greens, nationalists, racists, fundamentalists, cyclists, drivers, stamp collectors, etc., etc.) since our respective behavior itself is always simultaneously a part of natural selection that, during a long past, has formed all the members of our species, is still forming them today, and will continue to do so forever. Who then, must we ask, would want to abdicate willingly from his very own evolutionary influence? Moreover, the renunciation of giving good advice practiced here is only an apparent and perhaps even highly sophisticated suggestion of our genes, perhaps of the purely scientifically interested ones in our genome who do not want anything to do with any enervating moralistic ifs and buts.

Where does all this leave us with all this menacing genetics and molecular biology around us or, above all, in us with regard to the much-praised freedom of will one asks oneself at the very end. We need not be afraid, the theory of evolution has provided for this strongly human need since what it finally says is nothing other than that every living organism, from the most inconspicuous ameba to the self-appointed cleverest primate, possesses the freedom independent of all external influences to react to the environment in the way which according to his nature is the most sensible and up to the present day the most tried and tested. We do not want to be all too supercilious here at the end but anyone who has understood this will no longer consider intensified research into the phenomenon of life as merely materialist, unspiritual biology. From a supposed tyranny of anthropomorphic, malignant, apparently selfish genes which control us poor "vehicles" as if we were machine-like marionettes, in the present more extended evolutionary perspective, we have a basically autonomous, independent, self-determination of every living creature realized by a cognizant, self-organizing genome. It is rather a disgrace, but it was a theoretical physicist and not a biologist who predicted the consequences of the natural sciences for the interpretation of the human intellect more than fifty years ago— with a more than clear reference to all that what is now, some decades later, considered to be the newly created field of functional (Hieter and Boguski 1997) or, since starting after the mere sequencing, post-genomics (websites: www.sciencemag.org/feature/ plus/sfg/; www.nature.com/genomics/post-genomics/), a field which, if James Watson is right with his prediction that "a more important set of instruction books will never be found by human beings" (cited in Grayling 2000), will instigate a radically new era of biological research (Butler 2001, Spengler 2001):

In calling the structure of the chromosome fibers a code-script we mean that the all-penetrating mind, once conceived by Laplace, to which every causal connection lay immediately open, could tell from their structure whether the egg would develop, under suitable conditions, into a black cock or into a speckled hen, into a fly or a maize plant, a rhododendron, a beetle, a mouse or a woman. To which we may add, that the appearances of the egg cells are very often remarkably similar; and even when they are not, as in the case of the comparatively gigantic eggs of birds and reptiles, the difference is not so much in the relevant structures as in the nutritive material which in these cases is added for obvious reasons. But the term code-script is, of course, too narrow. The chromosome structures are at the same time instrumental in bringing about the development they foreshadow. They are law-code (today: "structural genomics"; Rowen, Mahairas and Hood 1997) and executive power (today: "functional genomics"; Henikoff et al. 1997, Tatusov, Koonin and Lipman 1997)—or, to use another simile, they are architect's plan and builder's craft—in one (Erwin Schrödinger 1944, p. 21/22).

Literature

Abel, T., Martin, K. C., Bartsch, D. & Kandel, E. R. (1998): Memory Suppressor Genes: Inhibitory Constraints on the Storage of Long-Term Memory. *Science* 279, 338—341.

Abelson, H. T. & Rabstein, L. S. (1970): Lymphosarcoma: virus-induced thymic-independent disease in mice. *Cancer Res* 30(8), 2213—2222.

Acheson, A. et al. (1995): A BDNF autocrine loop in adult sensory neurons prevents cell death. *Nature* 374, 450—453.

Agrawal, A. A. (2001): Phenotypic Plasticity in the Interactions and Evolution of Species. *Science* 294, 321—326.

Aguzzi, A. & Weissmann, C. (1997): Prion research: the next frontiers. *Nature* 389, 795—798.

Ahuja, A. (2000): He knows how we know. Gerald Edelman thinks he knows the secret of consciousness. *Times* July 6, 2000.

Alcock, J. (2001): *The Triumph of Sociobiology*. Oxford University Press, Oxford.

Alkon, D. L. (1983): Eine Meeresschnecke als Lernmodell. *Spektrum der Wissenschaft* 9, 38—49.

Ameisen, J. C. (1996): The Origin of Programmed Cell Death. *Science* 272, 1278—1279.

Anderson, J. R. & Gallup, G. G. Self-recognition in Saguinus? A critical essay. *Anim. Behav.* 54, 1562—1567 (1997).

Andersson, J. O., Doolittle, W. F. & Nesbø, C. L. (2001): Genomics: Enhanced: Are There Bugs in Our Genome? *Science* 292, 1848—1850.

Angermeier, W. F. (1984): *The Evolution of Operant Learning and Memory. A Comparative Etho-Psychology*. Karger, Basel.

Apter, M. & Wolpert, L. (1965): Cybernetics and Development. *J. theor. Biol.* 8, 244—257.

Archer, J. (1988): *The behavioral biology of aggression*. Cambridge University Press, Cambridge.

Arieli, A., Sterkin, A., Grinvald, A. & Aertsen, A. (1996): Dynamics of Ongoing Activity: Explanation of the Large Variability in Evoked Cortical Responses. *Science* 273, 1868—1871.

Arnold, M. L. (1997): *Natural Hybridization and Evolution*. Oxford University Press, Oxford.

Auerbach, C. (1976): *Mutation research. Problems, Results and Perspectives*. Chapman & Hall, London.

Avise, J. C. (2001): Evolving Genomic Metaphors: A New Look at the Language of DNA. *Science* 294, 86—87.

Axelrod, R. (1984): *The Evolution of Co-operation*. Penguin, Middlesex.

Baasner, R. (1997): *Georg Christoph Lichtenberg: In meinem Kopf des Nachts*. dtv, München.

Baldwin, J. (1994): *DNA Pioneer: James Watson and the Double Helix*. Walker & Co.

Bally, G. (1945): *Vom Ursprung und von den Grenzen der Freiheit, eine Deutung des Spieles bei Tier und Mensch*. Birkhäuser, Basel.

Barash, D. P. & Lipton, J. E. (1997): *Making Sense of Sex. How Genes and Gender Influence our Relationships*. Island Press.

Barbas III, C. et al. (1997): Immune Versus Natural Selection: Antibody Aldolases with Enzymic Rates But Broader Scope. *Science* 278, 2085—2091.

Barinaga, M. (1995): Receptors Find Work As Guides. *Science* 269, 1668—1670.

Bateson, P. (1978): Sexual imprinting and optimal outbreeding. *Nature* 273, 659—660.

— (1979): How do sensitive periods arise and what are they for? *Animal Behaviour* 27, 470—486.

— (1982): Preferences for cousins in Japanese quail. *Nature* 295, 236—237.

— (1983): *Mate choice*. Cambridge University Press, Cambridge.

— (1976b): Altruism, Egoism, and Genetic Fitness: Economics and Sociobiology. *Journal of Economic Literature* 14, 817—826.

Bekoff, M. & Jamieson, D. (1996): *Readings in Animal Cognition*. MIT Press, Cambridge, MA.

Bell, G. (1982): *The masterpiece of nature*. University of California Press, Berkeley.

Bennett, M. D. (1972): Nuclear DNA content and minimum generation time in plants. *Proc. R. Soc. Lond.* B 178, 259—275.

Berger, P. L. & Luckmann, T. (1982): *Die gesellschaftliche Konstruktion der Wirklichkeit*. Fischer, Frankfurt.

Berthoz, A. et al. (1995): Spatial Memory of Body Linear Displacement: What Is Being Stored? *Science*.

Biben, M. (1998): Squirrel monkey playfighting: making the case for a cognitive training function for play. In: *Animal Play. Evolutionary, Comparative, and Ecological Perspectives* (ed. M. Bekoff & J. A. Byers), pp. 161—182. Cambridge University Press, Cambridge.

Bickerton, D. (1990): *Language and Species*. University of Chicago Press, Chicago.

Bischof, N. (1985): *Das Rätsel Ödipus. Die biologischen Wurzeln des Urkonfliktes von Intimität und Autonomie*. Piper, München.

— (1991): *Gescheiter als alle die Laffen. Ein Psychogramm von Konrad Lorenz*. Rasch & Röhring, Hamburg.

Bischof-Köhler, D. (1988): Über den Zusammenhang von Empathie und der Fähigkeit, sich im Spiegel zu erkennen. *Schweizerische Zeitschrift für Psychologie* 47(2/3), 147—159.

Bishop, D. V. M., North, T. & Donlan, C. (1995): Genetic basis for specific language impairment: evidence from a twin study. *Dev. Med. Child Neurol.* 37, 56—71.

Bitterman, M. E. (1975): The comparative analysis of learning. *Science* 188, 699—709.

— (1988): Vertebrate-invertebrate comparisons. In: *Intelligence and evolutionary biology* (ed. H. J. Jerison & I. Jerison), pp. 251—276. Springer, Berlin.

Blaser, P. (1995): Mangels Darstellbarkeit nicht mehr präsent. *Univers* 6, 10—15.

Bloom, F. E. (2000): The Endless Pathways of Discovery. *Science* 287, 229.

Bornberg-Bauer, E. & Chan, H. S. (1999): Modeling evolutionary landscapes: Mutational stability, topology, and superfunnels in sequence space. *Proc. Natl. Acad. Sci. USA* 96(19), 10689—10694.

Bös, M. (1993): Ethnisierung des Rechts? Staatsbürgerschaft in Deutschland, Frankreich, Grossbritannien und den USA. *Kölner Zeitschrift für Soziologie und Sozialpsychologie* 45, 619—643.

Bouchard, T. J. et al. (1990): Sources of Human Psychological Differences: The Minnesota Study of Twins Reared Apart. *Science* 250, 223—228.

Bouchard, T. J. & McGue, M. (1981): Familial studies of intelligence: A review. *Science* 212, 1055—1059.

Bower, T. G. R. (1971): The object in the world of the infant. *Scientific American* 225, 30—38.

— (1977): *A primer of infant development.* Freeman, San Francisco.

— (1979): *Human Development.* Freeman, San Francisco.

Bowlby, J. (1969/97): *Attachment.* Pimlico.

Boyd, R. & Richerson, P. J. (1976): A simple dual inheritance model of the conflict between social and biological evolution. *Zygon* 11, 254—262.

— (1985): *Culture and the Evolutionary Process.* Chicago University Press, Chicago.

Bradshaw, J. L. & Rogers, L. J. (1993): *The Evolution of Lateral Asymmetries, Language, Tool Use, and Intellect.* Academic Press, San Diego.

Bragdon, A. D. (1996): *Can you pass these tests?* Barnes & Noble, New York.

Braitenberg, V. & Schüz, A. (1989): Cortex: hohe Ordnung oder grösstmögliches Durcheinander? *Spektrum der Wissenschaft* 5, 74—86.

Brand, C. (1996): *J. Biosocial Sci.* 28, 387—404.

Brown, P. (2001): Cinderella goes to the ball. *Nature* 410, 1018—1020.

Brownlee, A. (1954): Play in domestic cattle: an analysis of its nature. *Br. Vet. J.* 110, 48—68.

Bruner, J. S. (1964): The course of cognitive growth. *American Psychologist* 19, 1—15.

— (1968): *Processes of cognitive growth: Infancy.* Clark University Press, Worcester MA.

Bruner, J. S., Jolly, A. & Sylva, K. (ed., 1976): *Play: its role in development and evolution.* Basic Books, New York.

Brunswik, E. (1955): Ratiomorphic Models of Perception and Thinking. Acta *Psychologica* 11, 108—109.

Burgess, E. W. & Wallin, P. (1953): *Engagement and marriage.* Lippincott, New York.

Burghardt, D. & de al Motte, I. (1988): Big 'antlers' are favoured: female choice in stalk-eyed flies. *J. Comp. Physiol.* A162, 649—652.

Burghardt, G. M. (1998): The evolutionary origins of play revisited: lessons from turtles. In: *Animal Play. Evolutionary, Comparative, and Ecological Perspectives* (ed. M. Bekoff & J. A. Byers), pp. 1—26. Cambridge University Press, Cambridge.

Burkart, J. & Heschl, A. (2000): Marmosets and Mirrors. *6th Workshop of the European Marmoset Research Group,* Paris April 3—5, 2000.

Burke, T. et al. (1992): Molecular Variation and Ecological Problems. In: *Genes in Ecology* (ed. R. J. Berry et al.), Blackwell, Oxford.

Buss, D. M. & Barnes, M. (1986): Preferences in human mate selection. *Journal of Personality and Social Psychology* 50, 559—570.

Butler, D. (2001): Are you ready for the revolution? *Nature* 409, 758—760.

Callebaut, W. (1993): *Taking the Naturalistic Turn or How Real Philosophy of Science Is Done.* University of Chicago Press, Chicago.

Camargoa, A. A. et al. (2001): The contribution of 700,000 ORF sequence tags to the definition of the human transcriptome. *Proc. Natl. Acad. Sci.* USA 98(21), 12103—12108.

Campbell, D. T. (1960): Blind variation and selective retention in creative thought as in other knowledge processes. *Psych. Rev.* 67, 380—400.

— (1974): Evolutionary epistemology. In: *The Philosophy of Karl Popper* (ed. P. A. Schilpp), pp. 412—463. Open Court, LaSalle.

Campbell, R., De Gelder, B. & De Haan, E. (1996): The lateralization of lip-reading: a second look. *Neuropsychologia* 34(12), 1235—1240.

Cann, R. L. et al. (1987): Mitochondrial DNA and human evolution. *Nature* 325, 31—36.

Carpenter, C. R. (1964): *Naturalistic Behavior of Nonhuman Primates.* Pennsylvania State University Press, University Park.

Carroll, S. B., Grenier, J. K. & Weatherbee, S. D. (2001): *From DNA to Diversity. Molecular Genetics and the Evolution of Animal Design.* Blackwell, Malden, MA.

Cartwright, J. (2000): *Evolution and Human Behaviour.* Macmillan, London.

Cattell, R. B. (1968): Are IQ-tests intelligent? *Psychology today reprint series* p-16, March 68.

— (1966): *Handbook for the culture Fair Intelligence Test.* Champaign, Illinois.

Cavalier-Smith, T. (1978): Nuclear volume control by nucleoskeletal DNA, selection for cell volume and cell growth rate, and the solution of the DNA C-value paradox. *J. Cell Sci.* 34, 247—278.

— (1985): Cell volume and the evolution of eukaryote genome size. In: *The evolution of genome size* (ed. T. Cavalier-Smith), pp. 105—184. Wiley, Chichester UK.

Cavalli-Sforza, L. L. & Cavalli-Sforza, F. (1994): *Verschieden und doch gleich.* Knaur, München.

Cavalli-Sforza, L. L., Menozzi, P. & Piazza, A. (1994): *The History and Geography of Human Genes.* Princeton University Press, Princeton, NJ.

Chalmers, A. F. (1990): *What is this thing called Science?* Open University Press, Buckingham.

Chance, M. (1996): A socio-mental bimodality. In: *The Archeology of human Ancestry* (ed. J. Steele & S. Shennan), pp. 397—419. Routledge, New York.

Chao, L. & Tran, T. T. (1997): The advantage of sex in the RNA virus phi6. *Genetics* 147, 953—959.

Chazal, R. (1972): *Louis de Funès.* Denoel, Paris.

Cheverud, J. M. (1988): A comparison of genetic and phenotypic correlations. *Evolution* 42, 958—968.

Chomsky, N. (1965): *Aspects of the Theory of Syntax.* MIT Press, Cambridge MA.

— (1973): Linguistik und Politik. In: *Sprache und Geist* (N. Chomsky), pp. 165—189. Suhrkamp, Frankfurt.

— (1975): *Reflections on language.* The New Press, New York.

Clarke, B. (1976): The ecological genetics of host-parasite relationships. In: *Symposium of the British Society for Parasitology,* vol. 14 (ed. A. E. R. Taylor & R. Muller), pp. 87—103. Blackwell, Oxford.

Clausewitz, C. v. (1981): *Vom Kriege.* Reclam, Stuttgart.

Clutton-Brock, J. (1995): Origins of the dog: domestication and early history. In: *The Domestic Dog* (ed. J. Serpell), pp. 7—20. Cambridge University Press, Cambridge.

Colman, H., Nabekura, J. & Lichtman, J. W. (1997): Alterations in Synaptic Strength Preceding Axon Withdrawal. *Science* 275, 356—361.

Corballis, M. C. (1997): The genetics and evolution of handedness. *Psychological Review* 104, 714—727.

Coyne, J. (1996): Speciation in Action. *Science* 272, 700—701.

Crair, M., Gillespie, D. C. & Strykert, M. P. (1998): The Role of Visual Experience in the Development of Columns in Cat Visual Cortex. *Science* 279, 566—569.

Cranach, A. v. (2001): My Father of the Greylag Geese. In: *Konzepte der Verhaltensforschung: Konrad Lorenz und die Folgen* (ed. K. Kotrschal, G. B. Müller & H. Winkler). Filander, Fürth.

Crick, F. (1970): Central Dogma of Molecular Biology. *Nature* 227, 561—564.

Crick, F. & Koch, C. (1990): Towards a neurobiological theory of consciousness. *Sem. Neurosci.* 2, 263—275.

Dahlberg, G. (1893/1926): *Twin births and twins from a hereditary point of view.* Bokfèorlags, Stockholm.

Dale, P. S. et al. (1998): Genetic influence on language delay in two-year-old children. *Nature Neurosci.* 1, 324—328.

Damasio, A. R. & Damasio, H. (1992): Sprache und Gehirn. *Spektrum der Wissenschaft* 92, 80—92.

Dansky, J. L. & Silverman, I. W. (1973): Effects of play on associative fluency in preschool aged children. *Dev. Psychol.* 9, 38—44.

Darwin, C. (1859/1872 6): *On the Origin of Species by Means of Natural Selection, or the Preservation of Favoured Races in the Struggle for Life.* John Murray, London.

— (1871): *The Descent of Man, and Selection in Relation to Sex.* Murray, London.

— (1876): *The Effects of Cross and Self Fertilization in the Vegetable Kingdom.* Murray, London.

Darwin, F. (1887): *The Life and Letters of Charles Darwin.* Murray, London.

Dawkins, R. (1976): *The Selfish Gene.* Oxford University Press, Oxford.

— (1982): *The Extended Phenotype.* Oxford University Press, Oxford.

— (1994): The gene's eye view of creation. In: *Nature 125th Symposium,* London.

— (1996): *Climbing Mount Improbable.* W. W. Norton & Company.

— (2000): *The Blind Watchmaker.* Penguin, London.

Dawkins, R. & Krebs, J. (1979): Arms races between and within species. *Proc. Roy. Soc. B* 205, 489—511.

— (1981): Signale der Tiere: Information oder Manipulation? In: *Öko-Ethologie* (ed. J. Krebs & N. Davies), S. 222—242. Parey, Berlin.

Dayhoff, M. O. (1969): Computer Analysis of Protein Evolution. *Scientific American* 221, 86—97.

De Waal, F. (1997): *Good Natured : The Origins of Right and Wrong in Humans and Other Animals.* Harvard University Press, Cambridge.

— (2000): *Chimpanzee Politics: Power and Sex Among Apes.* Johns Hopkins University Press.

De Waal, F. de & Lanting, F. (1997): *Bonobo—The Forgotten Ape.* University of California Press, Berkeley.

Deacon, T. (1992): The brain and language. In: *Human Evolution* (ed. S. Jones, R. Martin & D. Pilbeam), p. 107—123. Cambridge University Press, Cambridge.

— (1997): *The Symbolic Species. The Co-Evolution of Language and the Human Brain.* Allen Lane, London.

Deary, I. J. et al. (in press): The stability of individual differences in mental ability from childhood to old age: follow-up of the 1932 Scottish Mental Survey. *Intelligence.*

Deichmann, U. (1995): *Biologen unter Hitler.* Fischer Taschenbuch, Frankfurt am Main.

DeLoache, J. S. (1987): Rapid Change in the Symbolic Functioning of Very Young Children. *Science* 238, 1556—1557.

Dempsey, P. W. et al. (1996): C3d of Complement as a Molecular Adjuvant: Bridging Innate and Acquired Immunity. *Science* 271, 348—350.

Dennett, D. C. (1993): *Consciousness Explained.* Penguin, London.

— (1996): *Darwin's Dangerous Idea: Evolution and the Meanings of Life.* Touchstone.

Devlin, B., Daniels, M. & Roeder, K. (1997): The heritability of IQ. *Nature* 388, 468—471.

Diamond, J. (1992): *The Third Chimpanzee. The Evolution and Future of the Human Animal.* HarperCollins, New York.

Dickman, S. (1997): Gene Mutation Provides More Meat on the Hoof. *Science* 277, 1922—1923.

Dolan, R. J. (1996): Dopaminergic modulation in the anterior cingulate cortex in schizophrenia. *Nature* 378, 180—182.

Doolittle, W. F. (1999): Phylogenetic Classification and the Universal Tree. *Science* 284, 2124—2128.

Dornhaus, A. & Chittka, L. (1999): Evolutionary origins of bee dances. *Nature* 401, 38.

Dose, K. (1982): Chemische Evolution und der Ursprung lebender Systeme. In: *Biophysik* (ed. W. Hoppe et al.), S. 948—949. Springer, Berlin.

Dragunow, M. & Faull, R. (1989): The use of c-fos as a metabolic marker in neuronal pathway tracing. *J. Neurosci. Meth.* 29, 261—265.

Dugatkin, L. A. (1997): *Cooperation Among Animals. An Evolutionary Perspective.* Oxford University Press, New York.

Dugatkin, L. A. & Reeve, H. K. (1994): Behavioral Ecology and Levels of Selection: Dissolving the Group Selection Controversy. *Advances in the Study of Behavior* 23, 101—133.

Dunbar, R. I. M. (1993): Coevolution of neocortical size, group size and language in humans. *Behavioral and Brain Sciences* 16, 681—735.

— (2000): Causal Reasoning, Mental Rehearsal, and the Evolution of Primate Cognition. In: *The Evolution of Cognition* (ed. C. Heyes & L. Huber), pp. 205—220. MIT Press, Cambridge, MA.

Duncan, J. et al. (2000): A Neural Basis for General Intelligence. *Science* 289, 457—460.

Du Pasquier, L. & Litman, G. W. (2000): *Origin and Evolution of the Vertebrate Immune System.* Springer, Berlin.

Dupré, J. (ed., 1987): *The Latest on the Best. Essays on Evolution and Optimality.* MIT Press, Cambridge MA.

Edelman, G. (1989): *Neural Darwinism: The theory of neuronal group selection.* Oxford University Press, Oxford.

Edelman, G. & Tononi, G. (2000): *A universe of consciousness. How matter becomes imagination.* Basic Books, New York.

Egid, K. & Brown, J. L. (1989): The major histocompatibility complex and female mating preferences in mice. *Animal Behaviour* 38, 548—550.

Ehrenstein, D. (1998): Immortality Gene Discovered. *Science* 279, 177.

Eibl-Eibesfeldt, I. (1973): *Der vorprogrammierte Mensch.* Molden, Wien.

— (1989): *Human Ethology.* De Gruyter, Berlin.

Eibl-Eibesfeldt, I. & Sütterlin, C. (1992): *Im Banne der Angst.* Piper, München.

Eigen, M. (1971): Selforganization of Matter and the Evolution of Biological Macromolecules. *Naturwissenschaften* 58, 465—522.

— (1987): *Stufen zum Leben.* Piper, München.

Eigen, M. & Schuster, P. (1977): The hypercycle. A principle of natural self-organization. *Naturwissenschaften* 64, 541—565.

Einstein, A. (1956): *Lettres à Maurice Solovine,* Paris.

Eizinger, A. & Sommer, R. J. (1997): The Homeotic Gene lin-39 and the Evolution of Nematode Epidermal Cell Fates. *Science* 278, 452—455.

Elbert, T. et al. (1995): Increased Cortical Representation of the Fingers of the Left Hand in String Players. *Science* 270, 305—307.

Elena, S. F., Cooper, V. S. & Lenski, R. E. (1996): Punctuated Evolution Caused by Selection of Rare Beneficial Mutations. *Science* 272, 1802—1804.

Elias, N. (2000): *The Civilizing Process.* Blackwell, Oxford.

Elkind, D. (1984): Zwei entwicklungspsychologische Ansätze: Piaget und Montessori. In: *Entwicklungspsychologie,* vol. 1 (ed. G. Steiner), S. 584—594. Beltz, Weinheim.

Ellis, R. E., Yuan, J. & Horvitz, R. H. (1991): *Annu. Rev. Cell Biol.* 7, 663.

Elman, J. L. et al. (1996): *Rethinking Innateness. A connectionist perspective on development.* MIT Press, Cambridge MA.

Elowson, A. M., Snowdon, C. T. & Lazaro-Perea, C. (1998): "Babbling" and social context in infant monkeys: Parallels to human infants. *Trends in Cognitive Science* 2, 35—43.

Encyclopedia (2000): "Immune System". *Microsoft® Encarta®* Online, http: //encarta.msn.com

Endler, J. (1991): Interactions between predators and prey. In: *Behavioural Ecology. An Evolutionary Approach* (ed. J. Krebs & N. Davies), pp. 169—196. Blackwell, Oxford.

Engels, E. M. (1989): Charles Darwin als Begründer der Evolutionstheorie menschlichen Erkennens. In: *Erkenntnis als Anpassung?* (ed. E. M. Engels), S. 63—129. Suhrkamp, Frankfurt.

— (1995): *Die Rezeption von Evolutionstheorien im 19. Jahrhundert.* Suhrkamp, Frankfurt.

Enquist, M. & Arak, A. (1994): Symmetry, beauty and evolution. *Nature* 372, 169—172.

Erckmann, W. J. (1983): The evolution of polyandry in shorebirds: an evaluation of hypotheses. In: *Social Behavior of Female Vertebrates* (ed. S. K. Wasser), Academic Press, New York.

Etienne, A. (1984): The meaning of object permanence at different zoological levels. *Hum. Dev.* 27, 309—320.

Etterson, J. R. & Shaw, R. G. (2001): Constraint to Adaptive Evolution in Response to Global Warming. *Science* 294, 151—154.

Evans, R. (1977): *Konrad Lorenz: Gespräche mit Richard Evans.* Ullstein, Frankfurt. [(1975): *Konrad Lorenz: The Man and his Ideas.* Harcourt, New York]

Eysenck, H. J. (1975): *Die Ungleichheit der Menschen.* Goldmann. [(1973): *The Inequality of Man.* TempleSmith, London]

— (1996): *Intelligenztest.* Weltbild Verlag, Augsburg.

Fábrega Jr., H. (1997): *Evolution of Sickness and Healing.* Univ. of Calif. Press, Berkeley.

Fagen, R. (1976): Exercise, play, and physical training in animals. In: *Perspectives in Ethology* (Vol. 2, ed. P. P. G. Bateson & P. H. Klopfer), pp. 189—219. Plenum, New York.

— (1981): *Animal Play Behavior.* Oxford University Press, New York.

Falk, D. (1983): Cerebral cortices of East African early hominids. *Science* 221, 1072—1074.

Farah, M. J. (1995): The Neural Bases of Mental Imagery. In: *The Cognitive Neurosciences* (ed. M. S. Gazzaniga), pp. 963—975. MIT Press, London.

Fearon, D. T. (1999): Innate immunity and the biological relevance of the acquired immune response. *QJ Med* 92, 235—237.

Fearon, D. T. & Locksley, R. M. (1996): The instructive Role of Innate Immunity in the Acquired Immune Response. *Science* 272, 50—54.

Ferguson, C. A. (1964): Baby talk in six languages. In: *Directions in Sociolinguistics. The Ethnography of Communication* (ed. J. L. Gumperz & D. H. Hymes), pp. 103—114. Holt, Rinehart & Winston, New York.

Fernald, A. (1985): Four month old infants prefer to listen to motherese. *Infant Behavior and Development* 8, 118—195.

Ferris, J. P. (1987): Prebiotic synthesis: problems and challenges. *Cold Spring Harbor Symp. Quant. Biol.* 52, 29—35.

Finch, C. E. & Tanzi, R. E. (1997): Genetics of Aging. *Science* 278, 407—411.

Fink, A. & Neubauer, A.C. (2001). Speed of information processing, psychometric intelligence and time estimation as an index of cognitive load. *Personality and Individual Differences* 30, 1009—1021.

Finkelhor, D. (1984): *Child Sexual Abuse*. Free Press, New York.

Finlay, B. L. & Darlington, R. B. (1995): Linked Regularities in the Development and Evolution of Mammalian Brain. *Science* 268, 1578—1584.

Fisher, J. A. (1996): The Myth of Anthropomorphism. In: *Readings in Animal Cognition* (M. Bekoff & D. Jamieson), pp. 3—16. MIT Press, Cambridge, MA.

Fisher, R. A. (1930): *The Genetical Theory of Natural Selection*. Oxford University Press, Oxford.

Fitch, W. T. (in press): Evolving Honesty: Kin Communication and "Mother Tongues". In: *The Evolution of Communication* (ed. K. Oller). MIT Press, Cambridge, MA.

Fleck, L. (1946/1983): Wissenschaftstheoretische Probleme. In: *Erfahrung und Tatsache* (L. Fleck), S. 128—146. Suhrkamp, Frankfurt.

Flint, J. et al. (1995): A Simple Genetic Basis for a Complex Psychological Trait in Laboratory Mice. *Science* 269, 1432—1434.

Flynn, J. R. (1998): IQ Gains Over Time: Toward Finding the Causes. In: *The Rising Curve. Long-Term Gains in IQ and Related Measures* (ed. U. Neisser), pp. 25—66. APA, Washington DC.

Foerster, H. v. & Pörksen, B. (2001): *Wahrheit ist die Erfindung eines Lügners. Gespräche für Skeptiker.* ISBN 3-89670-214-9.

Foley, R. A. (1991): Language origins. The silence of the past. *Nature* 353, 114—115.

Fontana, W. (1994): "The arrival of the fittest": toward a theory of biological organization. *Bulletin of Mathematical Biology* 56, 1—64.

Fontana, W. & Schuster, P. (1998): Continuity in Evolution: On the Nature of Transitions. *Science* 280, 1451—1455.

Forsthuber, T., Yip, H. C. & Lehmann, P. V. (1996): Induction of TH1 and TH2 Immunity in Neonatal Mice by a Murine Retrovirus. *Science* 271, 1728—1730.

Fox, R. (1967): *Kinship and Marriage*. Penguin, Middlesex.

Frank, E. (1997): Synapse Elimination: For Nerves It's All or Nothing. *Science* 275, 324—325.

Frank, R. H., Gilovich, T. & Regan, D. T. (1993): The Evolution of One-Shot Cooperation: An Experiment. *Ethology and Sociobiology* 14, 247—256.

Freeland, S. J., Knight R. D. & Landweber, L. F. (1999): Do Proteins Predate DNA? *Science* 286, 690—692.

Freeman, W. J. (1975): *Mass action in the nervous system.* Academic Press, New York.

— (1991): The physiology of perception. *Scientific American* 264, 78—85.

— (1997): Nonlinear Neurodynamics of Intentionality. *The Journal of Mind and Behavior* 18, 291—304.

Freud, S. (1900): *Die Traumdeutung.* Deuticke, Wien. [(1997): *The Interpretation of Dreams.* Wordsworth]

Frisch, K. v. (1965): *Tanzsprache und Orientierung der Bienen.* Springer, Berlin-Heidelberg-New York. [(1984): *Bees: Their Vision, Chemical Senses and Language.* Cape, London]

Fuchs, D., Gerhards, J. & Roller, E. (1993): Wir und die Anderen. Kölner *Z. f. Soz. u. Sozialpsychol.* 45, 238—253.

Furth, H. G. (1964): Research with the deaf: Implications for language and cognition. *Psychological Bulletin* 62, 145—164.

— (1971): Linguistic deficiency and thinking: Research with deaf subjects 1964—1969. *Psychological Bulletin* 74.

— (1972): *Denken ohne Sprache* [*Thinking without Language*]. Schwann, Düsseldorf.

Galison, P. & Stump, D. J. (eds., 1996): *The Disunity of Science. Boundaries, Contexts, and Power.* Stanford University Press, Stanford.

Gallagher, H. G. (1995): *By Trust Betrayed : Patients, Physicians, and the License to Kill in the Third Reich.* Vandamere.

Gallistel, C. R. (1990): *The organization of learning.* MIT Press, MA: Cambridge.

Gallup, G. G., Jr. (1970): Chimpanzees: Self-recognition. *Science* 167, 86—87.

Gannon, P. J., Holloway, R. L., Broadfield, D. C. & Braun, A. R. (1998): Asymmetry of Chimpanzee Planum Temporale: Humanlike Pattern of Wernicke's Brain Language Area Homolog. *Science* 279, 220—222.

Ganzberger, H. (1996): The Sun Illusion: limited speed of light and the perception of sunrise and sunset. Unpublished manuscript.

Garcia, J. (1991): Lorenz's impact on the psychology of learning. *Evolution & Cognition* 1, 31—41.

Garcia, J., Kimeldorf, D. J. & Koelling, R. A. (1955): A conditioned aversion towards saccharin resulting from exposure to gamma radiation. *Science* 122, 157—158.

Garcia, R. & Piaget, J. (1983): *Psychogenèse et histoire des sciences.* Flammarion, Paris.

Gardner, R. A. & Gardner, B. T. (1969): Teaching sign language to a chimpanzee. *Science* 165, 664—672.

Gavrilets, S. (2000): Rapid evolution of reproductive barriers driven by sexual conflict. *Nature* 403, 886—889.

Gegenfurtner, K. R. & Sharpe, L. T. (2001): *Color Vision. From Genes To Perception.* Cambridge University Press, Cambridge.

Gehring, W. J. & Ruddle, F. (1998): *Master Control Genes in Development and Evolution.* Yale University Press, New Haven.

Germain, R. N. (2001): The Art of the Probable: System Control in the Adaptive Immune System. *Science* 293, 240—245.

Gibbons, A. (1995): When It Comes to Evolution, Humans Are in the Slow Class. *Science* 267, 1907—1908.

— (1997a): Y Chromosome Shows That Adam Was an African. *Science* 278, 804.

— (1997b): Human Evolution: The Women's Movement. *Science* 278, 805.

— (1998): Which of Our Genes Make Us Human? *Science* 281, 1432—1434.

— (2001): Modern Men Trace Ancestry to African Migrants. *Science* 292, 1051—1052.

Gibson, J. J. (1982): *Wahrnehmung und Umwelt*. Urban & Schwarzenberg, München. [(1979): *The Ecological Approach to Visual Perception*. Houghton Mifflin, Boston]

Gigerenzer, G. (1991): From Tools to Theories: A Heuristic of Discovery in Cognitive Psychology. *Psychological Review* 98, 254—267.

Gigerenzer, G., Hoffrage, U. & Kleinbölting, H. (1991): Probabilistic Mental Models: A Brunswikian Theory of Confidence. *Psych. Rev.* 98, 506—528.

Gilbert, S. F. (2000): *Developmental Biology*. Sinauer, Sunderland.

Glasersfeld, E. v. (1996): *Radical Constructivism*. RoutledgeFalmer, New York.

Gnatt, A. L. et al. (2001): Structural Basis of Transcription: An RNA Polymerase II Elongation Complex at 3.3 Å Resolution. *Science* 292, 1876—1882.

Goedel, K. (1931): Über formal unentscheidbare Sätze der Principia Mathematica und verwandter Systeme. *Monatshefte für Mathematik und Physik* 38, 173—198.

— (1944): Russell's Mathematical Logic. In: *The Philosophy of Bertrand Russell* (ed. P. A. Schilpp), Evanston, Chicago.

Godfrey, J. (1995): The Pope and the ontogeny of persons. *Nature* 373, 100.

Goldenberg, G., Podreka, I. & Steiner, M. (1990): The cerebral localization of visual imagery. In: Imagery: *Current Developments* (eds. P. Hampson, D. F. Marx & J. Richardson), Routledge, London.

Goldin-Meadow, S. & Mylander, C. (1998): Spontaneous sign systems created by deaf children in two cultures. *Science* 391, 279—281.

Goodenough, O. R. & Dawkins, R. (1994): The "St. Jude" mind virus. *Nature* 371, 23—24.

Gopnik, A. & Meltzoff, A. N. (1997): *Words, Thoughts, and Theories*. MIT Press, Cambridge, Mass.

Gopnik, M. (1990): Feature-blind grammar and dysphasia. *Nature* 344, 715.

Gott III, J. R. (1993): Implications of the Copernican principle for our future prospects. *Nature* 363, 315—319.

Gottesman, I. I. (1997): Twins: En Route to QTLs for Cognition. *Science* 276, 1522—1523.

Gould, J. L. (1975): Honey bee communication: the dance-language controversy. *Science* 189, 685—693.

— (1979): Do honeybees know what they are doing? *Nat. Hist.* 88, 66—75.

Gould, J. L. & Marler, P. (1987): Learning by Instinct. *Scientific American* 255(1), 74—85.

Gould, S. J. (1979): Another look at Lamarck. *New Scientist* 38, 4 (4 October).

— (1991): *Zufall Mensch*. Hanser, München. [(2000): *Wonderful Life. The Burgess Shale and the Nature of History*. Vintage, New York]

Gould, S. J. & Eldredge, N. (1977): Punctuated equilibria: The tempo and mode of evolution reconsidered. *Paleobiology* 3, 115—151.

— (1993): Punctuated equilibrium comes of age. *Nature* 366, 223—227.

Gould, S. J. & Vrba, E. S. (1982): Exaptation—a missing term in the science of form. *Paleobiology* 8, 4—15.

Grafen, A. (1991): Modelling in behavioural ecology. In: *Behavioural Ecology. An Evolutionary Approach* (ed. J. R. Krebs & N. B. Davies), pp. 5—31. Blackwell, London.

Graham, A. (2000): Root and branch surgery. *Current Biology* 10(1), 36—38.

Grayling, A. C. (2000): Genes. *The Guardian*, June 10, 2000.

Greene, H. (1994): Homology and behavioral repertoires. In: *Homology. The hierarchical basis of comparative biology* (ed. B. K. Hall), pp. 369—386. Academic Press, San Diego.

Greenfield, P. M. (1991): Language, tools and brain: The ontogeny and phylogeny of hierarchically organized sequential behaviour. *Behavioural and Brain Sciences* 14, 531—595.

Griffin, D. (1991): *Wie Tiere denken*. DTV, München. [(1984): *Animal Thinking*. Harvard, Cambridge, MA]

Grodzinsky, Y. (2000): The neurology of syntax: Language use without Broca's area. *Behavioral and Brain Sciences* 23, 1—71.

Gruber, R. P. (1985): *Im Namen des Vaters*. Residenz, Salzburg.

Gruss, A. & Michel, B. (2001): The replication–recombination connection: insights from genomics. *Current Opinion in Microbiology* 4(5), 595—601.

Guttmann, G. (1966): Komplexe Ordnungstendenzen in Verhaltensabläufen und ihre differentialdiagnostische Bedeutung. *Z. f. exp. angew. Psychol.* 13, 19—30.

Güntürkün, O. (2001): Hemispheric Asymmetry in the Visual System of Birds, In: *Brain Asymmetry* (K. Hugdahl & R. J. Davidson, eds.), MIT Press, Cambridge, MA.

Güntürkün, O., Emmerton, J. & Delius, J. D. (1989): Neural asymmetries and visual behaviour of birds. In: *Biological Signal Processing: Cellular and Integrative Aspects* (H. C. Lüttgau & R. Necker, eds.), pp. 122—145. Verlag Chemie, Weinheim.

Güntürkün, O. et al. (2000): Asymmetry pays: Visual lateralization improves discrimination success in pigeons. *Current Biol.* 10, 1079—1081.

Haeckel, E. (1912): Die Fundamente des Monismus. In: *Der erste internationale Monisten-Kongress* (ed. W. Ostwald & C. Riess), S. 59—60. Kröner, Leipzig.

Haig, D. (1992): Intragenomic conflict and the evolution of eusociality. *Journal of Theoretical Biology* 156, 401—403.

— (1997): Parental antagonism, relatedness asymmetries, and genomic imprinting. *Proc. R. Soc. Lond.* B 264, 1657—1662.

— (1999): Asymmetric Relations: Internal Conflicts and the Horror of Incest. *Evolution and Human Behavior* 20, 83—98.

Haigh, J. (1978): The accumulation of deleterious genes in a population—Muller's ratchet. *Theoretical Population Biology* 14, 251—267.

Haken, H. (1989): *Information and Self-Organization. A Macroscopic Approach to Complex Systems*. Springer, Berlin.

Haken, H. & Haken-Krell, M. (1989): *Entstehung von biologischer Information und Ordnung*. Wissenschaftliche Buchgesellschaft, Darmstadt.

Haldane, J. B. S. (1949): Disease and evolution. *La Ricerca Scientifica* 19, 68—76.

Hall, J. C. (1994a): The Making of a Fly. *Science* 264, 1702—1714.

— (1994b): Pleiotropy of Behavioral Genes. In: *Flexibility and Constraint in Behavioral Systems* (ed. R. J. Greenspan & C. P. Kyriacou), John Wiley & Sons, Chicester.

Hamilton, W. D. (1964): The genetical evolution of social behaviour. *J. theor. Biol.* 7, 1—51.

— (1987): Discriminating nepotism: expectable, common, overlooked. In: *Kin recognition in animals* (ed. D. J. C. Fletcher & C. D. Michener), pp. 417—437. John Wiley & Sons, Chichester.

Hanken, J. & Richardson, M. K. (1998): Haeckel's Embryos. *Science* 279, 1283.

Hardin, G. (1968): The tragedy of the commons. *Science* 162, 1243—1248.

Harley, J. (1981): Evolutionarily stable strategies in learning games. *J. theor. Biol.* 89, 611—633.

Harper, D. G. C. (1991): Communication. In: *Behavioural Ecology. An Evolutionary Approach* (ed. J. R. Krebs & N. B. Davies), pp. 374—397. Blackwell, London.

Hartl, F. U., Hodlan, R. & Langer, T. (1994): Molecular chaperones in protein folding: the art of avoiding sticky situations. *Trends Biochem. Sci.* 19, 20—25.

Hartmann, N. (1964): *Der Aufbau der realen Welt.* de Gruyter, Berlin.

Hartwell, L. H. & Kastan, M. B. (1994): Cell cycle control and cancer. *Science* 266, 1821—8.

Hass, H. (1979): *Wie der Fisch zum Menschen wurde.* Bertelsmann, München.

Hassenstein, B. (1987): *Verhaltensbiologie des Kindes.* Piper, München.

Hauser, M. (1997): Artifactual kinds and functional design features: what a primate understands without language. *Cognition* 64, 285—308.

Hauser, M. et al. (1995): Self-recognition in primates: Phylogeny and the salience of species-typical features. *Proceedings of the National Academy of Sciences* (USA) 92, 10811—10814.

Hauser, M. D., Kralik, J. & Botto-Mahan, C. (1999): Problem solving and functional design features: experiments on cotton-top tamarins, *Saguinus oedipus. Anim. Behav.* 57, 565—582.

Heberer, G. (1968): *Der gerechtfertigte Haeckel.* Gustav Fischer, Stuttgart.

Hedges, S. B. (2000): Human evolution. A start for population genomics. *Nature* 408, 652—653.

Heidegger, M. (1978): *Being and Time.* Blackwell, London.

Hein, J. (1995): *Nestroy zum Vergnügen.* Reclam, Stuttgart.

Heisenberg, M. (1990): Über Universalien der Wahrnehmung und ihre genetischen Grundlagen. In: *Mannheimer Forum 89/90. Ein Panorama der Naturwissenschaften* (ed. H. v. Ditfurth & E. P. Fischer), S. 11—70. Piper, München.

Hell, W., Fiedler, K. & Gigerenzer, G. (1993): *Kognitive Täuschungen.* Spektrum Akademischer Verlag, Heidelberg.

Helmuth, L. (2001): From the Mouths (and Hands) of Babes. *Science* 293, 1758—1759.

Hemme, H. (1988): Spieglein, Spieglein an der Wand... *Bild der Wissenschaft* 8/88, 133—136.

Hengartner, M. O. & Horvitz, R. H. (1994): *Philos. Trans. R. Soc. London Ser.* B 345, 243.

Henikoff, S. et al. (1997): Gene Families: The Taxonomy of Protein Paralogs and Chimeras. *Science* 278, 609—614.

Herbert, A. & Rich, A. (1999): RNA processing and the evolution of eukaryotes. *Nature Genetics* 21(3), 265—269.

Herman, L. M., Richards, D. G. & Wolz, J. P. (1984): Comprehension of sentences by bottlenosed dolphins. *Cognition* 16, 129—219.

Herre, W. (1982): So kam der Wolf auf den Hund: Vom Wild- zum Haustier. *Das Tier* 12/82, 42—46.

Herre, W. & Röhrs, M. (1973): *Haustiere, zoologisch gesehen.* Fischer, Stuttgart.

Herrnstein, R. J. & Murray, C. (1996): *The Bell Curve. Intelligence and Class Structure in American Life.* Free Press Paperback, New York.

Herschel, J. F. W. (1861): *Physical Geography of the Globe.* Adam & Charles Block, Edinburgh.

Heschl, A. (1989): Integration of "Innate" and "Learned" Components within the IRME for Mussel Recognition in the European Bitterling *Rhodeus amarus*. *Ethology* 81, 193—208.

— (1990): L=C, A Simple Equation with Astonishing Consequences. *J. theor. Biol.* 145, 13—40.

— (1992a): Behaviour and the concept of "heritability": Axioms of an ethological refutation. *Acta Biotheoretica* 40, 23—30.

— (1992b): Das Erwachen des Bewusstseins beim Kinde. In: *Das Bewusstsein. Multidimensionale Entwürfe* (ed. G. Guttmann & G. Langer), S. 353—374. Springer, Wien.

— (1993a): Physiognomic Similarity and Political Cooperativeness: an Exploratory Investigation. *Politics and the Life Sciences* 14, 247—256.

— (1993b): On the ontogeny of seed harvesting techniques in free ranging ground squirrels (*Spermophilus citellus*). *Behaviour* 125 1/2, 39—50.

— (1993c): Epistemology as a natural science. *Evolution and Cognition* 2, 235—255.

— (1994a): Re-constructing the real unit of selection. *Behavioral and Brain Sciences* 17, 624—625.

— (1994b): Nature/nurture. *Nature* 369, 185.

— (1994c): Razor-blade of life. *Nature* 368, 93.

— (1996): Biological Determinism. *Science* 271, 743—744.

— (forthcoming): Natural selection and metaphors of "selection". Commentary on Hull, Langman & Glenn: A general account of selection. Behavioral and Brain Sciences.

Heschl, A. & Peschl, M. (1992): Natural versus artificial "intelligence": an axiomatic comparison. *Journal of Social and Evolutionary Systems* 15(1), 55—74.

Hess, B. (1997): Periodic patterns in biochemical reactions. *Quarterly Reviews of Biophysics* 30, 121—176.

Hess, B. & Markus, M. (1985): The Diversity of Biochemical Time Patterns. Ber. Bunsenges. *Phys. Chem.* 89, 642—651.

Heyes, C. M. (1998): Theory of mind in nonhuman primates. *Behavioral and Brain Sciences* 21(1), 101—114.

Hickok, G., Bellugi, U. & Klima, E. S. (1996): The neurobiology of sign language and its implications for the neural basis of language. *Nature* 381, 699—702.

Hieter, P. & Boguski, M. (1997): Functional Genomics: It's All How You Read It. *Science* 278, 601—602.

Hilgard, E. R. (ed., 1974): *Theories of Learning and Instruction*. University of Chicago Press, Chicago.

Hill, C. (1946): Playtime at the zoo. *Zoo-Life* 1, 24—26.

Hodgkin, J., Plasterk, R. H. A. & Waterston, R. H. (1995): The Nematode Caenorhabditis elegans and Its Genome. *Science* 270, 410—414.

Hoffmann, A. A. (1994): Behaviour genetics and evolution. In: *Behaviour and evolution* (ed. P. J. B. Slater & T. R. Halliday), pp. 7—42. Cambridge University Press, Cambridge.

Hoffmann, W. et al. (1978): Hybridization between gulls (*Larus glaucescens and Larus occidentalis*) in the Pacific Northwest. *Auk* 95, 441—458.

Hofmann, P. (1988): *The Viennese. Splendor, Twilight, and Exile*. Anchor, New York.

Hofstadter, D. R. (1979/2000): *Goedel, Escher, Bach: An Eternal Golden Braid*. Penguin, London.

Hoi, H. (1987): Brutaufteilung und Habitatnutzung beim Cassinschnäpper *Muscicapa cassini. J. Orn.* 128, 338—342.

Holden, C. (1995): Is It Time to Begin Ph.D. Population Control? *Science* 270, 123—128.

— (1996): The Vatican's Position Evolves. *Science* 274, 717.

Höllinger, S. (1992): *Universität ohne Heiligenschein.* Passagen, Wien.

Hol, T. et al. (1994): Consequences of short term isolation after weaning on later adult behavioural and neuroendocrine reaction to social stress. *Behav. Pharmacol.* 5, 88—89.

Holm, L. & Sander, C. (1996): Mapping the Protein Universe. *Science* 273, 595—602.

Honolka, K. (1976): *Weltgeschichte der Musik.* Rheingauer Verlagsgesellschaft, Eltville am Rhein.

Holzinger, K. J., Newman, H. H. & Freeman, F. N. (1875/1937): *Twins; a study of heredity and environment.* The University of Chicago Press, Chicago.

Horne, P. J. & Lowe, C. F. (1996): On the origins of naming and other symbolic behaviour. *J. exp. Anal. Behav.* 65, 185—241.

Horsfall, J. A. (1984): Brood reduction and brood division in coots. *Anim. Behav.* 32, 216—225.

Hösle, V. (1988): Tragweite und Grenzen der evolutionären Erkenntnistheorie. *Zeitschrift für allgemeine Wissenschaftstheorie* 19, 348—377.

Howard, R. S. & Lively, C. M. (1994): Parasitism, mutation accumulation and the maintainance of sex. *Nature* 367, 554—557.

Hubel, D. H. & Wiesel, T. (1962): Receptive fields of single neurons in the cat's striate cortex. *J. Physiol.* 148, 574—591.

Huff, T. E. (1993): *The rise of early modern science. Islam, China, and the West.* Cambridge University Press, Cambridge.

Huizinga, J. (1956): *Homo ludens.* Beacon, Boston.

Hull, D. L. (1981): The herd as means. In: *PSA 1980,* 2. East Lansing (ed. P. o. S. Association), MI.

Hull, D. L., Tessner, P. D. & Diamond, A. M. (1978): Planck's Principle. Do younger scientists accept new scientific ideas with greater alacrity than older scientists? *Science* 202, 717—723.

Hume, D. (1777/1975): *An Enquiry Concerning Human Understanding.* Clarendon, Oxford.

Humphrey, N. & Dennett, D. C. (1989): Speaking for Our Selves: An Assessment of Multiple Personality Disorder. *Raritan* 9, 68—98.

Hund, D. M. (1995): The Chemistry of John Dalton's Color Blindness. *Science* 267, 984—988.

Hunt, A. & Evan, G. (2001): Till Death Us Do Part. *Science* 293, 1784—1785.

Hurst, L. D. (1996): Segregation Distorter in fruitflies. *Genetics* 142, 641—643.

International Mouse Mutagenesis Consortium (2001): Functional Annotation of Mouse Genome Sequences. *Science* 291, 1251—1255.

Irwin, D. E., Bensch, S. & Price, T. D. (2001): Speciation in a ring. *Nature* 409, 333—337.

Ishai, A. & Sagi, D. (1995): Common Mechanisms of Visual Imagery and Perception. *Science* 268, 1772—1774.

Isles, A. R. & Wilkinson, L. S. (2000): Imprinted genes, cognition and behaviour. *Trends in Cognitive Sciences* 4(8), 309—318.

Jacob, F. (1998): Können, dürfen, sollen, müssen. *Die Presse* 14. 2. 1998, I—II.

Jacob, F. & Monod, J. (1961): Genetic regulatory mechanisms in the synthesis of proteins. *J. Mol. Biol.* 3, 318—356.

James, L. E. & Burke, D. M. (2000): Phonological priming effects on word retrieval and tip-of-the-tongue experiences in young and older adults. *J Exp Psychol Learn Mem Cogn* 26, 1378—1391.

Jänig, W. (1993): Vegetatives Nervensystem. In: *Physiologie des Menschen* (ed. R. F. Schmidt & G. Thews), S. 349—389. Springer, Berlin.

Jazwinski, S. M. (1996): Longevity, Genes, and Aging. *Science* 273, 54—59.

Jensen, A. R. (1969): Environment, heredity and intelligence. *Harvard Educational Review, Reprint Series* No. 2 .

— (1973): Wie sehr können wir Intelligenzquotient und schulische Leistung steigern? In: *Umwelt und Begabung* (ed. H. Skowronek), S. 63—155. Klett, Stuttgart.

— (1999): The g factor: the science of mental ability. *Psycoloquy*, 10(023).

Johnson-Laird, P. N. (1995): Mental Models, Deductive Reasoning, and the Brain. In: *The Cognitive Neurosciences* (ed. M. S. Gazzaniga), pp. 999—1008. MIT Press, London.

Johnstone, R. A. (1994): Female preferences for symmetrical males as a by-product of selection for mate recognition. *Nature* 372, 172—175.

Jones, S., Martin, R. & Pilbeam, D. (1996): *Human Evolution*. Cambridge University Press, Cambridge.

Jorde, L. B., Bamshad, M. & Rogers, A. R. (1998): Using mitochondrial and nuclear DNA markers to reconstruct human evolution. *BioEssays* 20(2), 126—136.

Joza, N. et al. (2001): Essential role of the mitochondrial apoptosis-inducing factor in programmed cell death. *Nature* 410, 549—54.

Juen, G. (1997): Ugandas gestohlene Generation. *ai-info* 10/97, 6—7.

Jusczyk, P. W. & Hohne, E. A. (1997): Infants' Memory for Spoken Words. *Science* 277, 1984—1985.

Kaas, J. H. (1995): The Reorganization of Sensory and Motor Maps in Adult Mammals. In: *The Cognitive Neurosciences* (ed. M. S. Gazzaniga), pp. 51—72. MIT Press, London.

Kaczmarek, L. (1993): Molecular biology of vertebrate learning: Is c-fos a new beginning? *J. Neurosci. Res.* 34, 377—381.

Kaessmann, H. et al. (1999): DNA sequence variation in a non-coding region of low recombination on the human X chromosome. *Nature Genetics* 22, 78—81.

Kaessmann, H., Wiebe, V. & Pääbo, S. (1999): Extensive Nuclear DNA Sequence Diversity Among Chimpanzees. *Science* 286, 1159—1162.

Kandel, E. R. & Hawkins, R. (1992): Molekulare Grundlagen des Lernens. *Spektrum der Wissenschaft* 11, 66—76.

Kandel, E. R., & Tauc, L. (1965): Heterosynaptic facilitation in neurons of the abdominal ganglion of *Aplysia depilans. Journal of Physiology* 181, 1—47.

Kanehisa, M. (2000): *Post-genome Informatics*. Oxford University Press, Oxford.

Kant, I. (1781/2000): *Critique of Pure Reason*. Cambridge University Press, Cambridge.

Katz, L. C. & Shatz, C. J. (1996): Synaptic Activity and the Construction of Cortical Circuits. *Science* 274, 1133—1138.

Keegan, L. P., Gallo, A. & O'Connell, M. A. (2000): Development. Survival is impossible without an editor. *Science* 290, 1707—1709.

Keith, C. T. & Schreiber, S. L. (1995): PIK-Related Kinases: DNA Repair, Recombination, and Cell Cycle Checkpoints. *Science* 270, 50—51.

Kenyon, L. & Moraes, C. T. (1997): Expanding the functional human mitochondrial DNA database by the establishment of primate xenomitochondrial cybrids. *Cell Biology* 94, 9131—9135.

Kerr, R. A. (2001): Evolutionary Pulse Found, But Complexity as Well. *Science* 293, 2377.

Kim, S. K. (1999): Poster Abstract: Developmental biology in the post-genome era: worms and chips. *Nature Genetics* The Microarray Meeting 1999.

Kimura, M. (1982): *The Neutral Theory of Molecular Evolution.* Cambridge University Press, Cambridge.

King, M.-C. & Wilson, A. C. (1975): *Science* 188, 107.

Kirchner, W. H. & Towne, W. F. (1994): The Sensory Basis of the Honeybee's Dance Language. *Scientific American* June 94.

Kirkpatrick, M. (1982): Sexual selection and the evolution of female choice. *Evolution* 36, 1—12.

Klicka, J. & Zink, R. M. (1998): The Importance of Recent Ice Ages in Speciation: A Failed Paradigm. *Science* 277, 1666—1669.

Knippers, R. (1997): *Molekulare Genetik.* Thieme, Stuttgart.

Knudson, A. (1971): The role for tumor suppressor in retinoblastoma. *Proc. Natl. Acad. Sci. U.S.A.* 68, 820.

Köchler, H. (1983): Erkenntnistheorie als biologische Anthropologie? *Veröff. d. Internat. Forschungszentrum Salzburg* 9, 43—63.

Kopp, A., Duncan, I. & Carroll, S. B. (2000): Genetic control and evolution of sexually dimorphic characters in Drosophila. *Nature* 408, 553—559.

Kornberg, A. & Baker, T. A. (1992): *DNA Replication.* Freeman, New York.

Kosslyn, S. M. et al. (1995): Topographical representations of mental images in primary visual cortex. *Nature* 378, 496—498.

Kosslyn, S. M. & Sussman, A. L. (1995): Roles of Imagery in Perception: Or, there is no such Thing as Immaculate Perception. In: *The Cognitive Neurosciences* (ed. M. S. Gazzaniga), pp. 1035—1042. MIT Press, London.

Krasnegor, N. A. (1991): *Biological and Behavioral Determinants of Language Development.* Lawrence Erlbaum, Hillsdale.

Krebs, J. (1996): Chewing it over. *Nature* 380, 304.

Krebs, J. & Davies, N. (1981): *Öko-Ethologie.* Parey, Berlin. [(1978[1]): *Behavioural Ecology—An Evolutionary Approach.* Blackwell, Oxford]

Krebs, J. & Davies, N. (1991[3]): *Behavioural Ecology. An Evolutionary Approach.* Blackwell, Oxford.

Krebs, J. & Dawkins, R. (1984): Animal signals: mind-reading and manipulation. In: *Behavioural Ecology: An Evolutionary Approach* (ed. J. Krebs & N. Davies), pp. 380—402. Blackwell, Oxford.

Krebs, U. (2001): *Erziehung in Traditionalen Kulturen. Quellen und Befunde aus Afrika, Amerika, Asien und Australien 1898—1983.* Reimer, Berlin.

Krechevsky, I. (1932): "Hypothesis" in rats. *Psychological Review* 39, 516—532.

Kreiman, G., Koch, C. & Fried, I. (2000): Imagery neurons in the human brain. *Nature* 408, 357—361.

Kretzschmar, H. A. et al. (1997): Cell death in prion disease. *J Neural Transm Suppl.* 50, 191—210.

Kuhl, P. K. et al. (1997): Cross-Language Analysis of Phonetic Units in Language Addressed to Infants. *Science* 277, 684—686.

Kuhn, H. & Waser, J. (1982): Selbstorganisation der Materie und Evolution früher Formen des Lebens. In: *Biophysik* (ed. W. Hoppe et al.), S. 860—906. Springer, Berlin.

Kuhn, T. (1959/92): Die grundlegende Spannung: Tradition und Neuerung in der wissenschaftlichen Forschung. In: *Die Entstehung des Neuen* (T. S. Kuhn), pp. 308—326. Suhrkamp, Frankfurt. [Engl.: The Essential Tension: Tradition and Innovation in Scientific Research. In: *The Third University of Utah Research Conference on the Identification of Scientific Talent* (ed. C. W. Taylor), pp. 162—174. University of Utah Press, Salt Lake City]

— (1962/96): *The Structure of Scientific Revolutions*. University of Chicago Press, Chicago.

— (1970/92): Logik oder Psychologie der Forschung? In: *Die Entstehung des Neuen* (T. S. Kuhn), pp. 308—326. Suhrkamp, Frankfurt. [Logic of Discovery or Psychology of Research? In: *Criticism and the Growth of Knowledge* (ed. I. Lakatos & A. Musgrave), pp. 1—23. Cambridge University Press, Cambridge]

— (1974/92): Neue Überlegungen zum Begriff des Paradigmas. In: *Die Entstehung des Neuen* (T. S. Kuhn), pp. 308—326. Suhrkamp, Frankfurt. [Second Thoughts on Paradigms. In: *The Structure of Scientific Theories* (ed. F. Suppe), pp. 459—482. University of Illinois Press, Urbana]

Kuro-o, M. (1997): Mutation of the mouse klotho gene leads to a syndrome resembling ageing. *Nature* 390, 45—51.

Lackner, J. R. & DiZio, P. A. (2000): Aspects of body self-calibration. *Trends in Cognitive Sciences* 4(7), 279—288.

LaFreniere, P. J. (1992): The ontogeny of tactical deception in humans. In: *Machiavellian Intelligence. Social Expertise and the Evolution of Intellect in Monkeys, Apes, and Humans* (ed. R. W. Byrne & A. Whiten), pp. 238—252.

Lai, C. S. L. et al. (2001): A forkhead-domain gene is mutated in a severe speech and language disorder. *Nature* 413, 519—523.

Lalouschek, W., Lang, W. & Deecke, L. (1995): Distinguishing between Randomness and Regularity—Evolutionary Epistemology and Neurophysiology. Evolution and Cognition 1, 86—94.

Lamarck de, J.-B. (1809/1909): *Zoologische Philosophie*. Kröner, Leipzig.

Lamberts, S. W. J., van den Beld, A. W. & van der Lely, A.-J. (1997): The Endocrinology of Aging. *Science* 278, 419—424.

Landau, B., Smith, L. & Jones, S. (1998): Object perception and object naming in early development. *Trends in Cognitive Sciences* 2(1), 19—24.

Lander, E. & Botstein, D. (1989): A new tool for investigating multigenic traits. *Science* 264, 1691.

Lander E. S. & Weinberg, R. A. (2000): Journey to the Center of Biology. *Science* 287, 1777—1782.

Law, L. E. & Lock, A. J. (1994): Do gorillas recognize themselves on television? In: *Self-awareness in animals and humans: Developmental perspectives* (ed. S. T. Parker, R. W. Mitchell & M. L. Bocchia), pp. 308—312. Cambridge University Press, Cambridge.

Lazell, J. D. jr. & Spitzer, N. C. (1977): Apparent play behavior in the American alligator. *Copeia* 13, 188.

Lee, D. H. (1996): A self-replicating peptide. *Nature* 382, 525—528.

Lee, R. & DeVore, I. (1968): *Man the hunter*. Aldine, Chicago.

Lehninger, A. L. (2000): *Principles of Biochemistry*. Worth Publishers.

Lehrman, D. S. (1974): Semantische und begriffliche Fragen beim Natur-Dressur-Problem. In: *Kritik der Verhaltensforschung* (ed. G. Roth), S. 72—117. Beck, München.

Leinfellner, W. (1976): Interne und externe Kriterien der Wissenschaften und der kybernetische Charakter des wissenschaftlichen Fortschritts. In: *Wissenschaftssteuerung* (ed. H. Strasser & K. D. Knorr), pp. 139—155.

— (1983): Das Konzept der Kausalität und der Spiele in der Evolutionstheorie. In: *Die Evolution des Denkens* (ed. K. Lorenz & F. M. Wuketits), S. 215—260. Piper, München.

Lenski, R. E. (2001): Come Fly, and Leave the Baggage Behind. *Science* 294, 533—534.

Lesch, K. P. et al. (1996): Association of anxiety-related traits with a polymorphism in the serotonin transporter gene regulatory region. *Science* 274, 1527—1531.

Leslie, A. M. et al. (1998): Indexing and the object concept: developing 'what' and 'where' systems. *Trends in Cognitive Sciences* 2(1), 10—18.

Levin, L. R. (1992): The Drosophila Learning and Memory Gene rutabaga Encodes a Ca2+/Calmodulin-Responsive Adenylyl Cyclase. *Cell* 68, 479—489.

Levitt, P. (1995): Experimental Approaches that Reveal Principles of Cerebral Cortical Development. In: *The Cognitive Neurosciences* (ed. M. S. Gazzaniga), pp. 147—180. MIT Press, London.

Lewin, B. (1985): *Genes*. Wiley & Sons, New York.

Lewin, R. (1987): Africa: cradle of modern humans. *Science* 237, 1292—1295.

Lewontin, R. C. (1981): Sleight of hand. Review of Genes, Mind, and Culture. *The Sciences* July/August, 23—26.

Lewontin, R. C., Rose, S. & Kamin, L. J. (1985): *Not in Our Genes: Biology, Ideology and Human Nature*. Pantheon, New York.

Lichtenberg, G. C. (1941): *Tag und Dämmerung. Aphorismen—Schriften—Briefe—Tagebücher*, Leipzig.

Lindh, A. G. (1993): Did Popper solve Hume's problem? *Nature* 366, 105—106.

Liu, F. G. et al. (2001): Molecular and morphological supertrees for eutherian (placental) mammals. *Science* 291, 1786—1789.

Lorenz, K. (1935/73): Companions as factors in the bird's environment. In: *Studies in animal and human behavior* (K. Lorenz), pp. 101—258. Methuen & Co, London.

— (1937): Über den Begriff der Instinkthandlung. *Folia biotheoretica* Serie B, 2, Instinctus 17—50.

— (1940a): Systematik und Entwicklungsgedanke im Unterricht. *Der Biologe* 9(1-2), 24—36.

— (1940b): Durch Domestikation verursachte Störungen arteigenen Verhaltens. *Z. angew. Psychologie u. Charakterkunde* 59, 2—81.

— (1941): Kants Lehre vom Apriorischen im Lichte gegenwärtiger Biologie. *Blätter für Deutsche Philosophie* 15, 94—125.

— (1943): Die angeborenen Formen möglicher Erfahrung. *Z. Tierpsychologie* 5, 235—409.

— (1961): Phylogenetische Anpassung und adaptive Modifikation des Verhaltens. *Z. f. Tierpsychologie* 18, 139—187.

— (1965): *Über tierisches und menschliches Verhalten (2 vols.)*. Piper, München.

— (1973/77): *Behind the Mirror. A search for a natural history of human knowledge.* Harcourt, New York.

— (1982): Das Gänsekind Martina: Vom Wunder der Vogelsprache. *Das Tier* 12,22.

— (1983): *Der Abbau des Menschlichen.* Piper, München.

— (1985): My Family and other animals. In: *Leaders in the Study of Animal Behavior* (ed. D. A. Dewsbury), pp. 258—287. Bucknell University Press, Lewisburg.

Lorenz, K. & Wuketits, F. M. (1983): *Die Evolution des Denkens.* Piper, München.

Losick, R. & Sonenshein, A. L. (2001): Turning Gene Regulation on its Head. *Science* 293, 2018—2019.

Löther, R. (1990): *Wegbereiter der Genetik: Gregor Johann Mendel und August Weismann.* Harri Deutsch, Leipzig.

Loveland, K. A. (1984): Learning about points of view: Spatial perspective and the acquisition of "I/you". *Journal of Child Language* 11, 535—556.

Löw, R. (1983): Evolution und Erkenntnis—Tragweite und Grenzen der evolutionären Erkenntnistheorie in philosophischer Absicht. In: *Die Evolution des Denkens* (ed. K. Lorenz & F. M. Wuketits), S. 331—360. Piper, München.

Lubinski, D. (2000): Scientific and social significance of assessing individual differences: "Sinking Shafts at a Few Critical Points". *Annual Review of Psychology* 2000.

Lumsden, C. J. & Wilson, E. O. (1981): *Genes, Mind, and Culture: The Coevolutionary Process.* Harvard University Press, Cambridge, Mass..

— (1983): *Promethean Fire: Reflections on the Origin of Mind.* Harvard University Press, Cambridge, Mass..

Luria, S. E. & Delbrück, M. (1943): Mutations of bacteria from virus sensitivity to virus resistance. *Genetics* 28, 491—511.

Lütterfelds, W. (1987): *Transzendentale oder evolutionäre Erkenntnistheorie.* Wissenschaftliche Buchgesellschaft, Darmstadt.

Lynch, A. (1996): *Thought Contagion.* Basic Books, New York.

Lynn, R. (1998): The Decline of Genotypic Intelligence. In: *The Rising Curve. Long-Term Gains in IQ and Related Measures* (ed. U. Neisser), pp. 335—364. APA, Washington DC.

Lyons, E., J. (1997): Sex and synergism. *Nature* 390, 19—20.

Mach, E. (1911): *Principien der Wärmelehre,* Leipzig.

Maclean, G. L. (1972): Clutch size and evolution in the Charadrii. *Auk* 89, 299—324.

Macphail, E. M. (1998): *The Evolution of Consciousness.* Oxford University Press, Oxford.

MacPhail, E. M. & Bolhuis, J. J. (2001): The evolution of intelligence: adaptive specializations versus general process. *Biological Reviews of the Cambridge Philosophical Society* 76, 341—364.

Maddox, J. (1997): Schizophrenia: The price of language? *Nature* 388, 424—425.

Makous, W. (2000): Limits to Our Knowledge. *Science* 287, 1399.

Mann, C. (1994): Behavioral Genetics in Transition. *Science* 264, 1686—1689.

Mantegna, R. N. et al. (1994): Linguistic Features of Noncoding DNA Sequences. *Physical Review Letters* 73(23), 3169—3172.

Maquet, P. et al. (2000): Experience-dependent changes in cerebral activation during human REM sleep. *Nature Neuroscience* 3, 831—836.

Marais, E. (1976): *Die Seele der weissen Ameise [The Soul of the White Ant].* Heyne, München.

Margulis, L. (1970): *Origin of Eukaryotic cells*. Yale University Press, New Haven.

Markl, H. (1997): Von der Mediengesellschaft zur Wissensgesellschaft: Unsere Zukunft im globalen Informationsnetzwerk. *Herbert Quandt Medien-Preis* Verleihung 1997, Frankfurt am Main.

Markus, M., Kuschmitz, D. & Hess, B. (1984): Chaotic dynamics in yeast glycolysis under periodic substrate input flux. *FEBS Lett.* 172, 235—238.

Marler, P. & Terrace, H. S. (1984): The Biology of Learning. In: *Reports of the Dahlem Workshop*. Springer, Berlin.

Marten, K. et al. (1996): Ring Bubbles of Dolphins. *Scientific American* August 1996.

Martin, N. G. et al. (1986): The transmission of social attitudes. *Proceedings of the National Academy of Sciences of the United States of America* 83, 4365—4368.

Marx, J. (1995): Helping Neurons Find Their Way. *Science* 268, 971—973.

Mascie-Taylor, C. G. N. (1987): Assortative mating in a contemporary British population. *Annals of Human Biology* 14, 59—68.

— (1989): Spouse Similarity for IQ and Personality and Convergence. *Behavior Genetics* 19, 223—227.

Mascie-Taylor, C. G. N. & Boyce, A. J. (1988): *Human mating patterns*. Cambridge University Press, Cambridge.

Massaro, D. W. (1999): Speechreading: illusion or window into pattern recognition. *Trends Cogn Sci* 3, 310—317.

Massaro, D. W. et al. (1998): Laterality in visual speech perception. *J Exp Psychol Hum Percept Perform* 24, 1232—42.

Maturana, H. R. & Varela, F. (1980): *Autopoiesis and Cognition: The Realization of the Living*. Reidel, Dordrecht.

May, R. M. (1995): The Cheetah Controversy. *Nature* 374, 309—310.

Mayer, L. (1997): Schutz für die Schwächsten. *ai-info* 10/97, 8—9.

Maynard Smith, J. (1983): Models of Evolution. *Proceedings of the Royal Society of London* B219, p. 315—325.

— (1989): *Evolutionary Genetics*. Oxford University Press, Oxford.

— (1991): *Evolution and the Theory of Games*. Cambridge University Press, Cambridge.

— (1998): *Shaping Life: Genes, Embryos, and Evolution*. Weidenfeld & Nicolson, London.

— (2000): The Concept of Information in Biology. *Philosophy of Science* 67, 177—194.

Maynard Smith, J., Burian, R., Kauffman, S., Alberch, P., Campbell, J., Goodwin, B., Lande, R., Raup, D. & Wolpert, L. (1985): Developmental constraints and evolution. *Q. Rev. Biol.* 60, 265—287.

Maynard Smith, J. & Warren, N. (1982): Models of cultural and genetic change. *Evolution* 36, 620—627.

Maynard Smith, J. & Price, G. R. (1973): The logic of animal conflict. *Nature* 246, 15—18.

Mayr, E. (1963): *Animal Species and Evolution*. Harvard University Press, Cambridge, Mass..

— (1988): *Toward a New Philosophy of Biology: Observations of an Evolutionist*. Harvard University Press, Cambridge, Mass..

McAdams, H. & Shapiro, L. (1995): Circuit Simulation of Genetic Networks. *Science* 269, 650—656.

McClearn, G. E. et al. (1997): Substantial Genetic Influence on Cognitive Abilities in Twins 80 or More Years Old. *Science* 276, 1560—1563.

McClintock, B. (1951): Genes and Mutations. *Cold Spring Harbor Symposium* XVI, Session I (Theory of the Gene).

McComb, K. & Semple, S. (1998): Are talkers the only thinkers? *Nature* 395, 656—657.

McGuffin, P., Riley, B. & Plomin, R. (2001): Genomics and Behavior: Toward Behavioral Genomics. *Science* 291, 1232—1249.

McKim, K. et al. (1998): Meiotic Synapsis in the Absense of Recombination. *Science* 279, 876—878.

McKim, K. S. & Hawley, R. S. (1995): Chromosomal Control of Meiotic Cell Division. *Science* 270, 1595—1601.

Meder, E. (1958): *Gnathonemus petersii* (Günther). *Z. Vivaristik* 4, 161—171.

Medicus, G. (1992): The Biogenetic Rule Has No Relevance for Behavioral Ontogeny. *Human Development* 35, 1—8.

Meyer-Holzapfel, M. (1960): Über das Spiel bei Fischen, insbesondere beim Tapirrüsselfisch (*Mormyrus kannume* Forskal). *Zool. Garten* 25, 189—202.

Miles, H. L. W. (1994): "Me Chantek!": The development of self-awareness in a signing orangutan. In: *Self-awareness in animals and humans: Developmental perspectives* (ed. S. T. Parker, R. W. Mitchell & M. L. Bocchia), pp. 254—272. Cambridge University Press, Cambridge.

Milinski, M. & Parker, G. A. (1991): Competition for resources. In: *Behavioural Ecology. An Evolutionary Approach* (ed. J. Krebs and N. Davies), S. 137—168. Blackwell, Oxford.

Milinski, M. & Wedekind, C. (2001): Evidence for MHC-correlated perfume preferences in humans. *Behavioral Ecology* 12(2), 140—149.

Miller, G. (2000): *The Mating Mind. How Sexual Choice Shaped the Evolution of Human Nature.* Heinemann, London.

Mirsky, A. E. & Ris, H. (1951): *J. Gen. Physiol.* 34, 451.

Mitchell, R. et al. (1993): Species concepts. *Nature* 364, 20.

Mittenecker, E. (1958): Die Analyse "zufälliger" Reaktionsfolgen. *Z. f. exp. angew. Psychol.* 5, 45—60.

— (1960): Die informationstheoretische Auswertung des Zeigeversuchs bei Psychotikern und Neurotikern. *Z. f. exp. angew. Psychol.* 7, 392—400.

Mittenecker, E. & Raab, E. (1973): *Informationstheorie für Psychologen.* Hogrefe, Göttingen.

Miyashita, Y. (1995): How the brain creates imagery. *Science* 268, 1719—1720.

Monod, J. (1970): *Le hasard et la nécessité. Essai sur la philosophie naturelle et la biologie moderne.* Seuil, Paris.

— (1975): De Homine. Rivista dell' Istituto di filosofia, *Roma* 53—56, p. 131.

— (1979): Discussion in Théories du langage, théories de l'apprentissage. In: *Le débat entre Jean Piaget et Noam Chomsky* (ed. Piattelli-Palmarini), p. 291. Éditions du Seuil, Paris.

Montgomery, H. E. et al. (1998): Human gene for physical performance. *Nature* 393, 221—222.

Morell, V. (1997): Sex Frees Viruses From Genetic 'Ratchet'. *Science* 278, 1562.

Morgan, J. I. & Curran, T. (1991): Stimulus transcription coupling in the nervous system: Involvement of the inducible proto-oncogenes fos and jun. *Ann. Rev. Neurosci.* 14, 421—451.

Morris, D. (1967): *The Naked Ape*. Vintage, New York.

— (1994): *Bodytalk, a World Guide to Gestures*. Jonathan Cape, London.

— (1995): Das Tier Mensch. *ORF-Nachlese* , 2—14.

Morrison, J. H. & Hof, P. R. (1997): Life and Death of Neurons in the Aging Brain. *Science* 278, 412—419.

Moseley, C. & Asher, R. E. (1994): *Atlas of the World's Languages*. Routledge, New York.

Moser, P. K. & vander Nat, A. (1987): *Human Knowledge. Classical and Contemporary Approaches*. Oxford University Press, Oxford.

Mountcastle, V. B. (1984): Central nervous mechanisms in mechanoreceptive sensibility. In: *Handbook of Physiology* (ed. I. Darian-Smith), Section I, Volume II. American Physiological Society, Bethesda.

Muller, H. J. (1932): Some genetic aspects of sex. *American Naturalist* 66, 118—138.

Müller, A., Müller, K. H. & Stadler, F. (eds., 2001): *Konstruktivismus und Kognitionswissenschaft*. Springer, Wien.

Müller, K. E., Müller, U. R. & Henning, C. (1999): *Soul of Africa. Magie eines Kontinents*. Könemann, Köln.

Murray, A. & Hunt, T. (1993): *The Cell Cycle, an Introduction*. Oxford University Press, Oxford.

Nagata, S. & Golstein, P. (1995): The FasL Death Factor. *Science* 267, 1449—1456.

Nagel, E., Newman, J. R. & Hofstadter, D. R. (2001): *Goedel's Proof*. New York University Press, New York.

Nagele, R., Freeman, T., McMorrow, L. & Lee, H.-y. (1995): Precise Spatial Positioning of Chromosomes During Prometaphase: Evidence for Chromosomal Order. *Science* 270, 1831-1835.

Naylor, G. J. & Brown, W. M. (1997): Structural biology and phylogenetic estimation. *Nature* 388, 527—528.

Nei, M. & Livshits, G. (1989): Genetic Relationships of Europeans, Asians and Africans and the Origin of Modern Homo sapiens. *Human Heredity* 39, 276—281.

Neitz, J., Neitz, M. & Kainz, P. M. (1996): Visual Pigment Gene Structure and the Severity of Color Vision Defects. *Science* 274, 801—804.

Nesse, R. M. & Williams, G. C. (1994): *Why we get sick*. Vintage, New York.

Neville, H. J. (1995): Developmental Specificity in Neurocognitive Development in Humans. In: *The Cognitive Neurosciences* (ed. M. S. Gazzaniga), pp. 219—231. MIT Press, London.

Nicolelis, M. A. et al. (1995): Sensorimotor encoding by synchronous neural ensemble activity at multiple levels of the somatosensory system. *Science* 268, 1353—1358.

Nicoll, R. A., Kauer, J. A. & Malenka, R. C. (1988): The current excitement in long-term potentiation. *Neuron* 1, 97—103.

Nooden, L. D. (1988): Whole plant senescence. In: *Senescence and aging in plants* (ed. K. V. Thiman & A. C. Leopold), Academic Press, San Diego.

Norris, S. (2000): Fatal attraction. *New Scientist* 17 June 2000.

Nudo, R. J. et al. (1996): Neural Substrates for the Effects of Rehabilitative Training on Motor Recovery After Ischemic Infarct. *Science* 272, 1791—1794.

Numbers, R. L. & Stenhouse, J. (eds., 1999): *Disseminating Darwinism. The Role of Place, Race, Religion, and Gender*. Cambridge University Press, Cambridge.

Nurse, P. (1997): Regulation of the eukaryotic cell cycle. *Eur J Cancer* 33(7), 1002—4.

O'Brien, S. J. (1994): Genetic and Phylogenetic Analyses of Endangered Species. *Annual Review of Genetics* 28, 467—489.

O'Brien, S. J. et al. (1999): The Promise of Comparative Genomics in Mammals. *Science* 286, 458—481.

Oeser, E. (1974a): *System, Klassifikation, Evolution.* Braumüller, Wien.

— (1974b): Der Tautologievorwurf und die Struktur der Darwinschen Selektionstheorie. In: *System, Klassifikation, Evolution* (ed. E. Oeser), S. 144—154. Braumüller, Wien.

Oller, D. K. (2000): *The Emergence of the Speech Capacity.* Lawrence Erlbaum Associates, Hillsdale.

Olsen, S. J. (1977): The Chinese wolf, ancestor of New World dogs. *Science* 197, 553—555.

— (1985): *Origins of the Domestic Dog: the Fossil Record.* University of Arizona Press, Tucson.

Osusky, M., Kissova, J. & Kovac, L. (1997): Interspecies transplacement of mitochondria in yeasts. *Curr. Genet.* 32, 24—26.

Pardoll, D. (2001): Immunology: T cells and tumours. *Nature* 411, 1010—1012.

Parker, S. T. & Mitchell, R. W. (1994): Evolving self-awareness. In: *Self-awareness in animals and humans: Developmental perspectives* (ed. S. T. Parker, R. W. Mitchell & M. L. Bocchia), pp. 413—428. Cambridge University Press, Cambridge.

Passingham, R. E. (1973): Anatomical differences between the cortex of man and other primates. *Brain, Behavior & Evolution* 7, 337—359.

— (1975): Changes in the size and organization of the brain in man and his ancestors. *Brain, Behavior & Evolution* 11, 73—90.

Patterson, F. G. P. & Cohn, R. H. (1994): Self-recognition and self-awareness in lowland gorillas. In: *Self-awareness in animals and humans: Developmental perspectives* (ed. S. T. Parker, R. W. Mitchell & M. L. Bocchia), pp. 273—290. Cambridge University Press, Cambridge.

Pääbo, S. (2001): The Human Genome and Our View of Ourselves. *Science* 291, 1219—1220.

Pease, A. & Pease, B. (2001): *Why Men Don't Listen And Women Can't Read Maps.* Orion Trade.

Pelkwijk, J. J. ter & Tinbergen, N. (1937): Eine reizbiologische Analyse einiger Verhaltensweisen von *Gasterosteus aculatus* L. *Z. Tierpsychologie* 1, 193—204.

Penfield, W. & Roberts, L. (1959): *Speech and brain mechanisms.* Princeton University Press, Princeton.

Pennisi, E. (1996): Seeking Life's Bare (Genetic) Necessities. *Science* 272, 1098—1099.

— (1997): New Developmental Clock Discovered. *Science* 278, 1564.

Pepperberg, I. (1990): Referential mapping: A technique for attaching functional significance to the innovative utterances of an African Grey parrot (*Psittacus erithacus*). *Applied Psycholinguistics* 11, 23—44.

Pepperberg, I. M., Willner, M. R. & Gravitz, L. B. (1997): Development of Piagetian object permanence in a grey parrot (*Psittacus erithacus*). *Journal of Comparative Psychology* 111, 63—75.

Perner, J., Baker, S. & Hutton, D. (1994): Prelief: The Conceptual Origins of Belief and Pretence. In: *Children's Early Understanding of Mind: Origins and Development* (ed. C. Lewis & P. Mitchell), pp. 261—286.

Perrett, D. I., May, K. A. & Yoshikawa, S. (1994): Facial shape and judgements of female attractiveness. *Nature* 368, 239—242.

Peschl, M. & Riegler, A. (1997): Does Representation Need Reality? *New Trends in Cognitive Science. Proceedings of the Austrian Society of Cognitive Science*, S. 5—17, Vienna.

Peskin, J. (1992): Ruse and representations. On children's ability to conceal information. *Developmental Psychology* 28, 84—89.

Petitto, L. A. et al. (2001): Language rhythms in baby hand movements. *Nature* 413, 35.

Petronis, A. & Kennedy, J. L. (1995): Unstable genes—unstable mind? *Am. J. Psychiatry* 152, 164—172.

Piaget, J. (1929): L'adaptation de la Limnea stagnalis aux milieux lacustres de la Suisse romande. *Revue suisse de Zoologie* 36, 263—531.

— (1936): *La naissance de l'intelligence chez l'enfant*. Delachaux & Niestlé, Neuchâtel.

— (1950/99): *The Construction of Reality in the Child*. Routledge, London.

— (1957): The child and modern physics. *Scientific American* 196, 46—51.

— (1967): *Biologie et Connaissance*. Gallimard, Paris.

— (1970): Piaget's theory. In: *Carmichael's manual of child psychology* (ed. P. H. Mussen), pp. 703—732. Wiley, New York.

— (1971): *Biology and knowledge : an essay on the relations between organic regulations and cognitive processes*. The University of Chicago Press, Chicago.

— (1976): *Le Centre International d'Epistemologie Génétique* (CIEG). Laboratoire audiovisuel universitaire, Geneva.

— (1989): *Psychogenesis and the History of Science*. Columbia University Press.

— (1997): *Selected Works*, Vol I—XI. Routledge,

Piaget, J. & Inhelder, B. (1948): *La représentation de l'espace chez l'enfant*. P.U.F., Paris.

Piaget, J. & Inhelder, B. (1969): *Die Entwicklung des physikalischen Mengenbegriffs*. Klett, Stuttgart.

Pilz, G. & Moesch, H. (1975): *Der Mensch und die Graugans. Eine Kritik an Konrad Lorenz*. Umschau, Frankfurt.

Pinker, S. (1991): Rules of Language. *Science* 253, 530—535.

— (1995): *The Language Instinct. How the Mind Creates Language*. Penguin, London.

Place, Ullin T. (2000): The Role of the Hand in the Evolution of Language. *Psycholoquy* 11(007).

Planck, M. (1928): *Wissenschaftliche Autobiographie*. Leipzig.

Pleijel, F. & Rouse, G. W. (2000): Least-inclusive taxonomic unit: a new taxonomic concept for biology. *Proceedings of the Royal Society London* 267 (1443), 627—630.

Plomin, R. (1994): *Genetics and experience: The interplay between nature and nurture*. Sage, London.

Plomin, R. & Bergeman, C. S. (1991): The nature of nurture: Genetic influence on "environmental" measures. *Behavioral and Brain Sciences* 14, 373—427.

Plomin, R. et al. (1998): A Quantitative Trait Locus Associated With Cognitive Ability in Children. *Psychological Science* 9(3).

Plotkin, H. (ed., 1982): *Learning, Development, and Culture: Essays in Evolutionary Epistemology*. Wiley, New York.

— (1994): *The Nature of Knowledge. Concerning Adaptations, Instinct and the Evolution of Intelligence.* Penguin, London.

Plowright, C. M. S., Reid, S. & Kilian, T. (1998): How mynah birds (*Gracula religiosa*) and pigeons (*Columba livia*) find hidden food: A study in comparative cognition. *Journal of Comparative Psychology* 112, 13—25.

Plutchik, R. & Kellerman, H. (1986): *Biological Foundations of Emotion.* Academic Press, Orlando.

Popper, K. R. (1935/59): *The Logic of Scientific Discovery.* Basic Books, New York.

— (1972): *Objective Knowledge. An Evolutionary Approach.* Clarendon, Oxford.

— (1988): Die erkenntnistheoretische Position der Evolutionären Erkenntnistheorie. In: *EE—Bedingungen, Lösungen, Kontroversen* (ed. R. Riedl & F. M. Wuketits), S. 29—37. Parey, Berlin.

— (1994): *Ausgangspunkte.* Hoffmann & Campe, Hamburg. [(1974): *Unended Quest. An Intellectual Autobiography.* Fontana/Collins, London]

Popper, K. R. & Eccles, J. C. (1976): *The Self and Its Brain.* Springer, Heidelberg-London.

Popper, K. R. & Lorenz, K. Z. (1994): *Die Zukunft ist offen.* Piper, München.

Portmann, A. (1967): *Zoologie aus vier Jahrzehnten.* Piper, München.

— (1971): *Entlässt die Natur den Menschen?* Piper, München.

Potts, W. K., Manning, C. J. & Wakeland, E. K. (1991): Mating patterns in seminatural populations of mice influenced by MHC genotype. *Nature* 352, 619—621.

Povinelli, D. J. (1987): Monkeys, apes, mirrors and minds: The evolution of self-awareness in primates. *Human Evolution* 2, 493—507.

Povinelli, D. J. & Cant, J. G. H. (1995): Arboreal clambering and the evolution of self-conception. *Quarterly Review of Biology* 70, 393—421.

Premack, D. & Woodruff, G. (1978): Does the chimpanzee have a theory of mind? *Behavioral and Brain Sciences* 1(4), 515—526.

Preston, S. H. (1998): Differential Fertility by IQ and the IQ Distribution of a Population. In: *The Rising Curve. Long-Term Gains in IQ and Related Measures* (ed. U. Neisser), pp. 377—388. APA, Washington DC.

Price, T. D. & Gibbs, H. C. (1987): Brood division in Darwin's ground finches. *Anim. Behav.* 35, 299—301.

Prusiner, S., Collinge, J., Powell, J. & Anderton, B. (1992): *Prion Diseases in Humans and Animals.* Ellis Horwood, New York.

Puca, A. A. et al. (2001): A genome-wide scan for linkage to human exceptional longevity identifies a locus on chromosome 4. *Proc. Natl. Acad. Sci.* USA 10.1073/pnas.181337598.

Pusey, A., Williams, J. & Goodall, J. (1997): The Influence of Dominance Rank on the Reproductive Success of Female Chimpanzees. *Science* 277, 828—831.

Puthalakath, H. et al. (2001): Bmf: A Proapoptotic BH3-Only Protein Regulated by Interaction with the Myosin V Actin Motor Complex, Activated by Anoikis. *Science* 293, 1829—1832.

Putnam, H. (1983): Why reason can't be naturalized. In: *Realism and Reason* (ed. H. Putnam), pp. 229—247, Cambridge University Press, Cambridge.

Queiroz, K. de & Gauthier, J. (1992): Phylogenetic taxonomy. *Annu. Rev. Ecol. Syst.* 23, 449—480.

Raff, R. A. (1996): *The Shape of Life. Genes, Development, and the Evolution of Animal Form.* University of Chicago Press, Chicago.

Rao, V. R., Cohen, G. B. & Oprian, D. D. (1994): Rhodopsin mutation G90D and a molecular mechanism for congenital night blindness. *Nature* 367, 639—642.

Recanzone, G. H. et al. (1992): Topographic reorganization of the hand representation in cortical area 3B of owl monkeys trained in a frequency-discrimination task. *J. Neurophysiol.* 67, 1031—1056.

Reichholf, J. (1979): Die Artabgrenzung im Tierreich, eine "Evolutionär Stabile Strategie"? *Spixiana* 2, 201—207.

Remmert, H. (1980): *Ecology.* Springer, Berlin.

— (1994): *Minimal Animal Populations.* Springer, Berlin.

Remschmidt, H. (1970): Experimentelle Untersuchungen zur sogenannten epileptischen Wesensveränderung. *Fortschr. Neurol. Psychiat.* 38, 524—540.

Rice, W. R. & Chippindale, A. K. (2001): Sexual Recombination and the Power of Natural Selection. *Science* 294, 555—559.

Richards, R. J. (1987): *Darwin and the emergence of evolutionary theories of mind and behavior.* University of Chicago Press, Chicago.

— (1997): The Foundations of Konrad Lorenz' Evolutionary Theory of Behavior. 2nd *Altenberg Workshop in Theoretical Biology* on "Evolutionary Naturalism", June 5, 1997.

Richardson, M. K. et al. (1997): There is no highly conserved embryonic stage in the vertebrates: implications for current theories of evolution and development. *Anat. Embryol.* 196, 91—106.

Ricken, E. (1984): *Lexikon der Erkenntnistheorie und Metaphysik.* Suhrkamp, Frankfurt.

Ridge, J. P., Fuchs, E. J. & Matzinger, P. (1996): Neonatal Tolerance Revisited: Turning on Newborn T Cells with Dendritic Cells. *Science* 271, 1723—1726.

Ridley, M. (1982): Coadaptation and the inadequacy of natural selection. British *Journal for the History of Science* 14, 45—68.

— (2001): *The Cooperative Gene. How Mendel's Demon Explains the Evolution of Complex Beings.* Free Press, New York.

Riedl, R. (1975): *Die Ordnung des Lebendigen.* Parey, Hamburg. [*Order in Living Organisms: A Systems Analysis of Evolution.* Wiley, New York]

— (1980): *Biologie der Erkenntnis.* Parey, Berlin. [*Biology of Knowledge : The Evolutionary Basis of Reason.* Wiley, New York]

— (1987a): *Begriff und Welt. Biologische Grundlagen des Erkennens und Begreifens.* Parey, Berlin.

— (1987b): *Kultur—Spätzündung der Evolution?* Piper, München.

Rieseberg, L. H. & Soltis, D. E. (1991): *Evol. Trends Plants* 5, 65—81.

Risch, N. J. (2000): Searching for genetic determinants in the new millennium. *Nature* 405, 847—856.

Rius (1983): *Marx für Anfänger.* Rowohlt, Reinbek b. Hamburg. [*Marx para Principiantes*]

Robin, N. & Holyoak, K. J. (1995): Relational Complexity and the Functions of Prefrontal Cortex. In: *The Cognitive Neurosciences* (ed. M. S. Gazzaniga), pp. 987—997. MIT Press, London.

Rose, M. (1991): *Evolutionary biology of aging.* Oxford University Press, Oxford.

Rosenbaum, D. E. (2000): *On Left-Handedness, Its Causes and Costs.* The New York Times on the Web, May 16, 2000.

Rosenzweig, M. L. (1998): Tempo and Mode of Speciation. *Science* 277, 1622—1623.

Roush, W. (1998): "Living Fossil" Fish Is Dethroned. *Science* 277, 1436.

Rowen, L., Mahairas, G. & Hood, L. (1997): Sequencing the human genome. *Science* 278, 605—607.

Rumbaugh, D. M., Beran, M. J. & Hillix, W. A. (2000): Cause-Effect Reasoning in Humans and Animals. In: *The Evolution of Cognition* (ed. C. Heyes & L. Huber), pp. 221—238. MIT Press, Cambridge, MA.

Rushton, J. P. (1989): Genetic similarity, human altruism, and group selection. *Behavioral and Brain Sciences* 12, 503—518.

Rushton, J. P., Russell, R. J. H. & Wells, P. A. (1984): Genetic Similarity Theory: Beyond Kin Selection. *Behavior Genetics* 14(3), 179—193.

Russell, D. E. H. (1986): *The Secret Trauma*. Basic Books, New York.

Ryle, G. (1949): *The Concept of Mind*. Hutchinson, London.

Sachidanandam, R. et al. (2001): A map of human genome sequence variation containing 1.42 million single nucleotide polymorphisms. *Nature* 409, 928—933.

Sacks, O. (1985): *The Man Who Mistook His Wife For a Hat*. Simon & Schuster, NY.

Saghatelian, A., Yokobayashi, Y., Soltani, K. & Ghadiri, M. R. (2001): A chiroselective peptide replicator. *Nature* 409, 797—801.

Salzberg, S. L., White, O., Peterson, J. & Eisen, J. A. (2001): Microbial Genes in the Human Genome: Lateral Transfer or Gene Loss? *Science* 292, 1903—1906.

Sanders, A. R. & Gershon, E. S. (1998): Clinical Genetics, VIII: From Genetics to Pathophysiology—Candidate Genes. *Am. J. Psychiatry* 155, 162.

Sarkar, S. (1998): The obsession with heritability. In: *Genetics and Reductionism* (S. Sarkar), pp. 71—100. Cambridge University Press, Cambridge.

Sarzotti, M., Robbins, D. S. & Hoffman, P. M. (1996): Induction of Protective CTL Responses in Newborn Mice by a Murine Retrovirus. *Science* 271, 1726—1728.

Savage-Rumbaugh, E. S. et al. (1986): Spontaneous Symbol Acquisition and Communicative Use by Pygmy Chimpanzees (*Pan paniscus*). *J. Exper. Psych.* 115(3), 211—235.

Savage-Rumbaugh, E. S. et al. (1993): Language comprehension in ape and child. *Monographs of the society for research in child development* 58, Nos. 3—4.

Savage-Rumbaugh, S. & Lewin, R. (1994): *Kanzi, the Ape at the Brink of the Human Mind*. Wiley, New York.

Saunders, W. B., Work, D. M. & Nikolaeva, S. V. (1999): Evolution of Complexity in Paleozoic Ammonoid Sutures. *Science* 286, 760—763.

Sawaguchi, T. & Kudo, H. (1990): Neocortical development and social structure in primates. *Primates* 31, 283—290.

Scheller, R. H. & Axel, R. (1984): Wie Gene ein angeborenes Verhalten steuern. *Spektrum der Wissenschaft* 5, 72—83.

Schiff, S. J. et al. (1994): Controlling chaos in the brain. *Nature* 370, 615—20.

Schleidt, W. (1992): Bewusstsein bei Tieren—Eine besondere Art der Wahrnehmung. In: *Das Bewusstsein. Multidimensionale Entwürfe* (ed. G. Guttmann & G. Langer), S. 309—329. Springer, Wien.

Schmidt, R. F. (ed., 1998): *Neuro- und Sinnesphysiologie*. Springer, Berlin-Heidelberg-New York.

Schmidt, R. F. & Struppler, A. (1983): *Der Schmerz. Ursachen, Diagnose, Therapie*. Piper, München.

Schmidt, R. F., Thews, G. & Lang, F. (ed., 2000): *Physiologie des Menschen* [*Human Physiology*]. Springer, Berlin-Heidelberg-New York.

Schneider, R. (1995): *Schlafes Bruder.* Reclam, Leipzig.

Scholl, B. J. & Tremoulet, P. D. (2000): Perceptual causality and animacy. *Trends in Cognitive Sciences* 4(8), 299—309.

Schröder, R. (2000): Funktion der Gene finden. *format* 27/00, 140—142.

Schrödinger, E. (1944/92): *What is Life?* Cambridge University Press, Cambridge.

Schuster, P. (1993): RNA-based evolutionary optimization. *Origins of Life* 23, 373—391.

Schusterman, R. J. (1990): Stimulus equivalence and cross-modal perception: a testable model for demonstrating symbolic representations in bottlenose dolphins. In: *Sensory Abilities of Cetaceans: Laboratory and Field Evidence* (ed. J. A. Thomas & R. A. Kastelein), pp. 677—683. Plenum, New York.

Schusterman, R. J. & Krieger, K. (1984): California sea lions are capable of semantic comprehension. *The Psychological Record* 34, 3—23.

Schweder, B. & Riedl, S. (1997): *Der kleine Unterschied [Why Men and Women think differently].* Deuticke, Wien.

Searle, J. (1995): Blurb written on a flyer for the Journal of Consciousness Studies. Imprint Academic, Thorverton.

Segal, N. (1988): Cooperation, Competition, and Altruism in Human Twinships: A Sociobiological Approach. In: *Sociobiological Perspectives on Human Development* (ed. K. B. MacDonald), pp. 168—206. Springer, New York.

Segerstrale, U. (2000): *Defenders of the truth. The battle for science in the sociobiology debate and beyond.* Oxford University Press, Oxford.

Service, R. F. (2000): Creation's Seventh Day. *Science* 289, 232—235.

Shadmehr, R. & Holcomb, H. H. (1997): Neural Correlates of Motor Memory Consolidation. *Science* 277, 821—825.

Shatz, C. (1992): Das sich entwickelnde Gehirn. *Spektrum der Wissenschaft* 11, 44—52.

Shettleworth, S. J. (1972): Constraints on learning. *Advances in the Study of Behavior* 4, 1—68.

— (1993a): Where is the comparison in comparative cognition? *Psychological Science* 4, 179—184.

— (1993b): Varieties of learning and memory in animals. *J. exp. Psychol.: Animal Behavior Processes* 19, 5—14.

— (1994): Biological Approaches to the Study of Learning. In: *Animal Learning and Cognition* (ed. N. J. Mackintosh), pp. 185—219. Academic Press, San Diego.

— (1998): *Cognition, Evolution, and Behavior.* Oxford University Press, New York.

Shields, A., Paskind, M., Otto, G. & Baltimore, D. (1979): Structure of the Abelson Murine Leukemia Virus Genome. *Cell* 18, 955—962.

Shubin, N., Tabin, C. & Carroll, S. (1997): Fossils, genes and the evolution of animals limbs. *Nature* 388, 639—648.

Shuter, B. J. et al. (1983): Phenotypic correlates of genomic DNA content in unicellular eukaryotes and other cells. *Am. Nat.* 122, 26—44.

Shyue, S.-K. et al. (1995): Adaptive Evolution of Color Vision Genes in Higher Primates. *Science* 269, 1265—1267.

Sibley, C. G. (1996): DNA-DNA hybridisation in the study of primate evolution. In: *Human Evolution* (ed. S. Jones, R. Martin & D. Pilbeam), p. 313—315. Cambridge University Press, Cambridge.

Siemens, H. W. (1891/1924): *Race hygiene and heredity.* Appleton, New York.

Sigmund, K. (1993): *Games of Life. Explorations in Ecology, Evolution, and Behaviour.* Oxford University Press, Oxford.

Sigmund, K. & Nowak, M. (2000): A Tale of Two Selves. *Science* 290, 949—950.

Silberzweig, D. A. et al. (1996): A functional anatomy of hallucinations in schizophrenia. *Nature* 378, 176—179.

Silver, L. (1997): *Remaking of Eden: Cloning and Beyond in a Brave New World.* Avon.

Singer, W. (1995): Time as Coding Space in Neocortical Processing: A Hypothesis. In: *The Cognitive Neurosciences* (ed. M. S. Gazzaniga), pp. 91—104. MIT Press, London.

Siviy, S. M. (1998): Neurobiological substrates of play behavior. In: *Animal Play. Evolutionary, Comparative, and Ecological Perspectives* (ed. M. Bekoff & J. A. Byers), pp. 221—242. Cambridge University Press, Cambridge.

Sjoelander, S. (1995): Some Cognitive Breakthroughs in the Evolution of Cognition and Consciousness. *Evolution and Cognition* 1(1), 3—11.

Skaggs, W. E. & McNaughton, B. L. (1996): Replay of Neuronal Firing Sequences in Rat Hippocampus During Sleep Following Spatial Experience. *Science* 271, 1870—1873.

Skoyles, J. R. (1997): Evolution's "missing link": A hypothesis upon neural plasticity, prefrontal working memory and the origins of modern cognition. *Medical Hypothesis* 48, 499—509.

Skuse, D. H. et al. (1997): Evidence from Turner's syndrome of an imprinted X-linked locus affecting cognitive function. *Nature* 387, 705—708.

Smith, J. D. et al. (1997): The uncertain response in humans and animals. *Cognition* 62, 75-97.

Sober, S. J. et al. (1997): Receptive Field Changes After Strokelike Cortical Ablation: A Role for Activation Dynamics. *J. Neurophysiol.* 78, 3438—3438 .

Sodian, B. (1994): Early Decpetion and the Conceptual Continuity Claim. In: *Children's Early Understanding of Mind: Origins and Development* (ed. C. Lewis & P. Mitchell), pp. 385—401.

Song, Z., McCall, K. & Steller, H. (1997): DCP-1, a Drosophila Cell Death Protease Essential for Development. *Science* 275, 536—540.

Spearman, C. (1904): "General Intelligence" objectively determined and measured. *American Journal of Psychology* 15, 201—293.

Spencer, H. (1893): Ways of Judging Conduct. In: *The Principles of Ethics* (H. Spencer), Williams and Norgate, London, pp. 47—83.

Spengler, S. J. (2001): Computers and Biology: Bioinformatics in the Information Age. *Science* 287, 1221—1223.

Stanford, C. B. (1999): *The Hunting Apes. Meat Eating and the Origins of Human Behavior.* Princeton University Press, Princeton.

Stearns, S. C. (1992): *The Evolution of Life Histories.* Oxford University Press, Oxford.

Stein, B. & Meredith, A. (1993): *The Merging of the Senses.* MIT Press, Cambridge.

Steller, H., Nagata, S., Golstein, P. & Thompsons, C. B. (1995): Apoptosis. *Science* 267, 1445—1462.

Stenseth, N. C. & Maynard Smith, J. (1984): Coevolution in ecosystems: red queen evolution or stasis? *Evolution* 38, 870—880.

Stern, D. (1971): A micro-analysis of mother-infant interaction behavior regulating social contact between a mother and her 3,5 months old twins. *J. Amer. Acad. Child Psychiatry* 10, 501—517.

— (1977): *The First Relationship, Mother and Infant*. Harvard University Press, Cambridge.

Stephens, G. & Yamazaki, T. (2000): *Legacy of the Cat*. Chronicle Books.

Stephens, J. C. et al. (2001): Haplotype Variation and Linkage Disequilibrium in 313 Human Genes. *Science* 293, 489—493.

Stone, L. J., Smith, H. T. & Murphy, L. B. (eds., 1973): *The competent infant: research and commentary*. Basic Books, New York.

Strauss, E. (1998): Getting a Handle on the Molecules That Guide Axons. *Science* 279, 481—482.

Strick, J. (ed., 2001): *Evolution and the Spontaneous Generation Debate*. Arizona State University.

Sylva, K. (1977): Play and learning. In: *Biology of Play* (ed. B. Tizard & D. Harvey), pp. 59—73. Heinemann, London.

Sylva, K., Bruner, J. S. & Genova, P. (1976): The role of play in the problem-solving of children 3-5 years old. In: *Play: its role in development and evolution* (ed. J. S. Bruner, A. Jolly & K. Sylva), pp. 244—260. Basic Books, New York.

Symons, D. (1978): The question of function: dominance and play. In: *Social play in Primates* (ed. E. O. Smith), pp. 193—230. Academic Press, New York.

Szagun, G. (1980): *Sprachentwicklung beim Kind*. Urban & Schwarzenberg.

Szathmary, E. (1989): The emergence, maintenance, and transitions of the earliest evolutionary units. *Oxford Surveys in Evolutionary Biology* 6, 169—205.

— (1999): The origin of the genetic code: amino acids as cofactors in an RNA world. *Trends in Genetics* 15(6), 223—229.

Tanabe, Y. & Jessell, T. M. (1996): Diversity and Pattern in the Developing Spinal Cord. *Science* 274, 1115—1123.

Tang, Y.-P. et al. (1999): Genetic enhancement of learning and memory in mice. *Nature* 401, 63—69.

Tatusov, R. L., Koonin, E. V. & Lipman, D. J. (1997): A Genomic Perspective on Protein Families. *Science* 278, 631—637.

Terrace, H. S. (1984): Animal cognition. In: *Animal cognition* (ed. H. L. Roitblat, T. G. Bever & H. S. Terrace), pp. 7—28. Lawrence Erlbaum, Hillsdale.

The MHC sequencing consortium (1999): Complete sequence and gene map of a human major histocompatibility complex. *Nature* 401, 921—923.

Theimer, W. (1983): *Das Rätsel des Alterns*. dtv, München.

Thiessen, D. D. & Gregg, B. (1980): Human assortative mating and genetic equilibrium. *Ethology and Sociobiology* 1, 111—140.

Thoenen, H. (1995): Neurotrophins and Neuronal Plasticity. *Science* 270, 593—598.

Thomas, J. H. (1994): The Mind of a Worm. *Science* 264, 1698—1699.

Tinbergen, N. (1939): On the analysis of social organization among vertebrates, with special reference to birds. *Amer. Midl. Natural*. 21, 210—234.

— (1948): Social releasers and the experimental method required for their study. *Wilson Bull*. 60, 6—52.

— (1951): *The Study of Instinct*. Oxford University Press, London.

— (1963): On aims and methods of ethology. *Z. Tierpsychol*. 20, 410—433.

Todt, D. (1975): Social Learning in the Grey Parrot. Zeitschrift für Tierpsychologie 51, 23—35.

Tomasello, M. (1999): *The Cultural Origins of Human Cognition*. Cambridge, Harvard University Press.

— (2000): Two Hypotheses About Primate Cognition. In: *The Evolution of Cognition* (ed. C. Heyes & L. Huber), pp. 165—184. MIT Press, Cambridge, MA.

Tomasello, M. & Call, J. (1997): *Primate cognition.* Oxford University Press, New York.

Tomasello, M., Kruger, A. C. & Ratner, H. H. (1993): Cultural learning. *Behavioral and Brain Sciences* 16, 495—552.

Tomblin, J. B. & Buckwalter, P. R. (1998): Heritability of poor language achievement among twins. *J. Speech Lang. Hear.* Res. 41, 188—199.

Tooby, J. & Cosmides, L. (1990): On the universality of human nature and the uniqueness of the individual: The role of genetics and adaptation. *Journal of Personality* 58, 17—67.

— (1992): The psychological foundations of culture. In: *The adapted mind: evolutionary psychology and the generation of culture* (ed. J. H. Barkow, L. Cosmides & J. Tooby), pp. 19—136. Oxford University Press, New York.

Topitsch, E. (1988): *Erkenntnis und Illusion.* Mohr, Tübingen.

Toth, N. (1985): Archeological evidence for preferential right-handedness in the Lower and Middle Pleistocene, and its possible implications. *Journal of Human Evolution* 14, 607—614.

Tough, D. F., Borrow, P. & Sprent, J. (1996): Induction of Bystander T Cell Proliferation by Viruses and Type I Interferon in Vivo. *Science* 272, 1947—1950.

Tracy, N. D. & Seaman, J. W. (1995): Properties of Evolutionarily Stable Learning Rules. *J. Theor. Biology* 177, 193—198.

Trappl, R. (1971): Die informationstheoretisch-statistische Behandlung sogenannter Zufallsfolgen in der Medizin. *Z. Nervenheilk.* 29, 143—176.

Treml, A. (1996): Das Zeigen. Funktion und Folgen der Zeigetechnik in der Kulturgeschichte aus pädagogischer Sicht. In: *Kulturethologische Aspekte der Technikentwicklung* (ed. M. Liedtke), S. 241—264. Austria Medien Service, Graz.

Trivers, R. (1985): *Social Evolution.* Benjamin/Cumming, Menlo Park.

— (1997): Genetic basis of intrapsychic conflict. In: *Uniting Psychology and Biology* (ed. N. Segal, G. Weisfeld & C. Weisfeld), pp. 385—395. American Psychological Association, Washington DC.

Turing, A. (1950): Computing Machines and Intelligence. *Mind* 59, 433—460.

Tversky, A. & Kahneman, D. (1983): Extensional versus intuitive reasoning: The conjunction fallacy in probability judgment. *Psychological Review* 90, 293—315.

Ullian, E. M. et al. (2001): Control of synapse number by glia. *Science* 291, 657—661.

U.S. Department of Energy (2001): *Genomes to Life.* Accelerating Biological Discovery. Program proposed by the Office of Advanced Scientific Computing Research, USA.

v. Glasersfeld, E. & Ackermann, E. (1997): Dialoge—Heinz von Foerster zum 85. Geburtstag. In: *Konstruktivismus und Kognitionswissenschaft* (ed. A. Müller, K. H. Müller & F. Stadler), S. 43—54. Springer, Wien.

Valbuzzi, A. & Yanofsky, C. (2001): Inhibition of the B. subtilis Regulatory Protein TRAP by the TRAP-Inhibitory Protein, AT. *Science* 293, 2057—2059.

Van den Berghe, P. L. (1983): Human inbreeding avoidance: culture in nature. *Behavioral and Brain Sciences* 6, 91—123.

Van Valen, L. (1973): A new evolutionary law. *Evolutionary Theory* 1, 1—30.

Vandenberg, S. G. (1972): Assortative mating or who marries whom? *Behavior Genetics* 2, 127—157.

Vandervert, L. (1997): *Understanding Tomorrow's Mind: Advances in Chaos Theory, Quantum Theory, and Consciousness in Psychology.* The Institute of Mind and Behavior, New York.

Velichkovsky, B. (1994): The levels approach in psychology and cognitive science. In: *International perspectives on psychological science: Leading themes* (ed. P. Bertelson, P. Eelen & G. d'Ydewalle), Lawrence Erlbaum Associates, Hillsdale, NJ.

Verhaag, B. (1995): *Blue Eyed.* eyetoeye, Marburg.

Viviani, P. & Terzuolo, C. (1980): Space-time invariance in learned motor skills. In: *Tutorials in Motor Behavior* (ed. G. E. Stelmach & J. Requin), pp. 525—533. North Holland Publishing Company.

Voelkl, B. & Huber, L. (2000): True imitation in marmosets. *Anim. Behav.* 60, 195—202.

Vogel, S. & Wainwright, S. A. (1969): *A Functional Bestiary.* Addison-Wesley, Reading.

Voland, E. (1995): Kalkül der Elternliebe—ein soziobiologischer Musterfall. *Spektrum der Wissenschaft* 6/95, 70—77.

— (1996): Konkurrenz in Evolution und Geschichte. *Ethik und Sozialwissenschaft* 7(1), 93—107.

Vollmer, G. (1975): *Evolutionäre Erkenntnistheorie.* Hirzel, Stuttgart.

— (1985): Über vermeintliche Zirkel in einer empirisch orientierten Erkenntnistheorie. In: *Was können wir wissen? Die Natur der Erkenntnis* (ed. G. Vollmer), pp. 217—267. Hirzel, Stuttgart.

— (1995): Was Evolutionäre Erkenntnistheorie nicht ist. In: Biophilosophie (G. Vollmer), S. 133—161. Reclam, Stuttgart.

Vroomen, J. & de Gelder, B. (2000): Sound enhances visual perception: cross-modal effects of auditory organization on vision. *J Exp Psychol Hum Percept Perform* 26, 1583—1590.

Waddington, C. H. (1968a): Does Evolution Depend on Random Search? In: *Towards a Theoretical Biology* (ed. C. H. Waddington), pp. 111—119. Edinburgh University Press, Edinburgh.

— (1968b): The basic ideas of biology. In: *Towards a Theoretical Biology* (ed. C. H. Waddington), pp. 1—32. Edinburgh University Press, Edinburgh.

Wagner, G. P. (1987): Der Passungsbegriff und die logische Struktur der evolutionären Erkenntnislehre. In: *Die Evolutionäre Erkenntnistheorie. Bedingungen, Lösungen, Kontroversen* (ed. R. Riedl & F. M. Wuketits), pp. 64—72. Parey, Berlin.

Wagner, G. P. & Altenberg, L. (1996): Complex adaptations and the evolution of evolvability. *Evolution* 50, 967—976.

Wagner, G. P., Booth, G. & Homayoun-Chaichian, H. (1997): A population genetic theory of canalization. *Evolution* 51, 329—347.

Wagner, G. P. & Gabriel, W. (1990): Quantitative variation in finite parthenogenetic populations: what stops Muller's ratchet in the absence of recombination? *Evolution* 44, 715—731.

Wahlsten, D. (1990): Insensitivity of the analysis of variance to heredity-environment interaction. *Behavioral and Brain Sciences* 13, 109—191.

Waldman, I. D. (1998): Problems in Inferring Dysgenic Trends for Intelligence. In: *The Rising Curve. Long-Term Gains in IQ and Related Measures* (ed. U. Neisser), pp. 365—376. APA, Washington DC.

Walhout, A. J. M. et al. (2000): Protein Interaction Mapping in C. elegans Using Proteins Involved in Vulval Development. *Science* 287, 116—122.

Wallace, A. R. (1855): On the law which has regulated the introduction of new species. *Annals and Magazine of Natural History* 16, 184—196.

— (1866): Letter to Charles Darwin dated 2 July. In: *Alfred Russel Wallace Letters and Reminiscences* (ed. J. Marchant, 1916), vol. 1, pp. 170—174. Cassell, London.

Watson, J. D. (1976): *Molecular Biology of the Gene.* Benjamin, Menlo Park.

Watzlawick, P., Beavin, J. H. & Jackson, D. D. (1982): *Menschliche Kommunikation.* Huber, Bern.

Waxman, D. & Peck, J. R. (1998): Pleiotropy and the Preservation of Perfection. *Science* 279, 1210—1213.

Weber, P. (1997): How do monkeys see the world? *Radio Broadcast,* Radio Austria One (20—23. September 1997).

— (1998): *Wie die Sprache in die Welt kam. Über den Ursprung der menschlichen Sprache.* Diplomarbeit, Grund- und Integrativwissenschaftliche Fakultät der Universität Wien.

Webster, R. (1995): *Why Freud was wrong: Sin, Science and Psychoanalysis.* Harper & Collins, New York.

Wedekind, C. et al. (1995): MHC-dependent mate preferences in humans. *Proc. R. Soc. London B Biol. Sci.* 260(1359), 245—249.

Wedemayer, G. J. (1997): Structural Insights into the Evolution of an Antibody Combining Site. *Science* 276, 1665—1669.

Weiner, J. (1995): Evolution Made Visible. *Science* 267, 30—33.

Weismann, A. (1889): *Essays Upon Heredity.* Clarendon, Oxford.

— (1892a): *Über die Vererbung* (2. Aufl.), Jena.

— (1892b): *Das Keimplasma. Eine Theorie der Vererbung.* Jena.

— (1895): *Neue Gedanken zur Vererbungsfrage.* Jena.

— (1904²): *Vorträge über Deszendenztheorie.* Jena.

— (1913³): *Vorträge über Deszendenztheorie.* Jena.

Weiss, P. A. (1971): The Basic Concept of Hierarchic Systems. In: *Hierarchically Organized Systems in Theory & Practice* (ed. P. A. Weiss), pp. 1—43. Hafner, New York.

Weiss, V. (1972): Empirische Untersuchung zu einer Hypothese über den autosomal-rezessiven Erbgang der mathematisch-technischen Begabung. *Biologisches Zentralblatt* 91, 429—535.

— (1982): *Psychogenetik: Humangenetik in Psychologie und Psychiatrie.* Fischer, Jena.

— (1988; als H. S. Weisman): Towards a quantum mechanics of intelligence. In: *Essays on the nature of intelligence and the analysis of racial differences in the performance of IQ tests* (ed. J. W. Jamieson), pp. 25—49. Cliveden, Washington.

— (1991): It could be neo-Lysenkoism, if there was ever a break in continuity. *ManKind Quarterly* 31(3), 231—253.

— (1992): Major genes of general intelligence. *Personality and Individual Differences* 13(10), 1115—1134.

— (1995): The Advent of a Molecular Genetics of General Intelligence. *Intelligence* 20, 115-124.

— (1996): Ungleiche Intelligenz und soziale Hierarchie als Ansatzpunkte unserer Evolution. *Ethik und Sozialwissenschaft* 7(1), 168—171.

Welch, A. M., Semlitsch, R. D. & Gerhardt, H. C. (1998): Call Duration as an Indicator of Genetic Quality in Male Gray Tree Frogs. *Science* 280, 1928—1930.

Werkmeister, W. H. (1990): *Nicolai Hartmann's New Ontology.* University Press of Florida.

Whalley, L. J. & Deary, I. J. (2001): Longitudinal cohort study of childhood IQ and survival up to age 76. *Brit. Med. J.* 322, 819.

White, B. L. (1970): *Human infants: Experience and psychological development.* Prentice Hall, New York.

Wilkinson, G. S., Presgraves, D. C. & Crymes, L. (1998): Male eye span in stalk-eyed flies indicates genetic quality by meiotic drive suppression. *Nature* 391, 276—279.

Willets, N. & Skurray, R. (1987): Structure and function of F factor and mechanisms of conjugation. In: *Escherichia coli and Salmonella typhimurium* (ed. F. C. Neidhardt), Cell. Mol. Biol., Amer. Soc. Microbiol., Washington.

Williams, G. C. (1957): Pleiotropy, natural selection, and the evolution of senescence. *Evolution* 11, 398—411.

— (1992): *Natural Selection: Domains, Levels, and Challenges.* Oxford University Press, New York.

Williams, G. C. (1993): Mother nature is a wicked old witch. In: *Evolutionary Ethics* (ed. M. H. Nitecki & D. V. Nitecki), pp. 217—231. State University of New York Press, Albany.

Williams, N. (1995): Chernobyl: Life Abounds Without People. *Science* 269, 304.

Wilson, A. C. & Cann, R. L. (1992): Afrikanischer Ursprung des modernen Menschen. *Spektrum der Wissenschaft* 6, 13—27.

Wilson, D. S. & Sober, E. (1994): Re-introducing group selection to the human behavioral sciences. *Behavioral and Brain Sciences* 17, 585—786.

Wilson, E. O. (1975): *Sociobiology: The New Synthesis.* Harvard University Press, Cambridge.

— (1978): The attempt to suppress human behavioral genetics. *The Journal of General Education* 29, 277—287.

Wilson, P. T. (1934): A Study of Twins with Special Reference to Heredity as a Factor Determining Differences in Environment. *Human Biology* 6, 324—354.

Wind, J. et al. (1992): *Language Origin: A Multidisciplinary Approach.* Kluwer, Dordrecht.

Wittgenstein, L. (1921/1963): *Tractatus logico-philosophicus.* Suhrkamp, Frankfurt.

— (1960): *Philosophische Untersuchungen.* Suhrkamp, Frankfurt.

Wolpert, C. R., Ghahramani, L. & Jordan, F. M. (1995): An Internal Model for Sensorimotor Integration. *Science* 269, 1880—1882.

Wolpert, L. (1968): The French Flag Problem: A Contribution to the Discussion on Pattern Development and Regulation. In: *Towards a Theoretical Biology* (ed. C. H. Waddington), pp. 125—133. Edinburgh University Press, Edinburgh.

Wolpert, L., Beddington, R., Brockes, J. & Jessell, T. (1997): *Principles of Development.* Oxford University Press, Oxford.

Wolpoff, M. H. et al. (2001): Modern human ancestry at the peripheries: a test of the replacement theory. *Science* 291, 293—297.

Wood, B. (1996): Human evolution. *Bioessays* 18 (12), 945—954.

— (1997): Ecce Homo—behold mankind. *Nature* 390, 120.

Wood, B. & Collard, M. (1999): The Human Genus. *Science* 284, 65-71.

Woodfield, A. (1976): *Teleology.* Cambridge University Press, London.

Woodworth, R. S. & Sherrington, C. S. (1904): A pseudoaffective reflex and its spinal path. *J. Physiol. (Lond)* 31, 234

Wu, K.-C. et al. (2001): The role of movement for embryonic development I: Embryonic motor activity and the influence of illumination intensity. *J. Experimental Zoology*, in press.

Wuketits, F. M. (1984): *Evolution, Erkenntnis, Ethik. Folgerungen aus der modernen Biologie.* Wissenschaftliche Buchgesellschaft, Darmstadt.

Wyllie, A. H. (1980): Glucocorticoid-induced thymocyte apoptosis is associated with endogenous endonuclease activation. *Nature* 284, 555—556.

Wynne-Edwards, V. C. (1962): *Animal dispersion in relation to social behavior.* Oliver & Boyd, Edinburgh.

Xerri, C. et al. (1998): Plasticity of Primary Somatosensory Cortex Paralleling Sensorimotor Skill Recovery From Stroke in Adult Monkeys. *J. Neurophysiol.* 79, 2119—2148.

Yamazaki, K. et al. (1976): Control of mating preferences in mice by genes in the major histocompatibility complex. *J. Exp. Med.* 144, 1324—1335.

Yamazaki, T. & Kojima, T. (1995): *Legacy of the Dog.* Chronicle Books.

Yu, C.-E. et al. (1996): Positional Cloning of the Werner's Syndrome Gene. *Science* 272, 258—262.

Zimen, E. (1988): *Der Hund. Abstammung, Verhalten, Mensch und Hund.* Bertelsmann, München.

Acknowledgement of sources

Introductory citation: A. Weismann 1892a, p. 15/17; Introduction: J. Monod 1975, p. 131; Chap. 1: W. Busch, Sämtliche Bildergeschichten (Prisma Verlag); Chap. 2: C. Darwin 1871, introduction; Chap. 3: C. Darwin 1859, chap. 3; Chap. 4: C. H. Waddington, 1968a, p. 111; J. Monod, 1970, p. 149; Chap. 5: C. Darwin 1872, chap. 2; Chap. 6: M. Eigen, 1987, p. 150/151; H. Haken & M. Haken-Krell, 1989, p. 51; Chap. 7: A. Weismann, 1904, p. 331; Chap. 8: D. Hume 1777/1992, p. 85; J. Garcia 1991, p. 40; Chap. 9: J. Nestroy, in Hein 1995, p. 97/105; J. Piaget, 1976 (C.I.E.G., Geneva); Chap. 10: C. H. Waddington 1968, p. 20; R. Plomin in *focus* 3/1998, S. 114; Chap. 11: J.-B. de Lamarck 1809, p. 73; Chap. 12: J. Diamond 1992, p. 56/141; Chap. 13: R. Trivers 1985, p. 315; J. Maynard Smith 1989, p. 244; Chap. 14: J. Searle 1995; J. Nestroy in Hein 1995, p. 145; Chap. 15: J. Monod 1973, p. 202; T. Kuhn 1962/96, p. 144; Chap. 16: J. Nestroy in Hein 1995, S. 147; H. Markl 1997, S. 19; Chap. 17: C. Darwin 1871, chap. 5; Chap. 18: A. Weismann 1904, Bd. 2, S. 125; Chap. 19: J. Maynard Smith 1983, p. 45;

Appendix: The Post-Genomic Era—Decoding the Metacode

Since February 2001, it has been known that the human genome comprises around 35,000 verifiable genes (*International Genome Sequencing Consortium*: 30,000—40,000 genes, *Nature* 409:813—958; *Celera Genomics*: 26,000—38,000 genes, *Science* 291:1304—1351). When one compares this number with previous expectations of more than 100,000 genes, it seems very low at first glance, since, with 300 additional genes, this is less than 1% more than that of the mouse (for comparison: *Drosophila* has 13,601, the nematode worm *Caenorhabditis* 19,909, and the inconspicuous plant thale cress *Arabidopsis thaliana* has 25,495 genes). In defense of the mouse, it must be said that the mouse represents a highly developed creature which possesses, for example, all the essential categories of intelligent behavior of the higher mammals (e.g., spatio-temporal orientation, learning and memory, exploration and inquisitiveness, social behavior, care of the young). This also explains the large number of similar and therefore comparable, and with great probability, homologous genes, which, in turn, is extremely helpful in the further clarification of the functions of human genes (*International Mouse Mutagenesis Consortium* 2001). With reference to ascertaining the real number of genes, it must also be mentioned that the now official number can in no way be definitive since recent estimates from various investigative methods for discovering genes, some of which seem to be very promising (e.g., retranscription of bits of mRNA found somewhere in the body into DNA called *expressed sequence tags* or EST; searching for new genes that do not resemble any known genes; and most important, searching for sequences that do not encode proteins but diverse kinds of RNA for regulatory functions) have resulted in figures which reach approximately 80,000 genes. An as-yet unpublished paper by a group of researchers at Ohio State University in Columbus even speaks of 850,000 gene segments for which evidence from proteins or RNA supposedly exists (*New Scientist* 12th May 2001) and a similar number (700,000) has been recently confirmed by the application of a new technique (ORESTES: "open reading frame expressed sequence tags") by a group of Brazilian geneticists (Camargoa et al. 2001). Concerning the real number of complete genes or, as Bo Yuan from Ohio State University prefers to say, "transcript clusters" (*The Scientist* 15th Oct. 2001), a new study based on a detailed comparison of 13 different public gene expression databases results in approx. 75,000 genes, more than 65,000 of which show mRNA splicing (www.labbook.com). Be that as it may, the essential point in this controversy is not just the absolute number of exons, i.e., directly peptide-, protein- or merely RNA-encoding gene sequences themselves, but something which could be called the "Metacode", or, under the restriction that, ultimately, the molecular semantics of the genome necessarily has to comprise Chomsky's "universal grammar", the "deep structure of the genome" (for a recent theoretical account, see Kanehisa 2000; for the suggestion to devise new "genomic metaphors", see Avise 2001). To understand this better, it is recommended that one imagines that genes represent nothing other than the simplest units of a kind of surface structure of the genome which are naturally of great significance in relation to the organism but are, however, only similar to single words which are the starting point for forming an unbelievably complex system of instructions which, in addition, can influence each other reciprocally. The first step to a better understanding of the genetic deep structure was taken by Francois Jacob and Jacob Monod as long ago as 1961 with their pioneering clarification of the regulation of gene expression by operons (Jacob and Monod 1961), but today a system of regulations and instructions different by several orders of magnitude is increasingly emerging (first hints in Losick and Sonenshein 2001 and Valbuzzi and Yanofsky 2001 for Bacteria). This is not astonishing in the light of the extremely long periods of time and the dimensions of the genetic resources which were available to the evolution of the multiplicity of various living forms. One therefore imagines the 3×10^9 base pairs of a human being as a written text which mirrors an actual content. The number of words (= genes) would of course still be limited (e.g., 35,000), but, through a hierarchically structured syntax similar to and at least as powerful as that of human language, an enormously compressed text with many layers of different significance would be created. In doing so, however, one would have to imagine, in addition, that everything which a human being—be it in an automatic, instinctive manner or not—could think of during speech about additional grammatical and semantic background knowledge for the analysis of a purely acoustic sequence of syllables (= linguistic "base triplets") must be laid down completely in the metacode in order to be realized in a particular situation. Because of the phenotypic complexity of our subjects of investigation, i.e., living organisms, this allows us to expect, as the theoretical biologist Rupert Riedl (1975) already forecast years ago, an extremely systemized, that is high-grade interdependent and hierarchically interrelated non-linear code laid down in a linear sequence of bases whose more detailed deciphering which has only recently been started is now displaying its first promising outlines. In the last few years, an increasingly well-defined synthesis from evolutionary biology, genetics, and developmental biology has emerged so that a complete hierarchy of intricately interlocked levels of the genetic code and diverse regulation mechanisms exists (Carroll, Grenier and Weatherbee 2001; Gehring and Ruddle 1998; Raff 1996; Stearns 1992; Wolpert, Beddington, Brockes and Jessell 1997). Sean Carroll, one of the leading pioneers in this field, rightly speaks of a proper molecular "genetic regulatory logic" which includes phenomena such as hierarchies, pathways, circuits, batteries, and networks of gene effects. Such a degree of functional networking which is always simultaneously accompanied by well-defined limits of adaptability caused by developmental constraints (Maynard Smith, Burian, Kauffman, Alberch, Campbell, Goodwin, Lande, Raup and Wolpert 1985) and thereby invariably leads to evolutionary canalizations (Wagner and Altenberg 1996; Wagner, Booth and Homayoun-Chaichian 1997) also explains why certain complex traits (e.g., eyes) often in very different groups of animals use the same homologous gene instructions (e.g., *eyeless* gene in man, mouse, and fruit fly). At the level of inheritance, this manifests itself in the course of evolution as only very slowly increasing complexity. For instance, it took at least 9 million years for the fins of fishes to develop into the locomotory appendages of the land-living tetrapods (Sordino, van der Hoeven and Duboule 1995) and not less than 360 million years to reach the present diversity of the limbs of four-legged animals (Shubin, Tabin and Carroll 1997). At the same time, this explains the fact that the great majority of characters of metazoans no longer follow the simple Mendelian laws based on a one gene/one phenotype-relationship according to the usual recessive/dominant pattern (Risch 2000). The general trend of scientific research with regard to the question of the genetic self-regulation of organisms is therefore easily identified. First, it seems, in principle, conceivable for a linear sequence of elementary chemical units, i.e., an instructive sequence, to evolve in a way that it can code both the realization of a basically unlimited diversity of complex, three-dimensionally folded structures of molecules (Schuster 1993; Fontana and Schuster 1998) as well as the regulation of spatio-temporally, highly differentiated and often incredibly robust developmental processes (Kim 1999, Walhout et al. 2000). Second, through many of these parallel advances in the field of developmental biology, it is becoming more and more clear that it is not the structure itself—as in the naive *homunculus*-picture—but that all of both the necessary and sufficient *information* for the embryonic development can be considered already present in the genome of a single cell, such as the fertilized egg. Or, as some important theorists have expressed it programmatically: "ex DNA omnia" (Lewis Wolpert 1991, p. 77), ten years ago, and, in more recent terms, "the relevant information resides in the genomic DNA" (John Maynard Smith 2000). Taken together these new empirical findings allow us nowadays to see the previously repeatedly controversial connection between morphological and genetic evolution in a much closer causal relationship than was still the case at the time of Darwin. As Sean Carrol so aptly expressed it recently: "If morphological diversity is all about development, and development results from genetic regulatory programs, then the evolution of diversity is directly related to the evolution of genetic regulatory programs" (Sean Carroll 2001, p. 13). The concluding step to a final synthesis of evolutionary theory and developmental biology consists quite simply of recognizing that from systems-theoretical reasons, i.e., the disadvantageousness of random somatic mutations, in contrast to Weismann's phylogenetic germ line, the whole of ontogeny must remain excluded from real evolutionary and hence, at the same time, cognitive gain. This also includes of necessity everything of behavioral patterns and cognitive abilities with which animals and humans are today equipped and therefore strictly rules out every Lamarckian instruction by the environment be it inanimate (geology, climate, physical forces) or animate (other organisms, conspecifics, "cultural milieu") as a matter of principle. Although still at its very beginnings if compared with morphology where the new synthesis is already taking a more concrete shape (cf. Kopp, Duncan and Carroll 2000), modern molecular behavioral genetics, which is increasingly changing to, in the original sense of the founders of comparative ethology (Tinbergen and Lorenz), a broadly based comparative behavioral genomics (McGuffin, Riley and Plomin 2001; psychiatric approaches in Sanders and Gershon 1998; Petronis and Kennedy 1995; for a general review, see O'Brien et al. 1999) has already proved that every single category of behavior (e.g., intelligence, aggression, sexuality, social behavior), every cognitive level of behavior no matter how complex (e.g., reflexes, associations, memory, mental respresentations, logical abstract thinking), as well as every individual variant (cf. Sachidanandam [2001]: A map of human genome sequence variation containing 1.42 million single nucleotide polymorphisms. *Nature* 409, 928—933) can be influenced, in detail as well as in general, by single genes but both influenced and altered even *more* so by complicated gene interactions. With regard to the main thesis of the book, i.e., 100% genetic instruction of behavior, the high degree of potential variability is particularly fascinating and Francis Collins, director of the National Human Genome Research Institute, has found the right words for it: "We have been talking a lot about how similar all of our genomes are, that we're 99.9% percent the same. That might tend to create an impression that it's a very static situation. But that 0.1 percent is still an awful lot of nucleotides" (from: [evol-psych] Huge Genetic Variation Found in Human Beings. 13.07. 2001). In fact, a recent study conducted by Gerald Vovis (Stephens et al. 2001) with 82 unrelated people from all over the world unveiled an average of 14 different versions for 313 human genes, that means 14^{313} theoretically possible variants. However, at a more general level, the great challenge for science in the years to come will be to bring these many still static information carriers called "genomes" to (real) life (cf. research program of the U.S. Department of Energy 2001) to be able to understand for the very first time how the endless process of life really works.

Index

Druck: Strauss Offsetdruck, Mörlenbach
Verarbeitung: Schäffer, Grünstadt